Lecture Notes in Mechanical Engineering

Lecture Notes in Mechanical Engineering (LNME) publishes the latest developments in Mechanical Engineering—quickly, informally and with high quality. Original research or contributions reported in proceedings and post-proceedings represents the core of LNME. Volumes published in LNME embrace all aspects, subfields and new challenges of mechanical engineering.

To submit a proposal or request further information, please contact the Springer Editor of your location:

Europe, USA, Africa: Leontina Di Cecco at Leontina.dicecco@springer.com
China: Ella Zhang at ella.zhang@cn.springernature.com
India: Priya Vyas at priya.vyas@springer.com
Rest of Asia, Australia, New Zealand: Swati Meherishi at swati.meherishi@springer.com

Topics in the series include:

- Engineering Design
- Machinery and Machine Elements
- Mechanical Structures and Stress Analysis
- Automotive Engineering
- Engine Technology
- Aerospace Technology and Astronautics
- Nanotechnology and Microengineering
- Control, Robotics, Mechatronics
- MEMS
- Theoretical and Applied Mechanics
- Dynamical Systems, Control
- Fluid Mechanics
- Engineering Thermodynamics, Heat and Mass Transfer
- Manufacturing Engineering and Smart Manufacturing
- Precision Engineering, Instrumentation, Measurement
- Materials Engineering
- Tribology and Surface Technology

Indexed by SCOPUS, EI Compendex, and INSPEC.

All books published in the series are evaluated by Web of Science for the Conference Proceedings Citation Index (CPCI).

To submit a proposal for a monograph, please check our Springer Tracts in Mechanical Engineering at https://link.springer.com/bookseries/11693.

Jan Kusiak · Łukasz Rauch · Krzysztof Regulski
Editors

Numerical Methods in Industrial Forming Processes

Numiform 2023

Springer

Editors
Jan Kusiak
AGH University of Science and Technology
Kraków, Poland

Łukasz Rauch
AGH University of Science and Technology
Kraków, Poland

Krzysztof Regulski
AGH University of Science and Technology
Kraków, Poland

ISSN 2195-4356   ISSN 2195-4364 (electronic)
Lecture Notes in Mechanical Engineering
ISBN 978-3-031-58005-5   ISBN 978-3-031-58006-2 (eBook)
https://doi.org/10.1007/978-3-031-58006-2

This Springer imprint is published by the registered company Springer Nature Switzerland AG
The registered company address is: Gewerbestrasse 11, 6330 Cham, Switzerland

If disposing of this product, please recycle the paper.

Preface

One of the vitally important branches of industry is forming materials, encompassing various efficient methods of shaping metals, alloys, polymers, etc. The final products of materials forming are present in all areas of our life, from aviation (satellites, aircrafts), marines (ships and yachts), our daily life (computers, household appliances) to healthcare (prostheses, medical equipment). Hence, any opportunity to discuss new matters and achievements in that vast industry attracts scientists and researchers worldwide. Mainly, nowadays, when we meet Industry 4.0, the numerical methods in modelling and simulation of all kinds of materials forming processes is a hot topic.

That was the motivation to organise the conference NUMIFORM 2023 (the 14th International Conference on Numerical Methods in Industrial Forming Processes) at the AGH University of Science and Technology in Kraków, Poland, on June 25–29, 2023. Additionally, this edition of the Conference was dedicated to Prof. O. C. Zienkiewicz, a great Polish scientist. One of the most important scientific achievements of Prof. O. C. Zienkiewicz was the transfer of the theoretical Finite Element Method of calculating the stresses to traditional structures as well as practical applications in solid mechanics, fluids, and other fields. Professor O. C. Zienkiewicz organised the first NUMIFORM conference in 1982 in Swansea, UK. Thus, this volume is a continuation of the celebrations of the 40th anniversary of the first Conference in this series.

The current volume of Lecture Notes in Mechanical Engineering (Springer series) contains chosen manuscripts presented during the NUMIFORM 2023 Conference. The published papers represent the state-of-the-art in the field of application of various numerical methods in the science and technology of materials forming, demonstrating the results of a broad spectrum of research, from micro- and nano-forming to digital materials modelling. The authors represent universities, research institutes, and industry.

Furthermore, we would like to express our gratitude to the reviewers of the papers submitted, principally to the Members Steering Committee of the Conference, for their work and critical but constructive remarks, which helped maintain the high scientific level of the Conference. We hope that the proceedings will become a source of valuable information and inspiration in the scientific work for academics, researchers, engineers, and students.

We want to thank all participants for their contributions to the Conference programme and for their contributions to the present volume, and we were pleased to welcome in Kraków every participant of the 14th NUMIFORM 2023.

On behalf of the Conference Committee.

Kraków, Poland   Jan Kusiak
                                                                       Łukasz Rauch
                                                                  Krzysztof Regulski

Contents

Analysis of Multiaxial Tensile Stresses in Cold Forging Tools

Martin Killmannⓘ **and Marion Merklein**ⓘ

Abstract Tools have a key role in cold forging as they directly influence part quality and process economics. High loads from the contact pressure exhibited by the work-piece are a major challenge regarding tool life. In complex geometries, tensile die stresses are concentrated in local areas and oriented both in axial and tangential directions. This superposition of loads leads to fatigue failure at the stress peaks. Since the interactions of multidirectional tensile stresses in cold forging tools have not been investigated in previous investigations, this paper aims to better understand the formation of local stress concentrations in axial and tensile directions as well as the possibilities to influence them. As tool loads in critical areas are difficult to measure during the process, simulation plays an important role in the evaluation of die stresses. Therefore, a model process with multiaxial tool stresses is simulated and the simulation results are compared to experimental values in order to validate the model. The simulation is then used to analyse the die load depending on the prestressing system.

Keywords Cold forming · Die · Stress

1 Introduction

Parts produced by cold forging have good mechanical properties and geometrical accuracy [1]. By omitting workpiece heating before forming, energy consumption and CO_2 emissions can be reduced. However, the large flow stress of materials formed at room temperature necessitates high forces. Cold forging tools are consequently subjected to high stresses. Preventing tool failure is essential for the economic viability of forging processes, since tooling costs amount to 30% or more for net shape applications [2]. The geometrical complexity of tools is increasing, as there

M. Killmann (✉) · M. Merklein
Institute of Manufacturing Technology, Friedrich-Alexander Universität Erlangen-Nürnberg, Egerlandstr. 13, 91058 Erlangen, Germany
e-mail: martin.killmann@fau.de

© The Rightsholder, under exclusive licence to [Springer Nature Switzerland AG], part of Springer Nature 2024
J. Kusiak et al. (eds.), *Numerical Methods in Industrial Forming Processes*, Lecture Notes in Mechanical Engineering, https://doi.org/10.1007/978-3-031-58006-2_1

1

is a demand for more intricate part geometries with functional integration [3]. For complex tool geometries, the main failure mechanism is fatigue [4], which occurs due to high alternating stresses. The high-strength steels or cemented carbides used for forging tools are especially sensitive towards tensile stresses [5], which are present in tangential direction for non-circular symmetrical part geometries [6] or in axial direction when there is a change in cross-section [7]. In increasingly complex tool geometries, local multiaxial tensile stresses in both tangential and axial directions are encountered as well. As fatigue depends on the multiaxial stress state [8], these types of loads are highly relevant, but have as of yet not been analysed in depth. Existing strategies to reduce tensile tool loads by compressive prestressing focus on one stress direction. For example, concepts for local prestressing by reinforcements with adapted interference fits exist both for axial [9] and for tensile stresses [10]. However, the influence on multiaxial tensile stresses is currently unknown. This paper therefore aims to provide a foundation for the analysis of multiaxial tensile tool stresses and methods to influence them. For this purpose, a model process with local multidirectional tensile stresses is designed. As stresses in forging dies cannot be measured experimentally, a numerical model is set up and validated. The model is then used to analyse the stress state and evaluate the influence of conventional prestressing systems on the multidirectional tool stresses.

2   Model Process

In this section, a model process is designed for the analysis of multiaxial tensile tool stresses. The process should be a typical cold forging operation and allow for flexible changes in tool geometry and prestressing system. Therefore, a forward extrusion process for parts with four functional elements is chosen as shown in Fig. 1. Compared to conventional forward extrusion, an additional slope is introduced, so that the functional elements are formed gradually in the tangential direction. This ensures multiaxial stresses with tangential and axial tensile loads at the same position as will be shown in the analysis of the stress state. Furthermore, angles of the extrusion shoulder and slope as well as the geometry and number of functional elements can be varied for future process analysis. Materials typical for cold forging applications are chosen for the process. The workpiece is made out of 16MnCrS5, and the active tool parts punch and die out of the powder metallurgical steel ASP2023. The material of the reinforcement is the hot-working steel 1.2344. The die is inserted into the reinforcement with an interference fit of 3% to induce a tangential prestress. The initial billet has a diameter of 25 mm and a height of 20 mm. In areas where no functional element is present, the diameter is reduced to 20 mm. To achieve this, the punch moves downwards until a fixed distance is reached creating a part height of 25 mm.

To analyse the stress state, a numerical model is set up using the software Simufact. Forming 15.0. The simulation is carried out in a decoupled approach, in which material flow and die load are calculated separately. The workpiece is modelled with

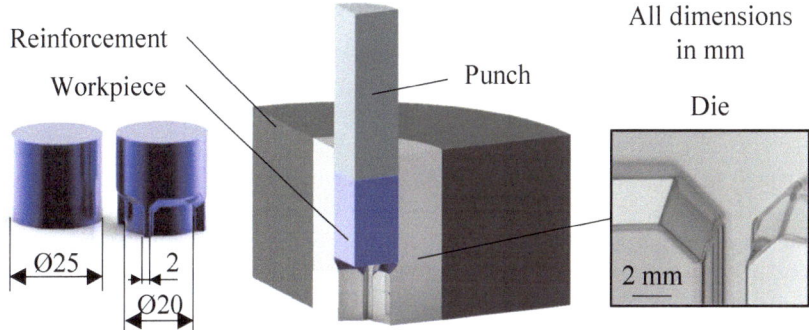

Fig. 1 Model process

the flow curve shown in Fig. 2, which was determined for the material 16MnCrS5 in previous research [6]. Furthermore, friction is taken into account with a friction factor of 0.04 according to Tresca. This factor was identified for the relevant material pairing of 16MnCrS5 and ASP2023 in a double cup extrusion test using a zinc-phosphate coating on the workpiece in combination with the lubricant soap [6].

Taking into account the cyclic symmetry of the process, a 45° segment is computed. The workpiece is meshed with hexahedral elements with an edge length of 0.5 mm and refinements to 0.25 mm below the extrusion shoulder and 0.125 mm at the extrusion shoulder and the functional elements. In the tool load simulation, die and reinforcement are represented using tetrahedral elements with a maximum geometrical deviation of 0.005 mm. The die's material ASP2023 is modelled elastically with a Young's modulus of 230 GPa. To validate the given assumptions, the numerical results are compared to experimental ones in the subsequent section.

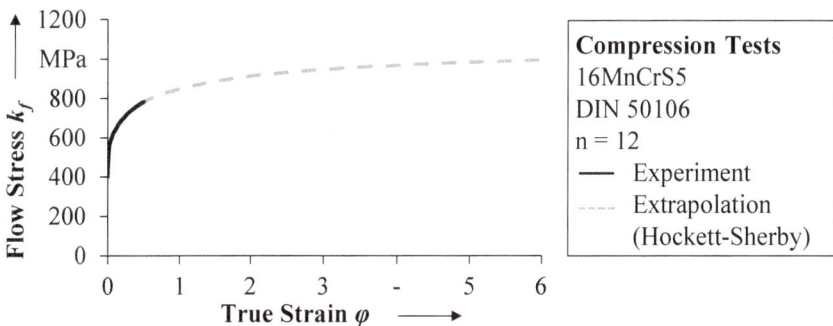

Fig. 2 Flow curve of 16MnCrS5 according to [6]

3   Validation of the Numerical Model

In order to qualify the numerical model for an evaluation of tool stresses, it needs to be validated by comparison with experimental results. For this purpose, force–displacement curves, part geometries as well as hardness, and true strain distributions are analysed. Figure 3 shows the force–displacement curves obtained in the experiment and simulation. After a slow rise in force, when elastic deformation takes place and transition radii are filled, the force rises steeply during the filling of the extrusion shoulder. Afterwards, the force increases with a less steep slope when the height of the functional elements increases. This type of force curve can be seen both in the simulation and the experiment and is similar to graphs obtained in the forward extrusion of gears [11]. While overall the curves are in good agreement, the increase in force is slightly higher for the simulation resulting in a higher maximum force compared to the experiment. The maximum values deviate by 8% with 334.3 kN in the simulation and 309.4 kN $\pm$ 2.2 kN in the experiment. The deviations at the beginning of the process can be explained by elastic tool deformation and varying friction conditions in the experiment, since rigid tools and a constant friction factor were used in the numerical model.

As second target value, part geometries are compared in Fig. 4. The experimentally formed parts were measured three dimensionally using a 3D sensor type Atos from the company GOM GmbH. A good agreement in part geometry can be seen in the graphs. The cross-section shows that the desired geometry with filled functional elements is reached both in simulation and experiment. Small deviations are present in the sections through the functional element and the extrusion shoulder. Here, the simulation slightly overestimates the material flow along the axis of symmetry. The height of the part in this area deviates by 3% with 24.75 mm for the simulation and 24.02 mm $\pm$ 0.04 mm for the experiment with n = 3 measured parts. This may be

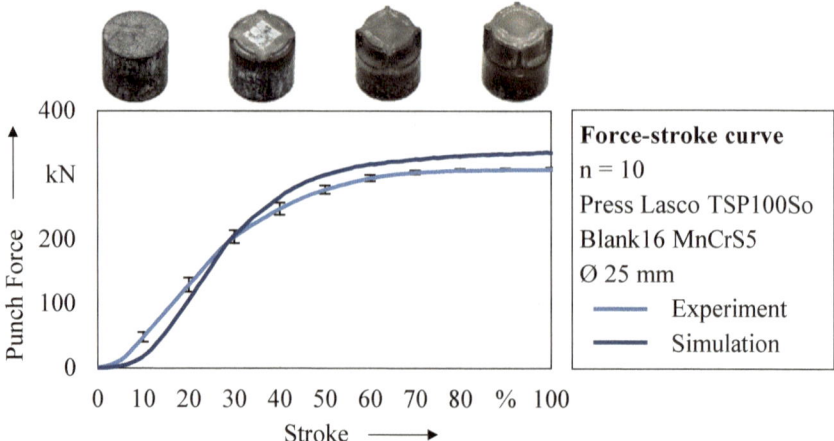

Fig. 3 Comparison of force–stroke curves in experiment and simulation

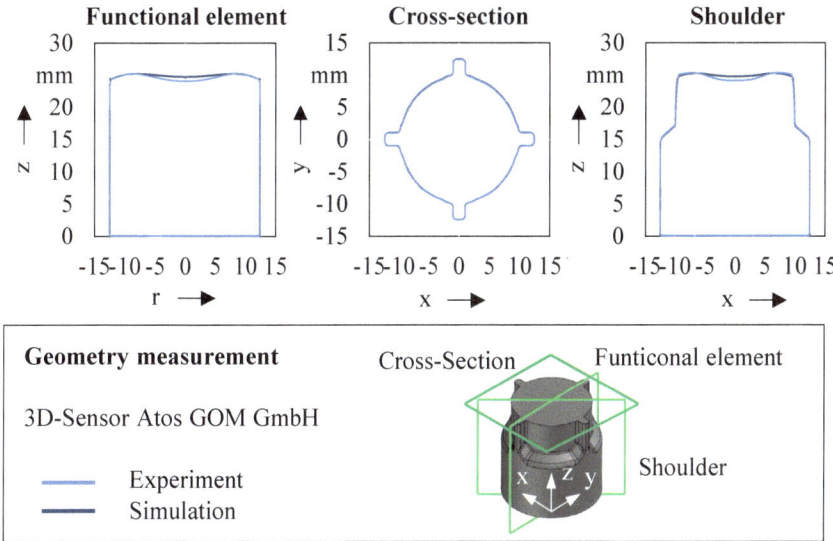

Fig. 4 Comparison of numerical and experimental workpiece geometry

due to locally deviating friction conditions between the double cup extrusion test and the model process.

The small deviation in part height is not expected to have a high influence on the tool load, since the critical die stresses occur at the extrusion shoulder. To verify the correct representation of material flow, hardness, and true strain distributions are compared in Fig. 5. As strain hardening of cold formed material induces a higher hardness, the mappings should be qualitatively similar. To achieve the results, micro hardness measurements were conducted in three different part sections as shown in Fig. 5.

The section through the shoulder and functional elements show increased hardness and true strains near the extrusion shoulder. Both hardness and true strains decrease towards the end of the functional elements and the part centre. The area below the extrusion shoulder shows no true strain and low hardness that corresponds to the initial hardness of the material. In the cross-section, it is evident that the highest strain hardening occurs near the functional elements, where the material is displaced to create the desired geometry. Furthermore, higher strains are present at the part edge than in the part middle, where close to no deformation takes place. Overall, the true strain and hardness distributions are in good agreement. Taking into account the force–displacement curves and part geometries as well, the predictive accuracy of the numerical model is evaluated as good. It is therefore qualified for an analysis of the die stresses in the following section.

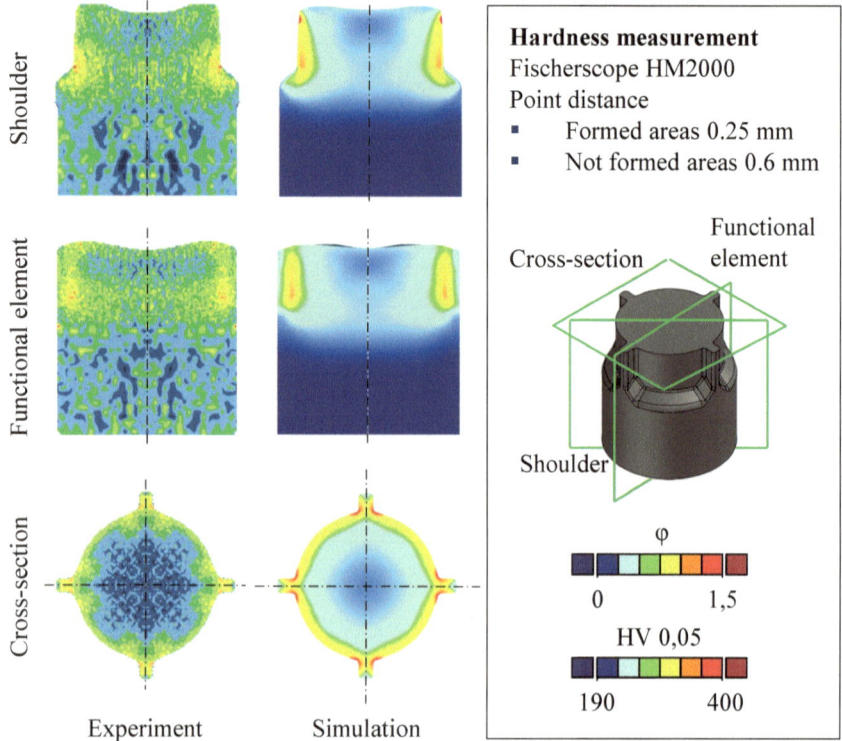

Fig. 5 Comparison of measured hardness and simulated true strain

4 Analysis of the Stress State

The critical tool stresses occur in the forming die, since it incorporates the negative of the part geometry with the transition to a non-circular symmetrical cross-section. The distribution of tangential, axial, and maximum principal stress is shown in Fig. 6. Axial and tangential stresses are normal stresses in the orientation of a cylindrical coordinate system corresponding to the outer die geometry.

As is typical for forward extrusion processes, axial stresses are present at the transition of the higher diameter to the extrusion shoulder. In this area, the pressure exhibited by the workpiece on the extrusion shoulder leads to tensile stresses. Furthermore, tangential stresses occur at the back of the functional element, which is in accordance with stresses obtained in dies for parts with functional elements and a constant cross-section [10]. Due to the slope between extrusion shoulder and functional elements, the two stress components overlap creating a multidirectional tensile stress. Consequently, the highest maximum principal stress occurs in this transition area because of the superposition of tangential and axial tensile loads. To illustrate the influence of conventional prestressing systems on this multidirectional tensile stress, axial, tangential, and maximum principal stresses are analysed depending on

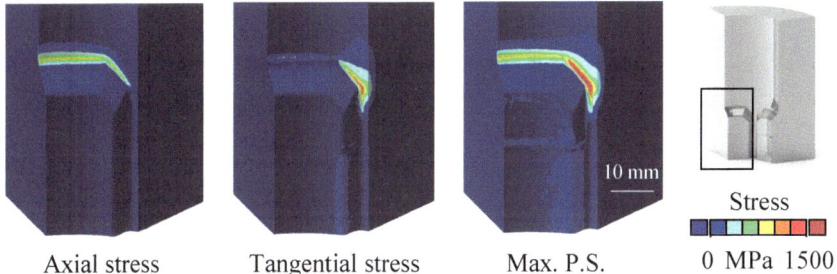

| Axial stress | Tangential stress | Max. P.S. | Stress
0 MPa 1500 |

Fig. 6 Multiaxial stress state in the die

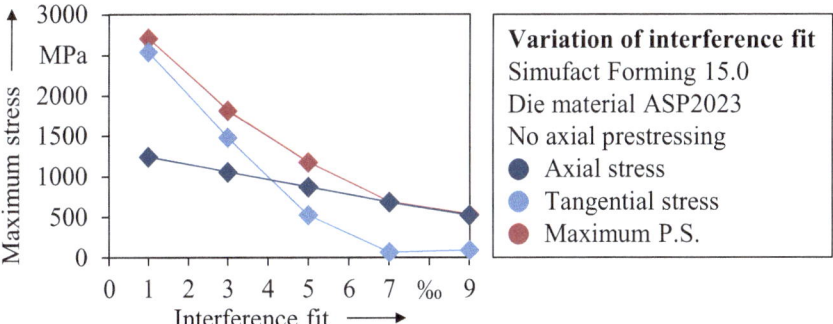

Fig. 7 Influence of interference fit on die stresses

the interference fit of the reinforcement in Fig. 7. The respective maximum stresses on the inner die wall were evaluated to obtain the graphs.

An increase in interference fit induces a higher prestress and therefore decreases all three stress types. However, the influence on the tangential direction is most pronounced with the tangential stress decreasing from 2536 MPa at 1% to 70 MPa at 7%. At the same interference fits, the axial stresses only decrease from 1244 to 683 MPa. From 7% upwards, no relevant tangential stresses are present in the die, because the reinforcement induced sufficient tangential compressive prestress. However, tensile stresses are still present in axial direction. The highest maximum principal stress is therefore equal to the axial stress at 7% and 9%. It is evident that reinforcements mainly influence tangential tensile stresses. For a reduction of axial loads, additional axial prestressing should be used. This is often implemented for horizontally divided dies by applying pressure in a hydraulic press and fixing the tool parts with a nut [1]. The axial prestressing is simulated using a rigid plate applying a force between 200 and 1000 kN on the die in axial direction. The interference fit of the reinforcement is kept constant at 3%. The results of the analysis are shown in Fig. 8.

While axial loads are influenced strongly by axial prestressing, tangential stresses decrease only slightly. From 0 to 1000 kN prestressing, the axial stress is reduced from

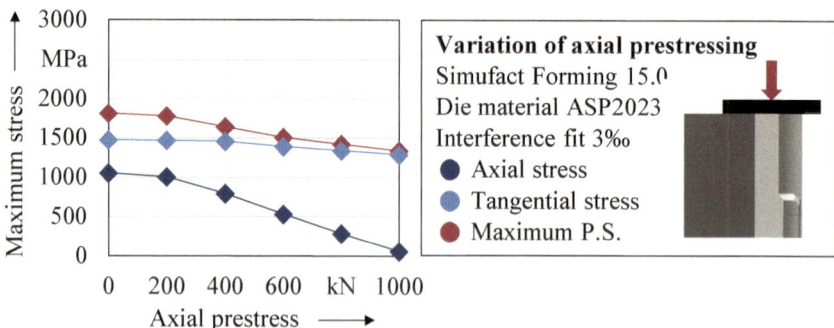

Fig. 8 Influence of axial prestressing on die stresses

1010 to 62 MPa, while the tangential stress only changes from 1478 to 1291 MPa. When axial loads are neutralised by a sufficient prestressing with 1000 kN, the critical tensile stress represented by the maximum principal stress occurs in tangential direction. While reinforcements are suitable for decreasing tensile loads, axial prestressing should be used for axial stresses. When dealing with multidirectional tensile stresses, a combination of measures is necessary to completely remove critical tensile die loads.

5   Summary and Outlook

Tool failure through fatigue is a major challenge when applying cold forming processes to produce intricate part geometries. As geometries with non-constant and non-circular symmetrical cross-sections create multiaxial tensile stresses, a model process for the research of this stress state was introduced in this paper. The corresponding numerical model was validated, as force–displacement curves, part geometry as well as true strain and hardness distributions are in good agreement between experiment and simulation. The analysis of the stress state showed that prestressing measures like reinforcements and axial prestressing are able to influence one stress direction, but have little influence on the other. A combination of measures is therefore recommended for multidirectional stress states. Further research should focus on the influence of novel local prestressing measures on the stress state. Furthermore, the fatigue behaviour should be analysed experimentally. Lastly, dies made of cemented carbide should be considered in addition to steel materials.

Acknowledgements The authors would like to thank the Bavarian Research Foundation (BFS) for supporting the project 1483_20 "Reduction of local tool stresses in cold forging dies". We further thank the supporting industrial partners Arnold Umformtechnik GmbH & Co. KG, Hoerbiger Antriebstechnik GmbH, Plansee SE, and ZF Friedrichshafen AG.

References

1. Lange K, Kammerer M, Pöhlandt K, Schöck J (2008) Fließpressen—Wirtschaftliche Fertigung metallischer Präzisionswerkstücke. Springer
2. International Cold Forging Group: Tool Life & Tool Quality in Cold Forging Part 1: General Aspects of Tool Life (2002) ICFG Document No. 14/02. Meisenbach
3. Engel U, Groenbaek J, Hinsel C, Kroiß T, Meidert M, Neher R, Räuchle F, Schrader T (2011) Tooling solutions for challenges in cold forging. In: UTF Science, Bd. 3, Meisenbach
4. Tekkaya A, Sonsöz A (1996) Life estimation of extrusion dies. Int J Mech Sci 38(5):527–538
5. International Cold Forging Group: General aspects of tool design and tool materials for cold and warm forging (1992) In: International cold forging group 1967–1992—objectives, history published documents. Meisenbach, pp 33–58
6. Killmann M, Merklein M (2021) Analysis of stress pins for the local prestressing of cold forging tools. Prod Eng Res Devel 15(1):119–131
7. Meidert M (2006) Beitrag zur deterministischen Lebensdauerabschätzung von Werkzeugen der Kaltmassivumformung. In: Geiger M (Hrsg) Fertigungstechnik—Erlangen, Bd. 172, Meisenbach
8. Lange K, Cser L, Geiger M, Kals JAG, Hänsel M (1992) Tool life and tool quality in bulk metal forming. CIRP Ann Manuf Technol 41(2):223–239
9. Dalbosco M, da Silva Lopes G, Schmitt PD, Pinotti L, Boing D (2021) Improving fatigue life of cold forging dies by finite element analysis: a case study. J Manuf Process 64:349–355
10. Killmann M, Merklein M (2022) Local prestressing of cold forging tools by reinforcements with adapted interference. Procedia Struct Integr 38:212–219
11. Rohrmoser A, Hagenah H, Merklein M (2021) Adapted tool design for the cold forging of gears from non-ferrous and light metals. Int J Adv Manuf Technol 113(7):1833–1848

Reduced Three-Dimensional Model for Cold Rolling Simulation of Aluminium Alloys

Florian Pachnek, Peter Simon, and Klaus Zeman

Abstract Mathematical modelling and computational analyses allow to deepen the understanding of manufacturing processes, to ensure high product quality and to reduce experimental efforts for the optimization of product properties and production processes. In order to predict the evolution of flat rolled aluminium product properties and their interdependence with process parameters, virtual representations of the entire process chain including process models of hot rolling, coiling, and cold rolling operations are highly beneficial. Starting from the solid foundation of a comprehensively validated 3D model of the hot rolling process of various aluminium alloys, the present work presents a 3D thermo-mechanical cold rolling simulation model using the commercial finite element code LS-DYNA. The material behaviour is considered by means of a user defined material model according to a modified dislocation density based flow model. The calculated microstructural and deformation history of the hot rolled strip is considered as input for the developed cold rolling model allowing for seamless prediction of product properties. For numerical studies, a modular modelling approach is pursued, comprising tailor-made simplifications depending on actual applications and research problems. For time efficient multi-pass simulations in an industrial environment, a reduced model with plane strain conditions is derived that extends the predictability of product properties and process parameters from hot rolling over coiling to cold rolling. Simulation results and computational costs are compared with detailed 3D models. Strip tensions can be adjusted showing the expected effects on roll separating forces and minimum rolled thickness. To ensure the desired strip tensions in the numerical model, a sensor-based control of strip tensions is implemented. For validation purposes, measured rolling forces from industrial pass schedules for 6xxx series alloys are compared against simulation results.

F. Pachnek (✉) · P. Simon
AMAG rolling GmbH, Lamprechtshausener Straße 61, 5282 Ranshofen, Austria
e-mail: florian.pachnek@amag.at

K. Zeman
Institute of Mechatronic Design and Production, Johannes Kepler University Linz, Altenberger Straße 69, 4040 Linz, Austria

© The Rightsholder, under exclusive licence to [Springer Nature Switzerland AG], part of Springer Nature 2024
J. Kusiak et al. (eds.), *Numerical Methods in Industrial Forming Processes*, Lecture Notes in Mechanical Engineering, https://doi.org/10.1007/978-3-031-58006-2_2

11

Keywords Multi-pass rolling of aluminium alloys · Hot and cold rolling simulation · Model verification and validation

1 Introduction

Virtual manufacturing is playing an increasingly important role in optimising manufacturing processes, product quality, and product properties. Flat rolled aluminium products undergo several production steps before reaching their final state, which includes hot rolling, coiling, cold rolling, and heat treatment processes. The development of numerical rolling models for aluminium was subject to numerous studies in the past. Olaogun et al. [1] presented a two-dimensional cold rolling model for AA8015 focussing on the heat transfer in the roll-sheet interface. Bátorfia et al. [2] compared different two-dimensional numerical approaches for identifying coefficients of friction in cold rolling of AA1050. Since two-dimensional models are not applicable in their studies, Gosh et al. [3] and Jiang et al. [4] used three-dimensional models to study edge cracking and strip flatness. However, in those studies, only single pass simulations were conducted. Simon and Falkinger [5] presented a simulation scheme for multi-pass hot rolling of aluminium including a microstructural material model and a comparison of two-dimensional and three-dimensional approaches. A similar approach was presented by Nemetz et al. [6] for multi-pass heavy plate rolling of steel. While single pass and multi-pass models are present in the literature, studies usually focus on either hot rolling or cold rolling. However, a reliable digital representation of the whole production process of flat rolled aluminium products requires modelling of the entire process chain and appropriate interfaces between different manufacturing processes. As the length of the virtual process chain representation increases, tailor-made simplifications become indispensable to answer specific questions within an acceptable effort.

The current work is based on complex numerical three-dimensional models for hot rolling of aluminium alloys that were established in the past [5]. They include elastic deformations of the work rolls and their thermal crowns, backup rolls, and stand deformations. In addition to roll separating forces, temperature, and microstructure evolution, these models provide reliable predictions of the strip profile and edge deformation. Due to the high level of detail, the computational costs of such models are immense, which makes them impractical for advanced numerical studies or multi-pass simulations in an industrial environment. Extending these models to include successive cold rolling passes increases the computational complexity even more. Therefore, the present work establishes a reduced thermo-mechanical cold rolling model using the commercial finite element code LS-DYNA to predict roll separating forces, temperatures, and microstructure evolution. A sensor-based control is introduced to manage strip tensions when the material is coiled. Once verified, this modelling approach is used to simulate a complete industrial pass schedule for an AA6016 aluminium alloy, including hot and cold rolling passes.

2 Modelling Approach

2.1 Modelling of the Cold Rolling Process

As a first step in extending the process chain to cold rolling, models predicting the roll separating forces and microstructural behaviour are beneficial. Using the finite element method, this can be done via two-dimensional models with plane strain conditions [1, 2] or three-dimensional models [3, 4] as shown in the introduction. However, for predicting the specific roll separating forces and the microstructural evolution, the actual width of the strip does not have to be considered. This allows for a reduction of the width dimension as the mentioned variables can be assumed as independent of the width and can therefore be evaluated at the half-width position of the strip. The commercial finite element code LS-DYNA used in this study utilises different solver versions (Shared Memory Parallel and Massively Parallel Processing) for two-dimensional elements and three-dimensional elements. Additionally, different keywords have to be used, e.g., for contact formulations, which can lead to inconsistent results [7]. Therefore, two-dimensional plane strain elements are not suitable for this current research.

Alternatively, a model containing three-dimensional solid elements is introduced in this work. However, it consists of only one row of elements in the width direction. The imposed boundary conditions limit the degrees of freedom of the nodes to translations only in length and thickness direction. Movements in width direction (z-direction, see Fig. 1) are suppressed, leading to plane strain conditions of the elements. The basic setup and boundary conditions are shown in Fig. 1. Both work rolls are modelled, and the symmetry is deliberately not used, since rolling conditions may occur where the strip centre in thickness direction is not horizontally aligned with the roll gap pass line.

To reach a steady state within a short period of computation time, the virtual length of the strip was set to 300 mm. Twelve elements are used in thickness direction with a fixed aspect ratio of 0.5 in length direction (see Fig. 1), to avoid large element distortions during the rolling process. The total number of elements therefore depends on the actual strip thickness and increases with thinner strips. The roll elements are larger in width than the strip elements to avoid line-on-line contact, which may lead to numerical problems.

The work rolls are treated as rigid bodies, however, heat transfer to the strip and the environment is taken into account. The elasto-plastic behaviour of the strip is implemented by a user defined material model according to a modified dislocation density based flow stress model of Kocks-Mecking type [8]. Tensile tests at usual cold rolling temperatures up to 150 °C were performed at the Fraunhofer Institute for Mechanics of Materials (IWM) and used to calibrate the model [9]. The extrapolated flow curves for 100 °C and different strain rates are shown in Fig. 2. Measured temperature dependencies of the Young's modulus, specific heat capacity, and thermal conductivity according to Table 1 are implemented, since considerable

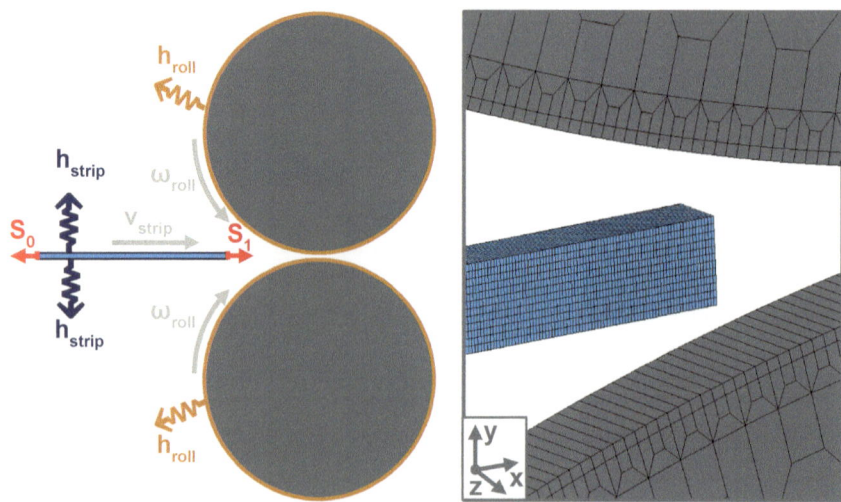

Fig. 1 Sketch of the basic simulation setup including the boundary conditions (left) and meshing for the reduced model (right): heat transfer coefficients h_{strip} and h_{roll}, translational strip velocity v_{strip}, rotational roll velocities ω_{roll} and strip tensions S_0 and S_1

temperature rises may occur during the deformation process. The thermal expansion coefficient and density are set to $2.31 \cdot 10^{-5}$ 1/K and 2700 kg/m³, respectively.

A surface-to-surface contact with heat transfer between the aluminium strip and the rolls is used, including a Coulomb friction model with a constant coefficient of friction. Based on measured roll separating forces, suitable coefficients of friction between 0.07 and 0.12 are used coinciding with earlier experimental research [10].

In addition to the restrictive boundary conditions, a prescribed motion is imposed on the strip, which depends on strip thickness, reduction, rolling speed, coefficient of friction, and roll diameter. A constant rolling speed is applied on both rolls.

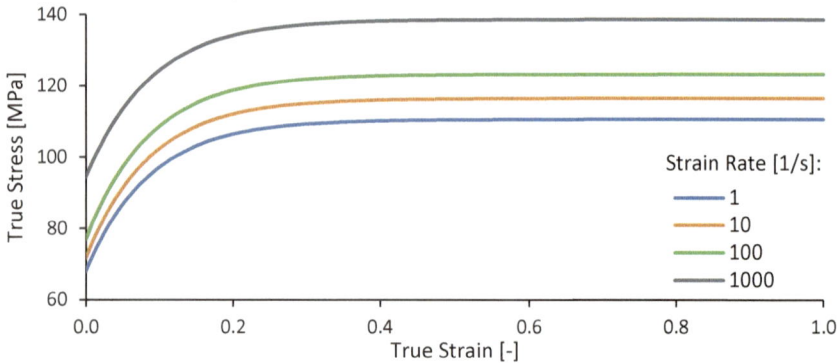

Fig. 2 Modelled flow stress curves for AA6016 at 100 °C for different strain rates [9]

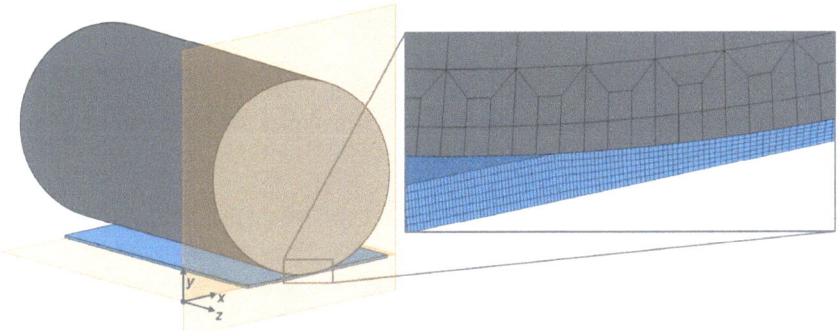

Fig. 4 Three-dimensional quarter model for the cold rolling process with x–y and x–z symmetry planes

in computational cost, a three-dimensional model considering the half-width of the strip is used as shown in Fig. 4. Only the upper work roll and the half-thickness of the strip are modelled resulting in a quarter model to keep the computation time reasonable. The identical physical boundary and contact conditions are used for both models. However, translational motion in width direction is only restricted in the mid of the strip where the x–y symmetry plane is located. Additionally, convective heat transfer is applied also at the lateral surface of the strip.

3 Model Verification and Validation

In the context of model development, model verification relates to reviews that are restricted to models exclusively, without using measurements from the model's original. Model verification thus includes checking the plausibility of its own results, i.e., comparing the results of the model against other results of the same model (model-internal verification), as well as checking the plausibility of the model's results against those of other models (model-external verification). In contrast, model validation relates to reviews that comprise the comparison of model results against measurements from suitable originals of the model. Model validation thus includes checking the validity of model results against measured data from some of the model's originals [14].

The first step is to verify the features of the presented model, namely the sensor-based tension control and the width reduction. The modelling approach is then used to simulate a complete industrial pass schedule for an AA6016 aluminium alloy including hot rolling and cold rolling.

3.1 Verification of Sensor-Based Strip Tensions

For the verification of the sensor-based strip tension control and its impact on the roll separating force, a single cold rolling pass with typical rolling parameters for an AA6016 aluminium alloy is used. The back and front tensions are set to 15 MPa and 25 MPa, respectively.

As the external strip tensions are not applied throughout the rolling process, their influence is visible on the transient roll separating force, as shown in Fig. 5. With no external loads applied, the force reaches its maximum at the start of the rolling process. After the first sensor reaches its critical value and both front and back tensions are applied, the roll separating force drops by 16% and the steady state begins. At the end of the process, the second sensor switch is activated, and the force increases again, but to a lower level than at the beginning, as the front tension remains.

The averaged stress in rolling direction over the cross section at the centre of the strip is shown in Fig. 6. The stress equals the applied back tension when the evaluated element row has not reached the roll gap yet. Due to the deformation in the roll gap, a compressive stress peak occurs for a limited time until the longitudinal stress reaches the applied front tension after the roll gap. These results verify the sensor-based strip tension approach.

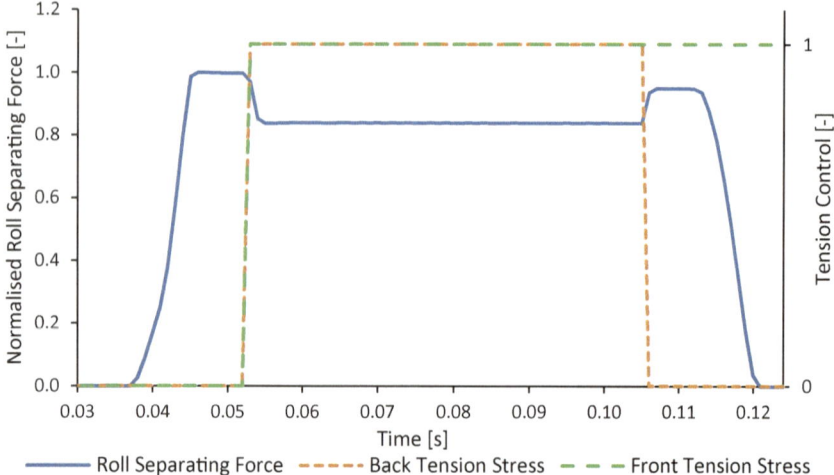

Fig. 5 Normalised roll separating force during a cold rolling pass with sensor-based strip tension control

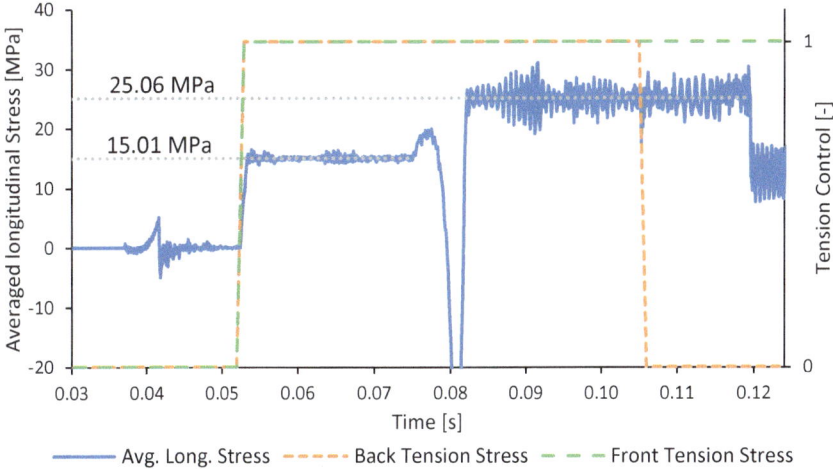

Fig. 6 Averaged longitudinal stress at the centre of the strip during a cold rolling pass with sensor-based strip tension control

3.2 Comparison of Reduced and Quarter Cold Rolling Model

To verify the effectiveness of the reduced modelling approach in predicting roll separating forces, temperatures, and microstructural behaviour, and to quantify the computational savings, three single pass simulations were performed using both the reduced model as well as the quarter model introduced in Sect. 2.3. The chosen rolling parameters are based on the actual industrial production. The initial strip thickness, starts at around 10 mm, and the pass reduction decreases from pass to pass, while the rolling speeds increase. This allows for model verification for different sets of rolling parameters used in industrial productions. Strip tensions are kept nearly constant for each pass. Identical numerical parameters and hardware settings were used to ensure comparability. Normalised roll separating forces, number of nodes, and resulting calculation times are shown in Fig. 7. The roll forces were normalised by the resulting roll separating force of the quarter model in the first pass. With a maximum deviation of 0.5% in the third pass, the roll separating forces can be considered identical in both models. As the initial strip thickness reduces from pass one to three, the total number of nodes increases drastically. The computation times of the three-dimensional quarter model exceed those of the reduced model by a factor of 8 to 10.

To evaluate the comparability of the material behaviour, Fig. 8 shows the transient temperature, strain, and dislocation density of an element at the surface and an element at the x–z symmetry plane for one exemplary pass. The temperature curves were normalised by the maximum occurring value at the surface of the quarter model. The evaluated elements are located at the centre of the strip in x-direction and lay in

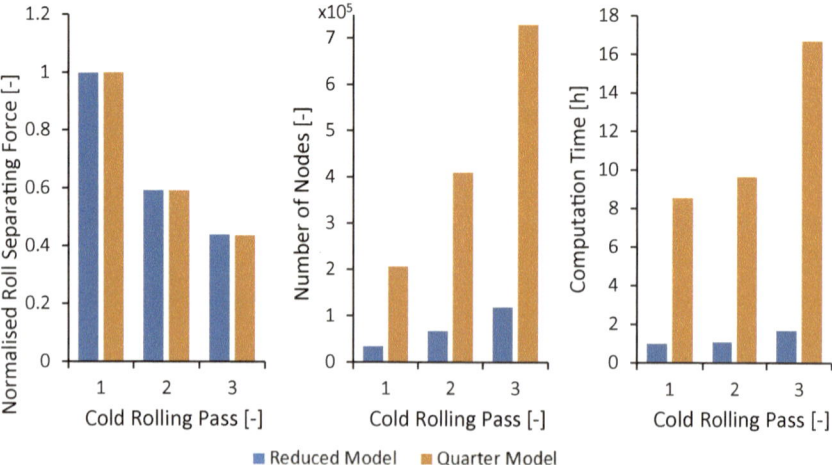

Fig. 7 Comparison of normalised roll separating forces, number of nodes, and computation time for three single pass simulations using the reduced and the three-dimensional quarter model

the x–y symmetry plane for the quarter model (see Fig. 4). All three material properties match exactly between both modelling approaches. In summary, the reduced model is suitable for the evaluation of width-independent variables, leading to the same kind and quality of information at a fraction of the computational costs of the quarter model.

3.3   Comparison with Industrial Pass Schedules

The reduced model is used to simulate an industrial pass schedule for an AA6016 aluminium alloy. The 600 mm thick cast slab is reduced to a 1 mm thin strip by 21 hot rolling passes followed by three cold rolling passes, which start at a strip thickness of around 10 mm. Strip tensions are applied during the last five passes in which the material is coiled. The coiling process itself is not modelled in this work. For the cold rolling passes, the actual initial strip temperatures for each pass vary due to different downtimes between successive passes and are implemented based on measurements. Between the first and second cold rolling pass the coiled strip is annealed to reduce the work hardening and to increase formability. As the heat treatment is not modelled, the material history is not transferred between these passes. Instead, a new mesh with initial material model variables for cold rolling is used as a simplification. For the contact zone, a speed-dependent friction coefficient in the range between 0.4 and 0.075 is used. For hot rolling, roll separating forces are almost independent on the coefficient of friction. Therefore, a nearly constant coefficient of friction is chosen there. Its value is determined from the bite condition, i.e., from the requirement that the entry of the strip into the roll gap is guaranteed for every

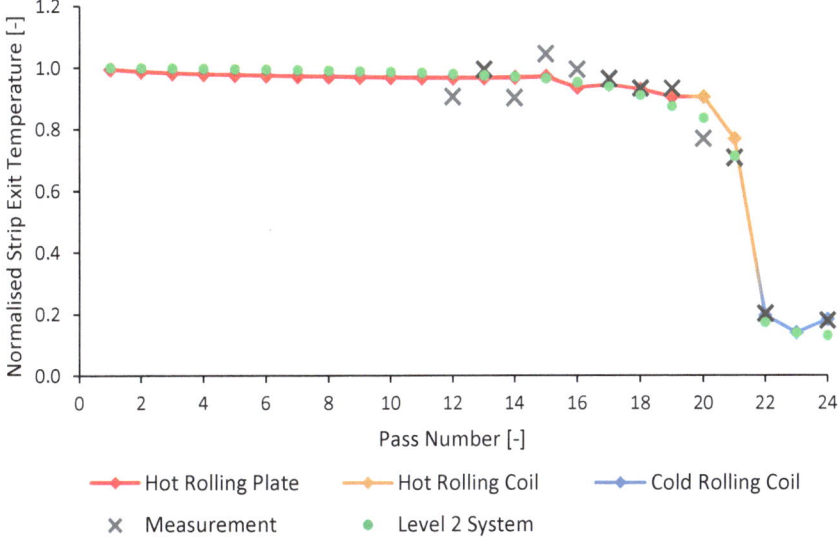

Fig. 10 Normalised strip exit temperatures along the entire process chain of hot and cold rolling for an AA6016 aluminium alloy strip

4   Conclusion and Outlook

The present work introduces a reduced three-dimensional model for rolling of aluminium alloys. Its aim is to reduce the computational effort with increasing length of the virtual process chain, by introducing tailor-made simplifications. An approach for sensor-based strip tensions for coiled material stocks is implemented and verified.

The results confirm that the reduced model concept is suitable for predicting width-independent variables like specific roll separating forces, temperatures, and the microstructural behaviour of the strip with acceptable accuracy compared to a conventional three-dimensional quarter model. Computational costs are reduced drastically compared to the quarter model considering the strip's half-width. Using the established model to simulate an entire industrial pass schedule of hot and cold rolling operations shows good agreement with measured roll separating forces.

Future work will focus on implementing more complex friction models based on rolling parameters. Additionally, the process chain will be extended by modelling the coiling and decoiling processes as well as heat treatments during or after rolling processes.

References

1. Olaogun O, Edberg J, Lindgren L-E, Oluwole L (2019) Heat transfer in cold rolling process of AA8015 alloy: a case study of 2-D FE simulation of coupled thermo-mechanical modeling. Int J Adv Manuf Technol 100:2617–2627
2. Bátorfia JG, Pál G, Chakravarty P, Sidora J (2022) Models for symmetric cold rolling of an aluminum sheet. Eng IT Solut 2(3):4–18
3. Gosh S, Li M, Gardiner D (2004) A computational and experimental study of cold rolling of aluminum alloys with edge cracking. J Manuf Sci Eng 126:74–82
4. Jiang ZY, Tieu AK, Zhang XM, Lu C, Sun WH (2003) Finite element simulation of cold rolling thin strip. J Mater Process Technol 140(1–3):542–547
5. Simon P, Falkinger G (2017) Hot rolling simulation of aluminium alloys using LS-Dyna. In: 11th European LS-DYNA conference, Salzburg
6. Nemetz AW, Parteder E, Reimer P, Kaltenbrunner T, Heise B, Lekue J, Gross T, Egger R, Zeman K (2022) Toward a 2D/3D finite-element–based digital shadow for the approximation of metrologically inaccessible dynamic fields of state variables during heavy plate rolling. Steel Res Int 94(5)
7. Livermore Software Technology Corporation (LST), an ANSYS Company: LS-DYNA Keyword User's Manual Vol I R13. Livermore (2021)
8. Kocks UF, Mecking H (2003) Physics and phenomenology of strain hardening: the FCC case. Prog Mater Sci 48(3):171–273
9. Cantergiani E, Falkinger G, Mitsche S, Theissing M, Klitschke S, Roters F (2022) Influence of strain rate sensitivity on cube texture evolution in aluminium alloys. Metall Mater Trans A 53:2832–2860
10. Tieu AK, Liu YJ (2004) Friction variation in the cold-rolling process. Tribol Int 37(2):177–183
11. Wusatowski Z (1969) Fundamentals of rolling. Pergamon Press, Oxford
12. Hallquist J (2006) Recent Developments in LS-Dyna. 5th German LS-DYNA Forum, Ulm
13. Li X, Schulte C, Abel D, Teller M, Hirt G, Lohmar J (2021) Modeling and exploiting the strip tension influence on surface imprinting. Adv Ind Manuf Eng 3
14. Vajna S, Weber C, Zeman K, Hehenberger P, Gerhard D, Wartzack S (2018) CAx für Ingenieure—Eine praxisbezogene Einführung. Springer Vieweg, Heidelberg

Numerical and Experimental Investigation of Multi-axial Forging of AA6082 Alloy

Srijan Prabhakar, D. Ravi Kumar, and S. Aravindan

Abstract Multi-axial forging is a useful technique for producing ultrafine-grained structures in bulk materials by means of severe plastic deformation. The workpiece is subjected to a specific plastic strain in the multi-axial forging process by repeatedly upsetting along all three axes by rotating the sample by 90° between the two passes; this leads to the accumulation of a large plastic strain in the material. The shape of the product does not change, as equal compressive strain is applied in all directions. Severe plastic deformation methods such as multi-axial forging can be used for producing lightweight high-strength Al alloys with ultrafine-grained structures. In this study, as-cast AA6082 has been multiaxially forged with a true strain of 0.1 in each direction, leading to a total effective strain of 0.3 in each cycle. The strain inhomogeneity from center to the surface has been predicted by a finite element simulation of the multi-axial forging with a Voce hardening model, and it has been correlated with an experimentally determined hardness variation. The peak loads in all of the passes have also been compared.

Keywords Multi-axial forging · Severe plastic deformation · AA6082 alloy

1 Introduction

Severe plastic deformation (SPD) is a technique in which a large plastic strain is imposed on a bulk metal to produce an ultrafine-grained structure. A finer grain size increases the strength and fracture toughness of the material [1, 2]. Due to the ability to produce grains with diameters of several hundred nanometers, SPD processes such as equal channel angular pressing (ECAP), constrained groove pressing (CGP), accumulative roll bonding (ARB), and high-pressure torsion (HPT) drew the interest of researchers in the 1990s. Among the SPD techniques, multi-axial forging (MAF)

S. Prabhakar (✉) · D. R. Kumar · S. Aravindan
Indian Institute of Technology Delhi, New Delhi, India
e-mail: srijan.prabhakar@mech.iitd.ac.in

© The Rightsholder, under exclusive licence to [Springer Nature Switzerland AG], part of Springer Nature 2024
J. Kusiak et al. (eds.), *Numerical Methods in Industrial Forming Processes*, Lecture Notes in Mechanical Engineering, https://doi.org/10.1007/978-3-031-58006-2_3

25

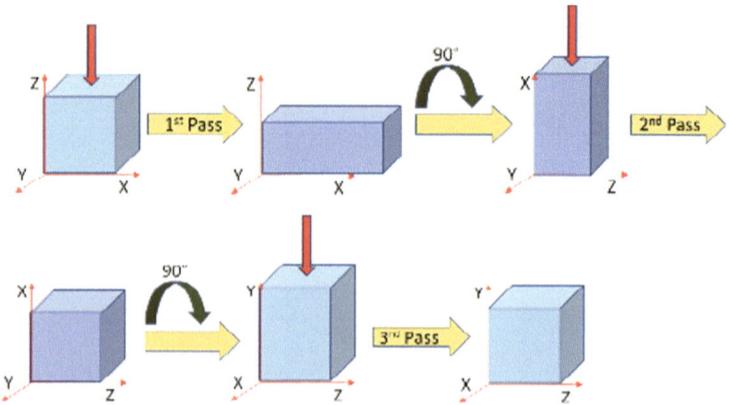

Fig. 1 Schematic diagram of multi-axial forging process

is of particular interest for its potential to be scaled up to relatively large samples [3, 4]. Figure 1 demonstrates the multi-axial forging principle.

In MAF, a workpiece is subjected to a specific strain in each pass several times on different axes using the same die; this leads to the accumulation of a large plastic strain in the material [5, 6]. The disadvantage of multi-axial forging is that the strain distribution along the specimen's cross-section is not uniform; however, this can be avoided through proper lubrication. Even though multi-axial forging produces less inhomogeneous strain than ECAP and HPT, it can be used to refine grain sizes for all metals [7, 8].

Qing Feng Zhu et al. studied the impact of forging passes on the refining of high purity aluminum during multi-axial forging [8]. The structural uniformity was focused due to deformation non-uniformity. When the forging passes reached 6, an X-shaped fine grain zone (Fig. 2) was initially formed. With further increases in the passes, this X-shaped zone started to spread the whole sample. Limitation in the structural refinement was observed with increasing strain during the multi-forging process at room temperature. The grain size in the center was refined to a certain extent as the forging passes reached 12, and there was no further grain refinement in the center after increasing the forging passes to 24. However, the sizes of the coarse grains near the surface continuously decreased with increasing forging passes [8].

Saeed Khani Moghanaki et al. [9] investigated the effect of initial heat treatment on mechanical properties and texture evolution during the multi-axial forging of an Al-Cu-Mg alloy. Under both solution-treated and over-aged conditions, the compressive stress during MAF increased up to the second pass; by further straining, the flow-stress decreased in the third pass. The deformation mode was a plane strain, and the equivalent strain per pass was approximately 0.4 (with a total strain of 1.2).

Multi-axial forging involves subjecting a material to complex loading conditions by applying sequential compression in all three principal directions. This process significantly alters the microstructure and enhances the mechanical properties of aluminum alloys, making them suitable for high-strength applications. A

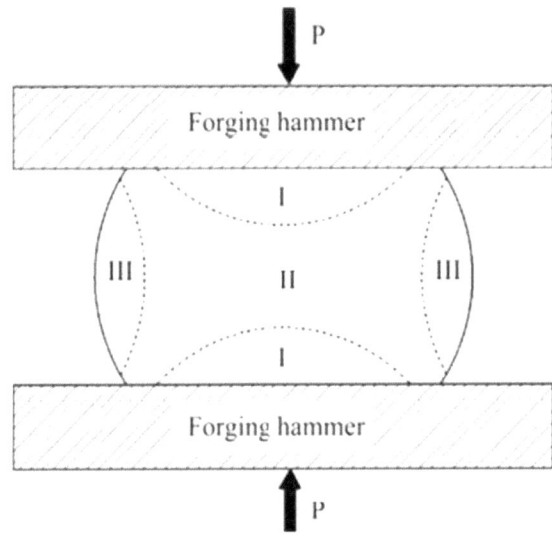

Fig. 2 Schematic diagram of three deformation zones in free forging process [8]

few researchers [10–13] have worked on microstructural changes during MAF and its effect on the mechanical properties of aluminum and its alloys; however, the published literature on the flow-stress modeling and finite element simulation of MAF on aluminum alloys is limited. Simulations that use suitable flow-stress models and their validation by using experimental results can lead to accurate predictions of required loads in MAF and the strain and stress distributions in the deformed samples. In view of this, finite element simulation of the multi-axial forging of the AA6082 alloy has been carried out in the present work. Reliable predictions from FE simulations could help engineers choose optimum choices of process parameters to reduce costly real-life experiments. A three-dimensional model of the tools and workpiece was created using CAD software. Isotropic hardening with a Voce flow-stress model has been used for numerical simulations. Also, the strain distributions that were obtained from our finite element simulations have been correlated with hardness distributions in order to validate the predicted variations through the cross-section.

Table 1 Chemical composition of AA6082 [15]

Element	Si	Mg	Mn	Cr	Fe	Cu	Zn	Ti	Al
Wt.%	0.7–1.3	0.6–1.2	0.4–1.0	0.25 max	0.5 max	0.1 max	0.2 max	0.1 max	Balance

2 Materials and Methods

2.1 Material

6XXX-series alloys have better strength and forgeability than other wrought Al alloys, and AA 6082 can provide better strength after heat treatment as compared to the other alloys of this series [14]. In this study, AA6082 has been used for multi-axial forging experiments. The nominal chemical composition of AA6082 is listed in Table 1 [15].

2.2 Experimental Work

Uniaxial Compression Tests

Uniaxial compression tests of AA6082 were carried out as per ASTM Standard E9 on an Instron 5900R/5582 machine (capacity—10 kN) at a constant cross-head speed of 1 mm/min. Specimens with an L/D ratio of 3 were generally used for determining the compressive properties of metallic materials as per ASTM E9. So, samples with a diameter of 13 mm and a height of 38 mm were prepared by wire electric discharge machining, and these samples were tested until the first load drop with the help of the load–displacement data that was obtained from the machine.

It is well-known that, during plastic deformation, a number of changes occur in the microstructure of a material that affects its properties. Work hardening (a phenomenon in which strength increases and plasticity decreases) is a consequence of increased deformation extent and cannot be ignored [16]; this must be incorporated into a flow-stress model to accurately predict the flow curve in finite element simulation.

The flow-stress curve of as-cast AA6082 that was obtained from uniaxial compression testing was compared to four flow-stress models (namely, Hollomon, Swift, Hocket-Sherby, and Voce) to obtain the best fit. The flow-stress models and the material constants that were obtained from our uniaxial compression tests are given in Table 2. The flow curves that were predicted by different models are shown in Fig. 3.

Table 2 Flow-stress models and corresponding material constants

Flow-stress model	Equation	Material constants
Hollomon	$\sigma = K\varepsilon^n$ [17]	$K = 379.95$ MPa, $n = 0.288$
Swift	$\sigma = K(\varepsilon_0 + \varepsilon)^n$ [18]	$K = 437.79$ MPa, $n = 0.368$
Hocket-Sherby	$\sigma = A - ((A - B)e^{-C\varepsilon^n})$ [19]	$A = -6.77$ MPa, $B = -2.80$ MPa, $C = -4.66$, $n = 0.073$
Voce	$\sigma = B(1 - Ce^{m\varepsilon})$ [20]	$B = 207.12$ MPa, $C = 0.672$, $m = -23.13$

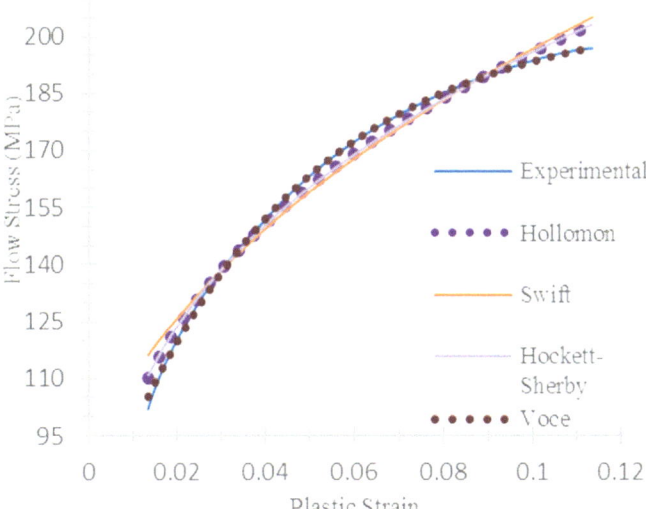

Fig. 3 Comparison of different flow-stress models with experimental data

Multi-axial Forging Experiments

The samples of the AA6082 that were produced by gravity die-casting were machined into cubes with sides of 30 ± 0.5 mm (Fig. 4a). The samples were polished on SiC emery paper with up to a 1500 grit size. The multi-axial forging was done at room temperature on a 60-ton (66-ton) hydraulic press at a constant ram speed of 0.2 mm/sec. The true strain in each pass is given by Eq. (1) [21].

$$\varepsilon_1 = \ln \frac{h_f}{h_0}, \tag{1}$$

where h_f and h_0 are the final and initial dimensions, respectively, of the sample in the direction of the compression.

$$\varepsilon_2 = \varepsilon_3 = -\frac{\varepsilon_1}{2} \text{ (considering material to be isotropic).} \tag{2}$$

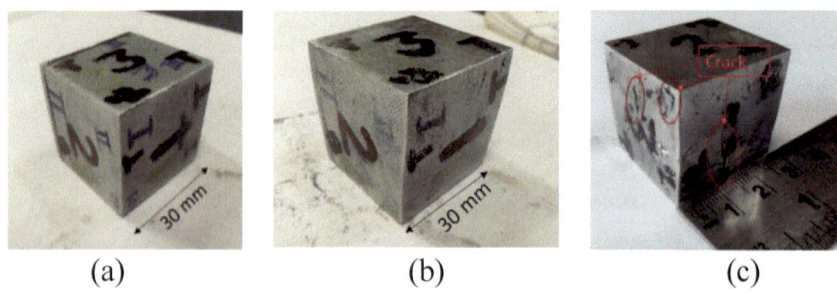

(a)   (b)   (c)

Fig. 4 **a** As-cast MAF sample; **b** MAFed sample after first cycle; **c** MAFed sample after second pass of second cycle

The effective strain after each pass is given by Eq. (3) [21].

$$\bar{\varepsilon} = \left[\frac{2}{3} \left(\varepsilon_1^2 + \varepsilon_2^2 + \varepsilon_3^2 \right) \right]^{1/2}, \tag{3}$$

where the total effective strain that was accumulated after one cycle (three passes) $= 3\,\bar{\varepsilon}$.

In the MAF experiment, a true strain of 0.095 ± 0.002 was given in each pass, so the total accumulated effective strain after one cycle (three passes) was 0.285 ± 0.002. Since there was a crack initiation in the second pass of the second cycle (Fig. 4b), the MAF study was done for one cycle only (Fig. 4b).

After three passes (Cycle 1) of MAF, the microhardness was measured at several points from the surface to the center of the specimen in the direction in which the last compression was done using a Vickers microhardness tester by applying a load of 0.5 kg for 10 s.

2.3 F.E. Simulation

The MAF process was simulated using QForm software. The 3-D model of the tools and MAF workpiece was prepared using CAD software separately and was imported into the software as a.step file. Triangular elements were used to mesh the tools and the block. The adaptive meshing technique was employed from a minimum mesh size of 1 to 3 mm with a multiplying factor of 0.5 mm. The Levanov model was used to define the friction between the tools and the workpiece with a friction factor (m) of 0.1 [22]. This is given by Eq. (4).

$$\tau = m \frac{\sigma_T}{\sqrt{3}} \left(1 - \exp\left(-1.25 \frac{\sigma_n}{\sigma_T} \right) \right), \tag{4}$$

where τ is the shear stress, σ_T is the flow-stress, and σ_n is the normal contact pressure.

In the material model, the von Mises yield function was used. Since the Voce hardening model fits well with the experimental flow curve, it was used in the simulation (as a subroutine) to define the flow curve. For the MAF process, the lower die was fixed, and the upper die was given the required displacement to deform the workpiece to a predetermined strain. From the simulations, the strain distribution was analyzed in the workpiece after each pass.

3 Results and Discussion

3.1 Strain Distribution

Figures 5a and d show that the strain distribution along the horizontal and diagonal axes was more homogeneous after the first pass as compared to the vertical direction due to the friction between the toll and the workpiece. In Fig. 7, it can also be seen that the inhomogeneity factor was higher in the vertical direction as compared to the other two directions. After Pass 2, there was the formation of an X-shaped zone (Fig. 5b), and the strain accumulation in this zone was greater as compared to the remaining region. This zone continued to expand in the third pass (Fig. 5c). This showed that the strain accumulated near the center as the number of passes increased. As the strain accumulated in the material, the dislocation became more piled-up to form shear bands that, in turn, formed sub-grains in the material and refined the grain sizes [23, 24]. The strain distribution after each pass is shown in Fig. 6 for comparisons among the passes.

There was a non-uniformity of the strain distribution due to the friction between the tools and the workpiece. So, an inhomogeneity factor (IF) was used to quantify the non-uniformity in the material [25]; this is defined in Eq. (5):

$$\text{IF} := \frac{\sqrt{\sum_{i=1}^{n} (S_i - S_{av})^2 / (n - 1)}}{S_{av}} \times 100, \tag{5}$$

where S_i is the value at the i-th point, S_{av} is the average value, and n is the number of measurements in a sample.

Since the IF in the horizontal direction was mainly influenced by friction, it was almost at its maximum in all passes (Fig. 7). The IF in the horizontal direction was the lowest after the first pass, but it increased after the second pass (because of the pre-strain from the first pass) and then decreased (due to the formation of the X-shaped zone).

The microhardness values and the plastic strain simulation results were correlated after the third pass along the last compression axis from the surface to the center (Fig. 8). There was an increasing trend in both cases due to the strain hardening of

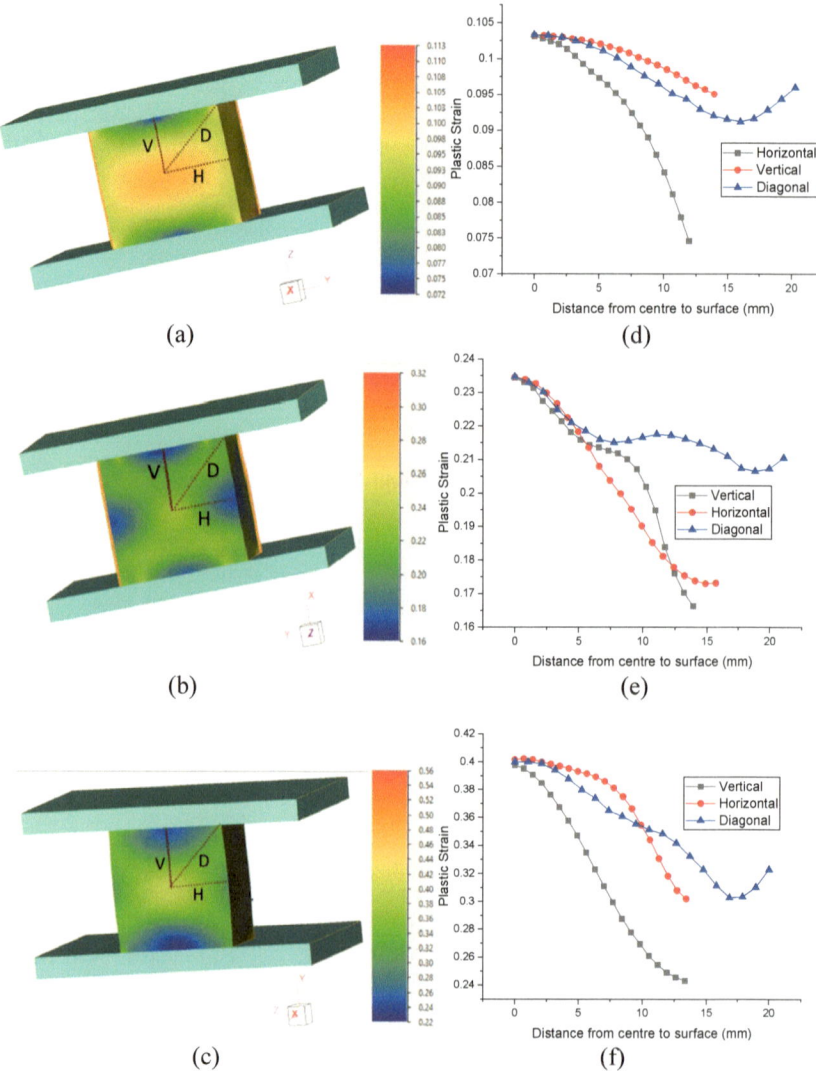

Fig. 5 Plastic strain distribution at mid-plane after **a** Pass 1, **b** Pass 2, and **c** Pass 3 and along the line from center to surface in vertical, horizontal, and diagonal directions after **d** Pass 1, **e** Pass 2, and **f** Pass 3

the material, as the strain accumulation near the center was greater as compared to the top surface.

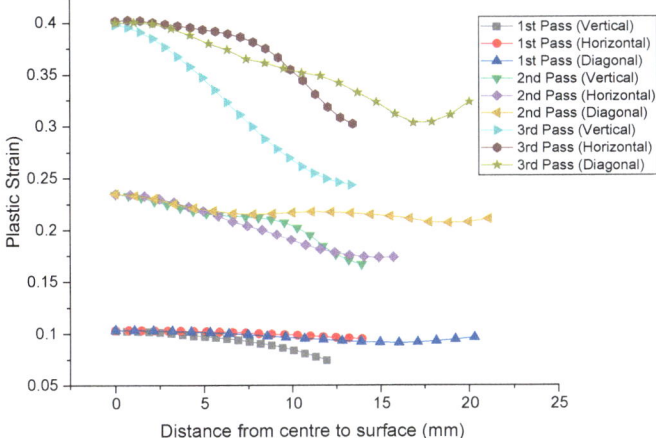

Fig. 6 Plastic strain distribution at mid-plane along the line from center to surface in vertical, horizontal, and diagonal directions after each pass

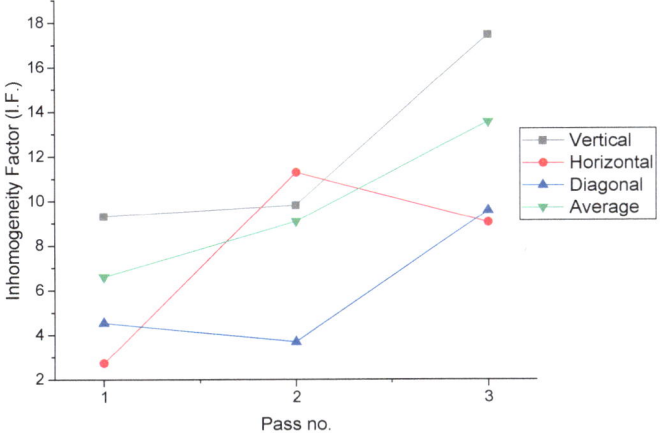

Fig. 7 Inhomogeneity factor after each pass along line from top to center in vertical, horizontal, and diagonal directions

3.2 Variation of Load in MAF

The load–displacement curve for the different passes of MAF was experimentally obtained from the N.I. data-acquisition system that was attached to the load cell and rotary displacement encoder in the hydraulic press. Figure 9b shows the jump in the required load in the initial portion of the load–displacement curve during the second pass due to the generated dislocation after the first pass. Also, the slope of the curve (after the initial load) decreased in the second and third passes; due to this,

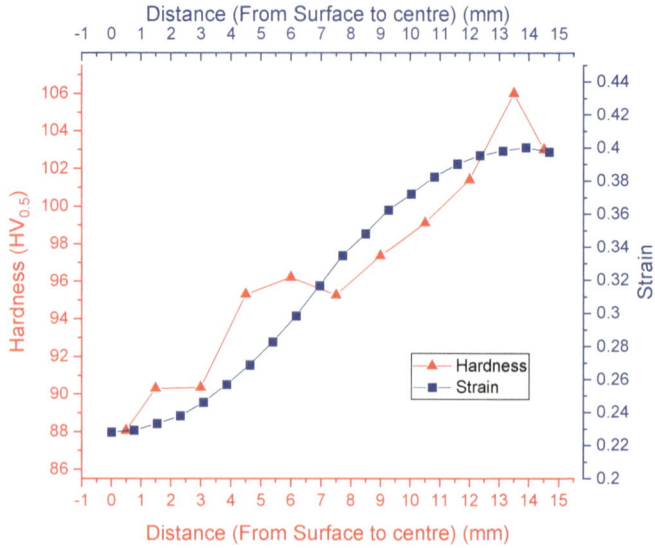

Fig. 8 Strain and hardness distribution along the line from surface to center in compression direction after Pass 3

the variation in the peak load in these two passes was lower. The load–displacement curve that was obtained from the QForm simulation is also presented in Fig. 9a.

In the experiments, the load that was predicted for the first pass was slightly greater than the experimental values, but the second and third passes correlated well with the experimental values (Table 3). The variation in the predicted and experimental loads was due to the isotropic hardening and structural change inside the material (such as dislocation interaction, texture evolution, and the possibility of sub-grain formation).

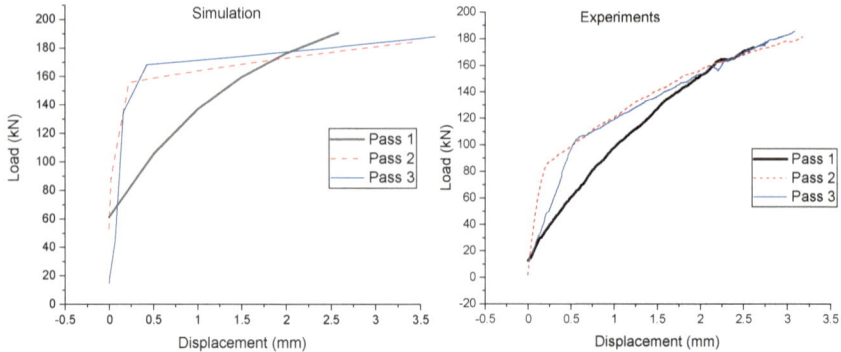

Fig. 9 Load–displacement curves in different passes of MAF that were obtained from **a** simulation and **b** experiments

Table 3 Peak load after each pass of MAF that was obtained from simulation and experiment

Condition	Peak load (kN)	
	Simulation	Experiment
After first pass	190.7	175.5
After second pass	184.3	181.1
After third pass	188.1	185.5

4 Conclusions

To study the behavior of AA6082 during multi-axial forging, a numerical simulation was successfully performed using QForm software. It was observed that there was a formation of an X-shaped zone after the second pass and that this zone expanded after the third pass. Although the accumulated strain was greater in this zone, there was more strain homogeneity here than could be found in the rest of the region. As there was a strain accumulation near the center of the workpiece, an increasing trend of the hardness and strain distributions could be observed after the third pass along the last compression axis from the surface to the center due to strain hardening.

The material models that were used for the FE simulation were the von Mises yield function and the Voce hardening law. Since the material model was defined for monotonic loading when a real experiment is cyclic in nature, the maximum predicted load by the model was slightly greater when compared to the experimental load values. Overall, this method can be an excellent resource for learning more about multi-axial forging; however, the material models must be re-examined in order to do this if we are to obtain a suitable response for the multidirectional load.

Acknowledgements Srijan Prabhakar acknowledges the assistantship that he received from the Ministry of Education, Government of India, during his research work.

References

1. Valiev RZ, Alexandrov IV (2002) Nanostructured materials from severe plastic deformation. Nanostructured Mater 12:35–40. https://doi.org/10.1016/s0965-9773(99)00061-6
2. Valiev RZ, Islamgaliev RK, Kuzmina NF, Li Y, Langdon TG (1998) Strengthening and grain refinement in an Al-6061 metal matrix composite through intense plastic straining. Scr Mater 40:117–122. https://doi.org/10.1016/S1359-6462(98)00398-4
3. Tang L, Liu C, Chen Z, Ji D, Xiao H (2013) Microstructures and tensile properties of Mg-Gd-Y-Zr alloy during multidirectional forging at 773K. Mater Des 50:587–596. https://doi.org/10.1016/j.matdes.2013.03.054
4. Kumar S, Venkatachalam S, Krishnaswamy H, Digavalli RK, Murthy HSN (2019) Influence of inhomogeneous deformation on tensile behavior of sheets processed through constrained groove pressing. ASME J Eng Mater Technol 141:1–10. https://doi.org/10.1115/1.4043492
5. Nie KB, Deng KK, Wang XJ, Gan WM, Xu FJ, Wu K, Zheng MY (2015) Microstructures and mechanical properties of SiCp/AZ91 magnesium matrix nanocomposites processed by

multidirectional forging. J Alloys Compd 622:1018–1026. https://doi.org/10.1016/j.jallcom. 2014.11.045

6. Miura H, Kobayashi M, Aoba T, Aoyama H, Benjanarasuth T (2018) An approach for room-temperature multi-directional forging of pure titanium for strengthening. Mater Sci Eng A 731:603–608. https://doi.org/10.1016/j.msea.2018.06.060

7. Liu WC, Chen MB, Yuan H (2011) Evolution of microstructures in severely deformed AA 3104 aluminum alloy by multiple constrained compression. Mater Sci Eng A 528:5405–5410. https://doi.org/10.1016/j.msea.2011.03.100

8. Zhu QF, Li L, Ban CY, Zhao ZH, Zuo YB, Cui JZ (2014) Structure uniformity and limits of grain refinement of high purity aluminum during multi-directional forging process at room temperature. Trans Nonferrous Met Soc China (English Ed.) 24:1301–1306. https://doi.org/ 10.1016/S1003-6326(14)63192-7

9. Khani Moghanaki S, Kazeminezhad M, Logé R (2018) Mechanical behavior and texture development of over-aged and solution treated Al-Cu-Mg alloy during multi-directional forging. Mater Charact 135:221–227. https://doi.org/10.1016/j.matchar.2017.11.048

10. Kumar N (2021) An exploration of microstructural in-homogeneity in the 6082 Al alloy processed through room temperature multi-axial forging. Mater Charact 176. https://doi.org/ 10.1016/j.matchar.2021.111134

11. Kumar N, Jayaganthan R, Owolabi GM (2022) Grain refinement mechanism in 6082 Al alloy fabricated by cryo-multiaxial forging. Mater Sci Eng A 833. https://doi.org/10.1016/j.msea. 2021.142518

12. Moghaddam M, Zarei-Hanzaki A, Pishbin MH, Shafieizad AH, Oliveira VB (2016) Characterization of the microstructure, texture and mechanical properties of 7075 aluminum alloy in early stage of severe plastic deformation. Mater Charact 119:137–147. https://doi.org/10.1016/ j.matchar.2016.07.026

13. Aoba T, Kobayashi M, Miura H (2017) Effects of aging on mechanical properties and microstructure of multi-directionally forged 7075 aluminum alloy. Mater Sci Eng A 700:220–225. https://doi.org/10.1016/j.msea.2017.06.017

14. Davis JR (1993) ASM handbook aluminum and aluminum alloys. ASM International, United States of America

15. Aalco—Ferrous and Non-Ferrous Metals Stockist: Aluminium Alloys—Aluminium 6082 Properties, Fabrication and Applications. https://www.azom.com/article.aspx?ArticleID= 2813. Accessed 10 May 2019

16. Aoba T, Kobayashi M, Miura H (2018) Microstructural evolution and enhancement of mechanical properties by multi-directional forging and aging of 6000 series aluminum alloy. J Japan Inst Light Met 67:277–283. https://doi.org/10.2320/matertrans.L-M2017856

17. Hollomon JH, Member J (1945) Tensile deformation. Met Technol 12:268–90

18. Swift HW (1952) Plastic instability under plane stress. J Mech Phys Solids 1:1–18. https://doi. org/10.1016/0022-5096(52)90002-1

19. Hockett JE, Sherby OD (1975) Large strain deformation of polycrystalline metals at low homologous temperatures. J Mech Phys Solids 23:87–98. https://doi.org/10.1016/0022-5096(75)900 18-6

20. Voce E (1948) The relationship between stress and strain for homogeneous deformation. J Instit Met 74:537–562

21. Dieter GE (1988) Mechanical metallurgy. In: Mechanical metallurgy, pp 69–99

22. Güner F, Cora ÖN, Sofuoğlu H (2018) Effects of friction models on the compaction behavior of copper powder. Tribol Int 122:125–132. https://doi.org/10.1016/j.triboint.2018.02.022

23. Bereczki P, Szombathelyi V, Krallics G (2014) Production of ultrafine grained aluminum by cyclic severe plastic deformation at ambient temperature. In: IOP conference series: materials science and engineering. Institute of Physics Publishing. https://doi.org/10.1088/1757-899X/ 63/1/012140

24. Kumar N, Owolabi GM, Jayaganthan R (2019) Al 6082 alloy strengthening through low strain multi-axial forging. Mater Charact 155. https://doi.org/10.1016/j.matchar.2019.06.003

25. Kumar S, Krishnaswamy H, Digavalli RK, Paul SK (2019) Accounting Bauschinger effect in the numerical simulation of constrained groove pressing process. J Manuf Process 38:49–62. https://doi.org/10.1016/j.jmapro.2018.12.013

Evaluation of Forming Factors for Titanium Gr. 2 Alloy Sheets with Multistage SPIF Process

Amrut Mulay and Vadher Sameer

Abstract In the vast prospects for automation, incremental sheet forming (ISF) is a reliable sector on which the industry may focus in the future. Due to spring back, poor surface finish, and production time, ISF has yet to apply in broad prospects in mainstream high-value manufacturing sectors. ISF performance is affected by process factors such as vertical step depth, feed rate, spindle speed, forming angle, tool path, intermetallic friction, and bending forces. One of the variants of ISF processes is Single Point Incremental Forming (SPIF). The formability of the SPIF process increases by incorporating certain intermediary phases known as Multistage Incremental Sheet Forming (MSPIF). The strain hardening effect that emerged in the distorted sheet alters its deformability. The integration of intermediate results into some new process parameters to address in this work includes the number of stages, angle interval between the stages, and tool diameter. The present research focuses on the influence of process parameters on section thickness, equivalent plastic strain, and spring back. A simulation study of various process parameters of multistage forming and their effects on multistage incremental forming performance were evaluated.

Keywords Titanium alloy · MSPIF · Thickness distribution · PEEQ · Multistage forming · MoS_2

1 Introduction

Conventional sheet metal forming processes require large batch sizes because these processes require large energy costs and very high investment in equipment and tooling. Single Point Incremental Forming (SPIF) is a relatively new metal forming process with a high potential economic payoff for rapid prototyping applications suitable for flexible and small quantity production, fulfilling this gap in metal

A. Mulay (✉) · V. Sameer
Department of Mechanical Engineering, Sardar Vallabhbhai National Institute of Technology, Surat 395007, India
e-mail: mas@med.svnit.ac.in

© The Rightsholder, under exclusive licence to [Springer Nature Switzerland AG], part of Springer Nature 2024
J. Kusiak et al. (eds.), *Numerical Methods in Industrial Forming Processes*, Lecture Notes in Mechanical Engineering, https://doi.org/10.1007/978-3-031-58006-2_4

forming processes. The processes differ as SPIF, two-point incremental forming (TPIF), double-sided incremental forming (DSIF), and hybrid incremental forming procedures based on the forming method. SPIF, also known as negative incremental forming, deforms the sheet with a single tool without a die. In contrast, the TPIF method uses a partial or full die to regulate the deformation of the blank. The SPIF process has several key limitations including geometric inaccuracy, sheet thinning, and forming constraints such as wall angle, step depth, tool diameter, temperature, etc. To increase geometric accuracy in understanding the many causes of faults in the SPIF process. The SPIF method may suffer from geometric inaccuracies due to spring back, unwanted plastic deformations, and non-uniform thickness distribution. Single-stage and multistage forming processes use forming will occur in a single-stage moment of a tool. Multistage point incremental forming (MSPIF) is a viable alternative than single-stage SPIF. As it reduces sheet thinning and results in more uniform thickness distributions. Geometrical imperfection, also known as stepped features, formed during the formation phase in the MSPIF process due to intermediary steps. With modeling and experimental data, studies of the development of stepped features at the formed parts are explained by Nirala et al. [1]. Material characterization, forming-limit curve, fracture forming-limit curve (FFLC) determination, numerical modeling, and experimentation, especially strain path and fracture strain evaluation, are multistage processes. The comparison of numerical modeling and experimental shows that the multistage forming sequence significantly impacts the strain sites' position in the major strain space. Strain routes are linear in the initial stage and significantly non-linear in the later phases of recognized forms by Skjoedt et al. [2]. A new specimen geometry offers to reduce the great amount of experimental work to determine the forming limits, concluding that formability is significantly higher in incremental forming than in conventional sheet forming. The process is very flexible and economical due to the low tool costs discussed by Tisza [3]. As Hirt and Bambach [4] depicted in Two Point Incremental Forming (TPIF), also termed positive incremental forming, the forces act on two points. Firstly, the interaction point between the tool and the sheet, and secondly, at the blank–die interface. Although the setup in TPIF is more complex than in SPIF, greater accuracy is obtained by reducing the spring back due to the presence of a partial or full die. Peng et al. [5] explains the Double-sided incremental sheet forming, two tools are present, i.e., one above the clamped sheet and the other below the sheet to support the blank and provide backup forces for better and controlled movement. DSIF provides extra support to the blank like TPIF but without the involvement of any die. Maqbool and Bambach [6] prove the beneficial influence of the dominating deformation modes on geometrical precision leftover moments by numerical simulations. A decrease in energy dissipation inside the bending deformation mode results in fewer residual moments by increasing pitch and lowering tool diameter. The diminished contribution of the bending deformation mode and lower residual moments, rising pitch, and reduced tool diameter increase geometrical precision. Increased geometrical precision results in a reduced contribution of the bending deformation mode and smaller residual moments. Carette et al. [7] refers to the inaccuracy that may lead to the difference in the formed part from that of the ideal part, such as obtaining fillet instead of a sharp corner, inappropriate height, non-uniform thickness throughout the part, obtaining a smaller size than the actual requirements, etc. Many ways suggest improving geometrical accuracy. They were studied by altering physical process parameters like wall angle, step size, speed rate, feed rate, and lubrication or by varying tool paths. Najafabady and Ghaei [8] explain that Ti-6Al-4V sheets deform into three elementary shapes cone-shaped solid, variable wall angle cone-shaped solid, and pointed solid. It also determines the impact of various method parameters on dimensional accuracy, surface quality, and work hardening. The results discovered that the roughness of the inner surface wherever the tool meets the workpiece is larger than the roughness of the outer surface of a workpiece. The micro-hardness check findings conjointly discovered that the hardness varies from the flange to the vertex of the workpiece. SPIF process indicates additional plastic deformation, and therefore work hardening is observed. Thyssen et al. [9] focus on robot-based incremental sheet metal forming. Experimentally identifies the effect of various parameters. The test with decided parameters demonstrated a functional link between process parameters such as forming velocity, the temperature of the forming zone, the step depth, and the wall angle can be defined as the largest actuating variables of the forming process and measured part deviation. The robot controller has designed an online compensating method based on this functional relationship. The compensatory technique reveals a favorable trend in the dimensional accuracy obtained. Ren et al. [10] discuss a constraint agglomerative hierarchical clustering algorithm approach, divided into three parts: determining critical control points to address geometric complexity, simplified simulation models for anticipating spring back offline, and in-situ tool path change during forming. Experiment results demonstrate that the approach delivers an efficient and resilient solution for many geometries at a low setup cost. Akrichi et al. [11] investigated that deep learning as a valuable tool for geometric accuracy prediction in single point incremental forming. Otherwise, deep learning outperforms shallow learning in performance prediction. Furthermore, compared to the stack, the deep belief network model outperforms it in predicting roundness and position deviation. Raju and Sathiya [12] use a hybrid optimization technique combining Taguchi grey relational analysis (TGRA) and response surface methodology (RSM). Combining TGRA and RSM determines an optimal combination of input process parameters, such as the number of sheets, the tool diameter, the feed rate, the spindle speed, and the vertical step depth. The optimization yielded promising findings confirmed by a confirmatory experiment, suggesting that the process may improve and that the technique is reliable. Mohanty et al. [13] explain that the sheet thickness decreases as the wall angle increases and theoretically becomes zero at the maximum wall angle. MSPIF can be used to obtain a large wall angle by redistributing the material. Several types of research have attempted to find an optimal number of stages to achieve the maximum possible wall angles with satisfactory geometric accuracy. In addition, the actual tool path also significantly influences the successful forming of sheet metal with maximum achievable wall angles. Gonzalez et al. [14] discuss the geometries of truncated cones created using two multistage approaches and compared them to the same geometry created using a single-stage method. Digital image correlation assesses each geometric accuracy and thickness distribution. The results show that multistage forming considerably impacts the geometric correctness of the treated sheets compared to single-stage forming. Kumar et al. [15] considered the tool path is a contour along which the forming tool moves to achieve the object's targeted shape. The tool moves layer by layer from one contour to subsequent contour. The toolpath is a critical parameter in the ISF process since it specifies the component's dimensional accuracy. The right selection of the tool path improves the formability, surface finish, and dimensional accuracy, as well as the processing time of the resulting part. Blaga and Oleksik [16] aim to find the best forming strategy by adjusting the press position of the punch and the path it takes to form a truncated cone using single point incremental forming. As selected, three different paths to create a truncated cone for the cone. After each vertical press, the punch covers a circular path with a constant step in the first and second variations. The differences show that the following circular trajectory can begin at the same point as the previous press point or shift at an angle of 90 degrees from the previous press point. The punch in the final variation follows a spatial spiral trajectory. Thibaud [17] developed a single point incremental sheet forming process and a numerical toolbox to carry out fully parametric simulations using finite element software.

A thorough literature analysis reveals several limitations in the single-stage SPIF process in terms of forming ability, thickness distribution. The product can be prepared in better way with the help of multistage forming technique. It is necessary to understand the many reasons for such flaws in the SPIF process and to enhance geometric precision. The geometrical inaccuracy, undesirable plastic deformations, and non-uniform thickness distribution can cause geometric inaccuracies in the process. Multistage forming is an alternative to the limitations of single-stage SPIF since it decreases sheet thinning and produces a more uniform thickness distribution. The main goal is to use FEA simulation to investigate the effect of various multistage forming process parameters such as the number of stages (n), angle interval between successive stages (Da), and tool diameter (D) on section thickness (STH), Equivalent Plastic Strain (PEEQ), and Spring back effect.

2 Methodology

The research activity begins with selecting materials and investigating various process parameters, the first phase of the research development. Figure 1 depicts a project plan. As described above in the introduction chapter in the application section, titanium material has many uses in the biomedical industry. Because of its excellent strength-to-weight ratio, titanium has a wide range of services in the aerospace and marine sectors. Due to its superior corrosion resistance to aluminum and steel is employed in orthopedic and dental applications like prosthetic limbs or knees. Furthermore, titanium has good biocompatibility and corrosion resistance because the oxidation of the outer layer in the presence of oxygen produces a highly stable oxide known as Titanium dioxide (TiO_2). The oxidized titanium layer protects

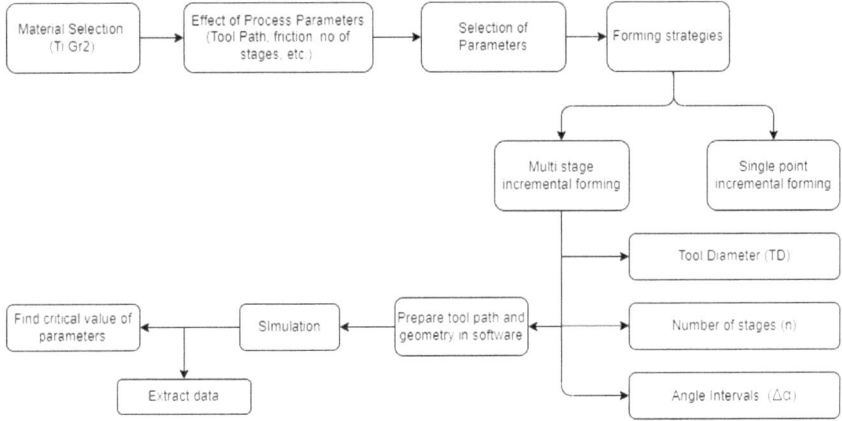

Fig. 1 Plan for execution

the surface or structure of implants against environmental conditions such as fatigue, stress, etc. Titanium's complete integration with the human body and its durable and flexible nature make it an excellent biocompatible metal referred by Kumar et al. [18]. Various researchers have recently studied process factors such as step depth, tool path, spindle speed, etc. Furthermore, creating the sheet in several phases introduces other characteristics and the ones listed above. Thus, it is necessary to investigate these parameters. Hence, the parameters chosen for the current research activity are the number of intermediate stages, angle intervals between successive stages, and tool diameter.

The Abaqus explicit program creates a square-shaped blank with 222 × 222 * 0.5 mm dimension. The blank creates the 2-D deformable model. The tool was modeled as a rigid analytical body with a hemispherical end because the diameter of the tool was one of the criteria. The diameter tools for each experiment were 8mm, 10mm, and 12mm. Furthermore, the size and form of the blank remained consistent. The tool and blank geometry, as shown in Fig. 2.

Material characteristics of Ti Gr 2 (commercially pure titanium) by Yoganjaneyulu and Sathiys [19] were acquired from numerous research articles and entered into the Abaqus explicit software's property manager. Table 1 displays the mechanical qualities. A scaling factor was also applied to shorten the simulation duration without affecting the results.

A spiral tool path with a down technique runs for the numerical study to shape the sheet into a truncated cone shape. The step depth is usually between 0.5 and 1 mm. Furthermore, the shorter the step depth, the greater the precision, and the better the outcomes explained by Sornsuwit et al. [20]. As a result, MATLAB software by Yeshiwas [21] operates to build the tool path with 0.5 mm step depth. The total time required to produce the sheet framed using a feed rate of 40 mm/sec. A total of 18001 steps operate to obtain the desired results.

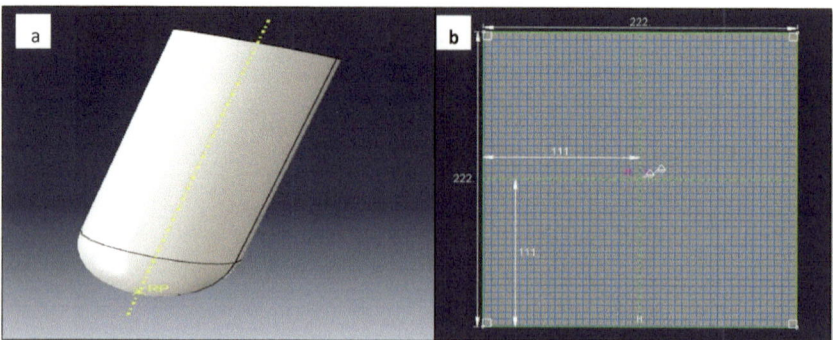

Fig. 2 a Geometry of tool, **b** Dimensions for sheet

Table 1 Mechanical properties of titanium grade 2 [19]

Sr no.	Material	Yield strength (MPa)	Tensile strength (MPa)	Strain hardening exponent (n value)	Strength coefficient (MPa) (K value)
1	Titanium grade 2	284	420	0.17	495

Furthermore, the distance traveled in each step time was computed using the equation. Again, the total time necessary for each step was determined using total length and feed rate. The total length (L), feed rate (Fr), and total duration (T) are necessary for the forming process calculated using the total length to feed rate ratio. Also, step time is the ratio of total time to complete steps. Interaction attributes such as touch and friction assign the sheet tool interfaces. The coulomb's friction law establishes the sheet blank and the forming tool with a lubricant with a friction coefficient of 0.045. The sheet has meshed with linear quadrilateral components of the S4R type with a mesh size of 0.8 mm. Sketch the interaction view between the blank and the tool. Later in a multistage procedure, importing the distorted sheet from the previous step will further deform it. The results of the previously produced sheet were added to the load manager using the preset field option. The results achieved in the previous stage were considered the beginning condition for the following stage. As a result, the sheet that deforms in the following step already possesses the circumstances achieved in the previous stage.

A total of sixteen experimental tests were required to simulate using Abaqus explicit. Table 2 shows the number of experiments.

Table 2 Design matrix for the experiments

Sr No.	Number of stages (n)	Angle intervals (Da)	Tool diameter (D)
1	2	5	10
2	4	5	10
3	2	15	10
4	2	10	8
5	4	10	8
6	2	10	12
7	4	10	12
8	3	5	8
9	3	15	8
10	3	5	12
11	3	15	12
12	3	10	10
13	3	10	10
14	3	10	10
15	3	10	10
16	3	10	10

3 Result and Discussions

On commercially pure Titanium (Ti Gr 2) of $222 \times 222 \times 0.5$ mm dimension, check the influence of simulation analysis on forming parameters (number of steps, angle intervals, and tool diameter) on the sheet thickness, spring back, and equivalent plastic strain. The forming process utilizes a hemispherical tool. MATLAB creates a spiral tool path to shape the sheet into a cup shape. Abaqus creates a 2-D deformable model of Ti Gr 2 sheet. The tool recognizes a rigid analytical body with a diameter of 10 mm. The sheet was held together by four edges and meshed with linear quadrilateral elements of the S4R type with a mesh size of 0.8mm. Furthermore, the coulomb's friction law applies between the sheet blank and the forming tool, with a lubricant with a friction coefficient of 0.045.

Figures 3a, b, and c show the simulation results of a cup made at 45° with a 10 mm tool diameter from a 0.5 mm thick Ti Gr 2 sheet When compared to two-stage, three-stage, and four-stage forming, the sheet created in three stages has a more consistent sheet thickness compared to two other. Furthermore, progressive and homogeneous deformation reflects the rise in intermediate phases.

According to Table 3, the highest sheet thickness of 0.300949 mm after deformation reflects in experiment 10. The sheet deforms in three stages with an angle interval of 5° between each stage and a tool diameter of 12 mm. When deforming with the same angle interval is impossible, the sheet can operate using a variety of angle intervals. A similar simulation for the experiment in which deforming in the

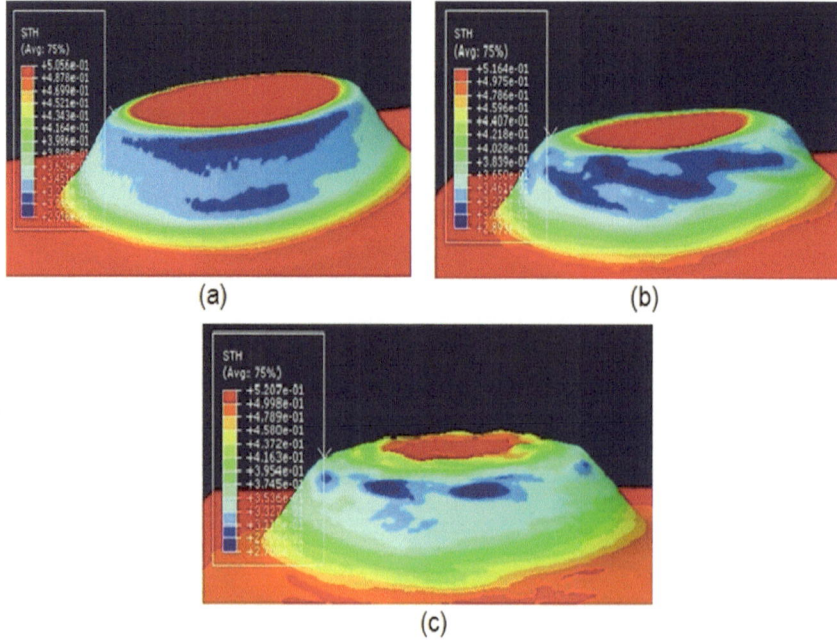

Fig. 3 Simulation for **a** 2 Stage incremental forming. **b** 3 Stage incremental forming. **c** 4 Stage incremental forming

same angle interval was impossible. The first stage had a 5° interval for such studies, and the final stage to complete with a predetermined angle interval. However, in experiment 1(2 Stage, Da = 5°, and TD = 10mm), the greatest equivalent plastic strain of 2.22942 and a minor inaccuracy at the bottom corner of the produced cup were attained. Furthermore, the influence of various responses (Section thickness, Equivalent plastic strain (e), and spring back effect) is depicted in the following subsections using various graphs and tables.

3.1 Section Thickness Distribution of Multistage Incremental Forming

A detailed analysis was complete by Li et al. [22] to study the influence of the number of forming stages and the angle interval between the consecutive stages on section thickness (STH). The MSPIF process uses the DC06 sheet as material and finds that increasing the number of stages resulted in a more uniform thickness distribution. However, as the number of steps increases, the minimum sheet thickness does. Furthermore, when the angle interval (Da) grows, the thickness initially increases but decreases. The maximum uniform sheet thickness is achievable with a 10° angle

Table 3 Responses data for various parameters

Parameters				Responses		
Sr No.	Number of stages (n)	Angle intervals (Da)	Tool diameter (D)	Section thickness (STH)	Equivalent plastic strain (PEEQ)	Spring back
1	2	5	10	0.294	1.740	0.976
2	4	5	10	0.297	1.938	2.839
3	2	15	10	0.298	1.697	1.621
4	2	10	8	0.293	1.695	1.528
5	4	10	8	0.283	2.295	3.309
6	2	10	12	0.285	1.745	1.379
7	4	10	12	0.285	2.071	2.648
8	3	5	8	0.262	2.229	1.653
9	3	15	8	0.285	1.997	2.487
10	3	5	12	0.300	1.755	1.467
11	3	15	12	0.292	1.871	1.678
12	3	10	10	0.294	1.821	2.256
13	3	10	10	0.294	1.821	2.256
14	3	10	10	0.294	1.821	2.256
15	3	10	10	0.294	1.821	2.256
16	3	10	10	0.294	1.821	2.256

separation between successive steps. The minimum section thickness was 0.262133 mm for experiment 8 (3 stages, Da = 5°, and TD = 8 mm), whereas the highest thickness was 0.300949 mm for experiment 10 (3 stages, Da = 5°, and TD = 12 mm). Figure 4a depicts the influence of angle interval and tool diameter on section thickness in a two-stage forming process. However, experiment 1 (2 stage, Da = 5°, and TD = 10mm) exhibits more consistent wall thickness, but it also shows early thinning of the sheet compared to the other three experiments. Experiment 3 (2 stages, Da = 15°, and TD = 10mm) reveals an immediate increase and decrease in thickness at the bottom. The increase and decrease could be due to the previously formed part and large angle interval. Performing repetitions for experiments 4, (2 stages, Da = 10°, and TD = 8mm) and 6, (2 stages, Da = 10°, and TD = 12mm) can be found with the same angle interval but different tool diameter. However, the change was smooth. As can be seen, the influence of tool diameter was minor or non-existent compared to the effect of angle interval. Figure 4b shows the outcome of three stages of shaping and adjusting the angle interval and tool diameter. As a result of an experiment with a Da of 5°, the sheet's final thickness is more uniform and maximum. However, as the Da between subsequent phases rises, the homogeneity declines.

Furthermore, among 5° Da trials, a sheet created with an 8mm tool diameter had a somewhat better thickness distribution than a sheet formed with a 12mm tool diameter. When the Da grows, so does the unevenness of the thickness distribution.

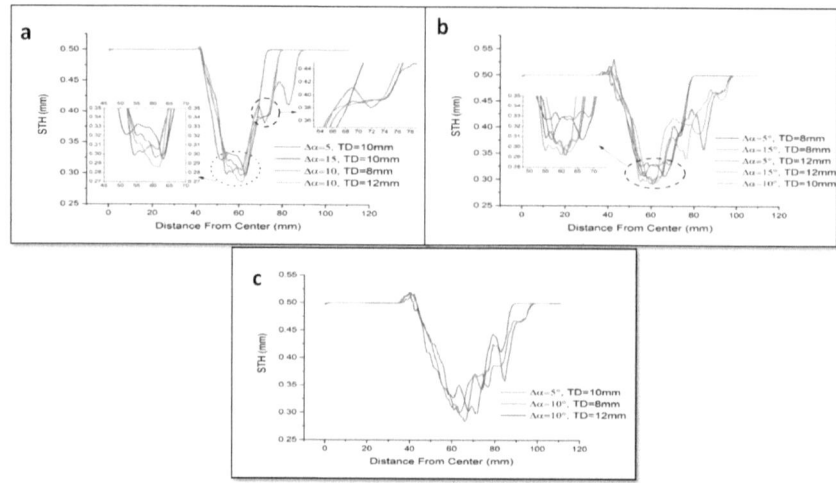

Fig. 4 Forming in **a** Two-stage, **b** Three-stage & **c** four stages with varying angle intervals and tool diameter

Figure 4c depicts the influence of angle interval and tool diameter on the section thickness created in four stages. Experiment with Da of 10° and 12 mm tool diameters to see if you can get a more consistent thickness at the start. On the other hand, the trial with 15° Da and 10 mm tool diameter exhibits higher consistency towards the bottom of the cup than the other experiments.

The influence of the number of intermediate phases on section thickness reflects in Fig. 5. As seen in graph 4, stage formation exhibits a delay in thinning, with the most negligible thickness measured at around 70 mm from the fixed edge (clamp). However, as the number of steps grows, the unevenness increases, as the sheet generated with two stages exhibits a smooth thickness distribution. The formation of the previously formed cup would cause unevenness (formed in the previous stage). Furthermore, three steps forming demonstrate the maximum sheet thickness in the produced section. Similarly, when the number of steps grows, the bending of the sheet at the clamps increases.

Figure 5 depicts the combined influence of the number of steps and angle interval on sheet thickness. It clearly shows that the minimum sheet thickness value increases initially as the number of steps grows but decreases after the threshold. Initially, the sheet is 0.5 mm thick and is reduced by 39.81 to 47.57%, from 0.300949 to 0.262133 mm.

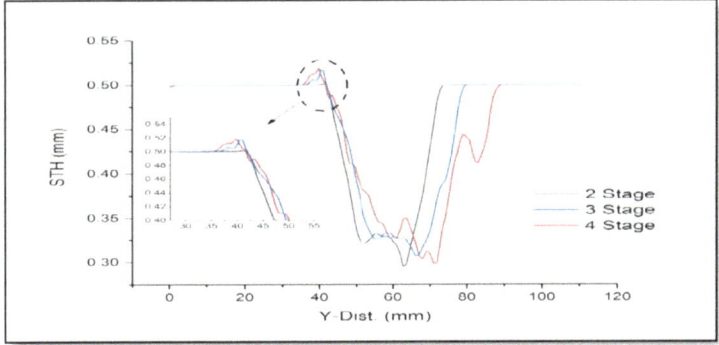

Fig. 5 Effect of the number of intermediate stages on section thickness

3.2 Spring Back of Multistage Incremental Forming

Another crucial impact noticed in the formed portion after the forming process is Spring back. It is an inaccuracy or deviation detected in the produced part due to the material's yield strength. The displayed graph in this section depicts the influence of the number of steps and tool diameter on the spring back. The impact of angle interval and tool diameter on a sheet created with a 5° angle interval reflects in Fig. 6a. The inaccuracy in the created cup raises as the number of steps increases, as the spring back effect incorporates each iteration.

Furthermore, the sheet created with four steps has a higher cup height than necessary due to sidewall stretching during the forming process. The mistake overlap in the former top section also has a mistake overlap. Furthermore, the 12 mm tool diameter cup exhibits less stretching than the sheet formed with a 10 mm tool diameter. The component created in the second step had higher accuracy among the plotted experiments than the part formed in the first stage. However, the pillow effect reflects in all four experiments. The sheets created at 10° angle intervals with varied numbers of stages and tool diameter yielded a similar outcome. However, the stretching phenomenon found in the 3 and 4 stages forming was relatively more than in the two stages. Figure 6b depicts the influence of the number of intermediate stages and tool diameter on the spring back effect with a constant angle interval of 10°. The extension of the wall rises as the number of phases increases.

Moreover, as shown in Fig. 6c for the exact stage forming (3 stage forming), the sheet formed using a 12 mm tool diameter shows a flat bottom compared to the sheet formed using an 8 mm tool diameter. The more contact area of the tool with a 12 mm diameter would result from less or no pillow effect in experiment 11 (3 stages, Da = 15°, TD = 12 mm), wherein the tool with an 8 mm diameter has a more negligible contact area sheet observed pillow effect. However, the bending effect was more for 8 mm tool diameter.

As shown in Fig. 7, the deviation in the sidewall of the formed cup is negligible with varying angle intervals. However, the height of the formed sheet increases by

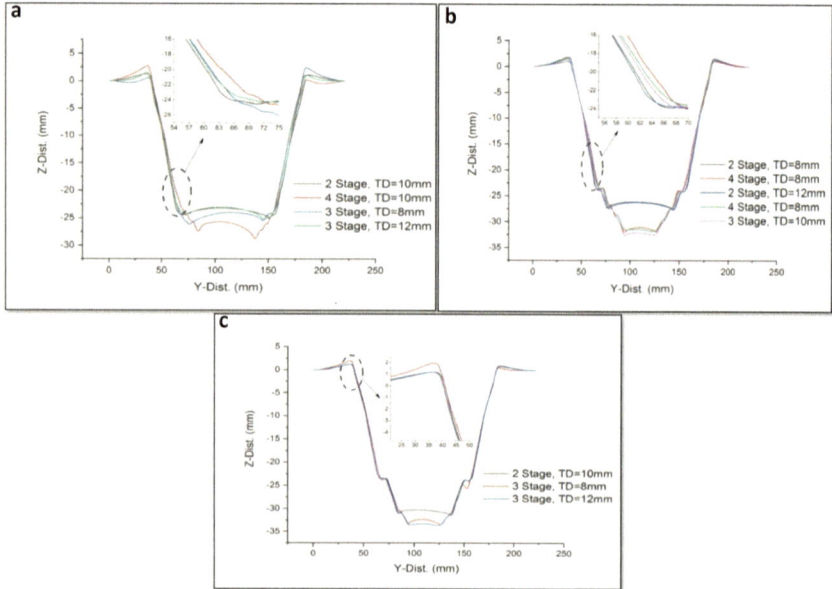

Fig. 6. The effect of the number of stages and tool diameter **a** with 5° angle interval of spring back **b** with 10° angle interval of spring back **c** with 15° angle interval of spring back

increasing the angle interval between the stages. In this case, the cup wall increases with angle interval, and the sheet formed at 15° shows the greatest stretching.

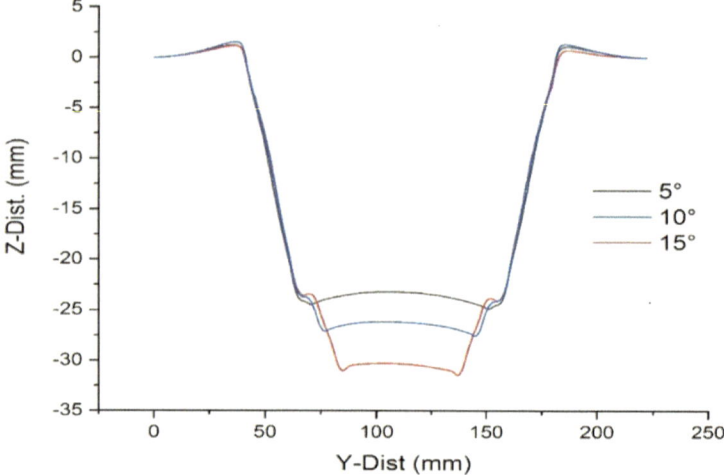

Fig. 7 The effect of the angle interval (Da) on spring back

3.3   Equivalent Plastic Strain (PEEQ) of Multistage Incremental Forming

The material's formability generated by the SPIF process reflects in equivalent plastic strain. The equivalent plastic strain reduces as the number of phases increases. Furthermore, increasing the angled gap between successive steps reduces the equivalent plastic strain. Shen et al. [23] performed a simulation analysis on a DC56 sheet with a dimension of $380 \times 380 \times 1$ mm to investigate the effect of the number of forming stages (n) and the angle interval (Da) between the consecutive stages on the equivalent plastic strain and spring back. As the number of forming stages increases, so do the equivalent plastic strain and the maximum total strain.

Furthermore, as the forming processes progress, more uniform deformation is noticed. Similarly, when the angle interval (Da) between steps rises, the equivalent plastic strain and the ultimate total strain Fig. 8a depict the sheet's two-stage shaping with different angle intervals and tool sizes. Figure 8b shows that when the angle interval increases from $5°$ to $15°$, the PEEQ value decreases. However, the final PEEQ and total maximum strain in the experiment $(Da = 5°, TD = 10$ mm) and experiment $(Da = 10°, TD = 12$ mm) were high compared to the other remaining experiment made with two stages. The sheet generated with a 2-stage, $10°$ angle interval and variable tool diameter $(TD = 8$ mm and 12 mm) had a more excellent value of PEEQ. The three-stage forming process with changing angle intervals and tool diameters appear in Fig. 8b. The figure shows that the value of PEEQ after the second stage is nearly the same for the experiment $(Da = 5°, TD = 8$ mm) and experiment $(Da = 5°, TD = 12$ mm), as well as the experiment $(Da = 10°, TD = 10$ mm) and experiment $(Da = 15°, TD = 8$ mm). However, the maximum value of PEEQ appears after the third stage sheet approves with fewer angle gaps between stages, which decreases as the angle interval grows. Similar results appear for the sheet generated in four stages, as shown in Fig. 8c. After each step, the sheet with a low angle interval exhibits a high PEEQ value.

Figure 9 shows the effect on an equivalent plastic strain by varying the number of stages. Three-stage forming shows the highest equivalent plastic strain (PEEQ) value. However, after the first stage, the sheet formed with the two stages shows the maximum PEEQ value, while for the three stages, the value of PEEQ was the least. Moreover, after every stage, the value of PEEQ decreases by increasing the number of intermediate stages. Thus sheet shows high formability when formed with a minimum or an optimal number of stages and angle intervals.

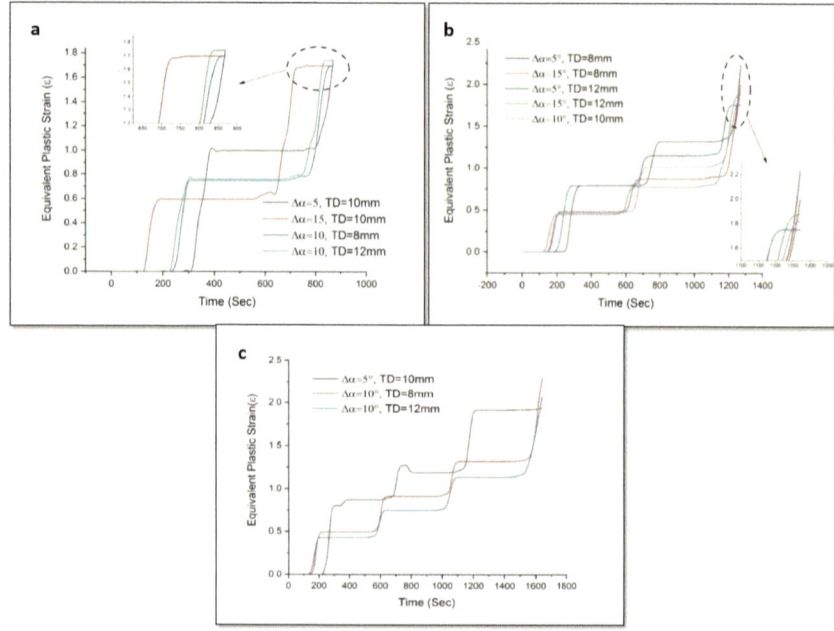

Fig. 8 **a** Two-stage **b** Three-stage & **c** Four-stage forming with variation in angle intervals and tool diameter

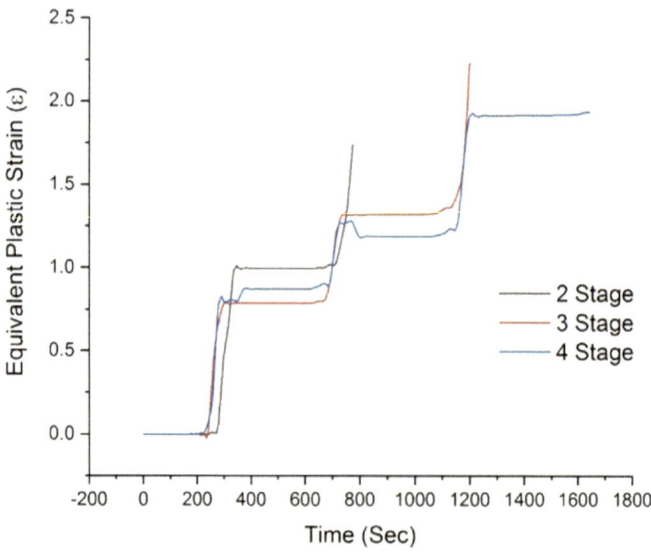

Fig. 9 Effect of the number of intermediate stages on equivalent plastic strain (PEEQ)

4 Conclusions

In the present work, FE Simulation work is carried through by deforming Titanium Gr. 2 sheet at different parameters. Section Thickness (STH), Equivalent plastic strain (PEEQ), and spring back were evaluated to understand the process mechanism. Below is a summary of some of the findings:

- The angle interval between phases does not influence spring back. However, the corresponding plastic strain and the ultimate total strain decrease when the angle interval rises.
- The number of stages doesn't change the thickness of a section. But as the number of stages increases, so does the amount of plastic strain and the spring.
- Using the right tool diameter and forming in the transverse direction may improve formability.
- The simulation findings show that section thickness decreases in multistage forming compared to single-stage forming.
- The core part of the cup has a consistent section thickness and the appropriate angle interval between the stages.
- With an increasing number of steps, the minimum sheet thickness increases initially but decreases over time. In the beginning, the sheet had a thickness of 0.5 mm, reduced by 39.81 to 47.57%, from 0.30 to 0.26 mm.

Acknowledgement The authors acknowledge the Science and Engineering Research Board, Department of Science and Technology, India, for funding the present research work [grant number SRG/2020/001355]

Conflict of Interest The authors declare that they have no conflict of interest.

References

1. Nirala HK, Jain PK, Roy JJ, Samal MK, Tandon P (2017) An approach to eliminate stepped features in multistage incremental sheet forming process: experimental and FEA analysis. J Mech Sci Technol 31(2):599–604. https://doi.org/10.1007/s12206-017-0112-6
2. Skjoedt M, Silva MB, Martins PAF, Bay N (2010) Strategies and limits in multistage single-point incremental forming. J Strain Anal Eng Des 45(1):33–44. https://doi.org/10.1243/030 93247JSA574
3. Tisza M (2012) General overview of sheet incremental forming. Manuf Eng 55(1):113–120. [Online]. http://www.journalamme.org/papers_vol55_1/55114.pdf.
4. Hirt G, Bambach M (2020) Incremental sheet forming. In: Sheet metal forming, pp 273–287. https://doi.org/10.31399/asm.tb.smfpa.t53500273
5. Peng W, Ou H, Becker A (2019) Double-sided incremental forming: a review. J Manuf Sci Eng Trans ASME 141(5):1–12. https://doi.org/10.1115/1.4043173
6. Maqbool F, Bambach M (2018) Dominant deformation mechanisms in single point incremental forming (SPIF) and their effect on geometrical accuracy. Int J Mech Sci 136(2018):279–292. https://doi.org/10.1016/j.ijmecsci.2017.12.053

7. Carette Y, Vanhove H, Duflou J (2019) Multi-step incremental forming using local feature based toolpaths. Procedia Manuf 29:28–35. https://doi.org/10.1016/j.promfg.2019.02.102
8. Najafabady SA, Ghaei A (2016) An experimental study on dimensional accuracy, surface quality, and hardness of Ti-6Al-4 V titanium alloy sheet in hot incremental forming. Int J Adv Manuf Technol 87(9–12):3579–3588. https://doi.org/10.1007/s00170-016-8712-3
9. Thyssen L, Magnus CS, Störkle DD, Kuhlenkötter B (2017) Compensating geometric inaccuracies in incremental sheet forming at elevated temperatures. Procedia Eng 207:860–865. https://doi.org/10.1016/j.proeng.2017.10.842
10. Ren H, Xie J, Liao S, Leem D, Ehmann K, Cao J (2019) In-situ springback compensation in incremental sheet forming. CIRP Ann 68(1):317–320. https://doi.org/10.1016/j.cirp.2019.04.042
11. Akrichi S, Abbassi A, Abid S, Ben yahia N (2019) Roundness and positioning deviation prediction in single point incremental forming using deep learning approaches. Adv Mech Eng 11(7):1–15. https://doi.org/10.1177/1687814019864465
12. Raju C, Narayanan CS (2016) Application of a hybrid optimization technique in a multiple sheet single point incremental forming process. Meas J Int Meas Confed 78:296–308. https://doi.org/10.1016/j.measurement.2015.10.025
13. Mohanty S, Regalla SP, Rao YVD (2015) Multistage and robot assisted incremental sheet metal forming, pp 1–5. https://doi.org/10.1145/2783449.2783457
14. Gonzalez MM et al (2019) Costing models for capacity optimization in analysis of geometric accuracy and thickness reduction in multistage analysis of geometric accuracy and thickness reduction in multistage incremental sheet. Procedia Manuf 34:950–960. https://doi.org/10.1016/j.promfg.2019.06.105
15. Kumar A, Gulati V, Kumar P, Singh H (2019) Forming force in incremental sheet forming : a comparative analysis of the state of the art, vol 41, no 6. Springer, Berlin, Heidelberg
16. Blaga A, Oleksik V (2013) A study on the influence of the forming strategy on the main strains, thickness reduction, and forces in a single point incremental forming process 2013
17. Thibaud S, Ben Hmida R, Richard F, Malécot P (2012) Simulation modelling practice and theory a fully parametric toolbox for the simulation of single point incremental sheet forming process : numerical feasibility and experimental validation. Simul Model Pract Theory 29:32–43. https://doi.org/10.1016/j.simpat.2012.07.004
18. Kumar R, Kumar G, Singh A (2019) An assessment of residual stresses and micro-structure during single point incremental forming of commercially pure titanium used in biomedical applications. Mater Today Proc 28:1261–1266. https://doi.org/10.1016/j.matpr.2020.04.147
19. Yoganjaneyulu G, Narayanan CS (2019) A comparison of fracture limit analysis on titanium grade 2 and titanium grade 4 sheets during single point incremental forming. J Fail Anal Prev 19(5):1286–1296. https://doi.org/10.1007/s11668-019-00721-y
20. Sornsuwit N, Sittisakuljaroen S, Sangsai N, Suwankan P (2019) Effect of heat treatment on single point incremental forming for titanium grade 2 sheet. In: ESIT 2018—3rd international journal of engineering and technology innovation, pp 6–9. https://doi.org/10.1109/ESIT.2018.8665124
21. Yeshiwas Z (2020) Spiral toolpath definition and G-code generation for single point incremental forming, January
22. Li JC, Yang FF, Zhou ZQ (2015) Thickness distribution of multistage incremental forming with different forming stages and angle intervals. J Cent South Univ 22(3):842–848. https://doi.org/10.1007/s11771-015-2591-x
23. Li J, Geng P, Shen J (2018) Numerical simulation and experimental investigation of multistage incremental sheet forming, October 2013. https://doi.org/10.1007/s00170-013-4870-8

Integration of Material Simulation to Extend the Accuracy of FEM Results for Metallic Forming Processes

Kristin Helas and Nikolay Biba

Abstract Nowadays, almost all metallic forming processes can be mapped by FEM simulation. Demands on FEM simulation are becoming increasingly higher with regard to the representation of very fine tolerances, whether in terms of temperature or dimensional accuracy. The combination of a FEM software suitable for the forming process and a connected materials database is a great potential to enlarge these precision figures. In the article, real simulation examples with QForm UK are used to show how this combination positively affects the accuracy of simulation results. For this purpose, several material data—such as flow curves, temperature-dependent properties, or CCT diagrams—and material models—such as the recrystallization or phase transformation model—from the materials database MatILDa® are applied. Hereby, the possibilities of material simulation are demonstrated and practical methods to improve individual simulation projects are given.

Keywords Material simulation · FEM simulation · Flow curves · Recrystallization model · Phase transformation model

1 Interconnection of Material Data and Forming Simulation

During hot forming processes, metallic materials are formed with tools like rolls or dies to get final dimensions and properties and undergo fundamental microstructural changes. Each forming process possesses unique challenges, which is why the representation of individualized industrial processes within a FEM simulation is becoming increasingly attractive. In recent years, considerable effort was invested to

K. Helas (✉)
GMT Gesellschaft für metallurgische Technologie- und Softwareentwicklung mbH, Heinersdorfer Str. 12, 13086 Berlin, Germany
e-mail: kristin.helas@gmt-berlin.com

N. Biba
Micas Simulations Limited, Oxford, UK

© The Rightsholder, under exclusive licence to [Springer Nature Switzerland AG], part of Springer Nature 2024
J. Kusiak et al. (eds.), *Numerical Methods in Industrial Forming Processes*, Lecture Notes in Mechanical Engineering, https://doi.org/10.1007/978-3-031-58006-2_5

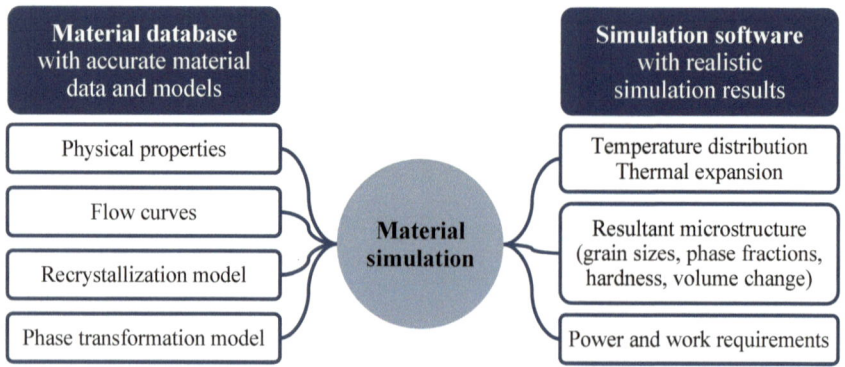

Fig. 1 The use of material data and models in material and FEM simulation

reach realistic simulation results and FEM software has been significantly improved. But often the stored material data is not questioned. In practice, similar alloys or standard values are applied without knowing the experimental test setting or the validity range. However, there is great potential in the use of accurate material data and models tailored to the industrial process (see Fig. 1). Thus, the influence of material data sets developed to be included into FEM simulation, for example, temperature-dependent properties, flow curves, microstructural models and CCT diagrams or phase transformation models, is described in the following sections.

At best, these material datasets are implemented in FEM or individual process software through a direct interface. In the simulation software, the user simply selects the alloy for the workpiece and tool, and the material data and models are loaded from the materials database in the background.

The application of material datasets within a FEM simulation can be demonstrated using the example of simulating the temperature distribution. The material datasets are imported from the materials database into the simulation software. The boundary conditions of the forming process are defined in the simulation software. The software contains a temperature model for simulating the time–temperature curve during forming. The required temperature-dependent material properties are chosen from the initially provided material datasets. These properties are applied according to the condition of each node during the FEM simulation. This results in the temperature distribution in the component or die.

2   Material Data for the Calculation of the Flow Behavior

The flow behavior of a metal alloy is essential to simulate a forming process. Therefore, flow curves are often used to include the flow behavior in a FEM simulation. Flow curves can be affected by the chemical composition, the experimental setting (such as tensile, compressive, or torsion test), forming parameters (such as temperature, true strain, and strain rate) as well as the heat treatment condition (such as as-rolled or heat-treated) [1]. Validated material datasets obtained from practical material investigations tailored to match the forming parameters and the main stress state in the forming process can enhance the accuracy of simulation results. For example, the effects of huge deformation on the dynamic and static recrystallization behavior of Inconel 718 were investigated by Borowikow et al. [2] by using flow curves from torsion experiments. Significant improvements in the accuracy of simulation results can be achieved on the basis of accurate material data regarding the calculation of temperature distribution, thermal expansion, force and work requirements, microstructure as well as phase fractions, and resulting final properties.

The next question at hand is: what occurs if you unintentionally or even unknowingly use flow curves that are not applicable to the specific forming process? How does it affect the accuracy of FEM results?

To investigate this, a FEM simulation was conducted using QForm UK to map a standard compressive test for the metallic alloys Ti6AlV4 (ASTM grade 5), 41Cr4 (AISI 5140), S420N (StE 420) and 100Cr6 (AISI 52,100) with a sample size of d_0 = 8 mm, h_0 = 16 mm and h_1 = 6.5 mm (see Fig. 2). Based on this, the forming parameters true strain $\varphi = 0.9$ and strain rate $\dot{\varphi} = 0.1 \text{ s}^{-1}$ were calculated.

The process parameters explained above were kept constant for all FEM simulation projects. But, for the same process parameters, flow curves with different validity ranges were used to see the impact on the resulting force (see Fig. 3). For example, the compressive test in Fig. 3a was conducted at a simulation temperature T_{Sim} of 1000 °C but two different flow curves were used: the first one is valid for the temperature range of 200 to 500 °C, whereas the second one is valid in between 800 and 1200 °C. Even though it is obvious that the temperature range of the flow curve should match the temperature range of the forming process, the temperature results are the most remarkable. However, it is important to note that information on the temperature range's validity is crucial.

In the next example, two flow curves for one steel alloy were available: for a lower and an upper analysis limit (see Fig. 3b). Also, the difference in the calculated force is visible for variations in the chemical composition, even though they belong to the same steel grade. From this result, it can be inferred how significant the difference in resulting force can be when using a flow curve for a completely different alloy.

Fig. 2 Parameters for the compressive test within the FEM simulation

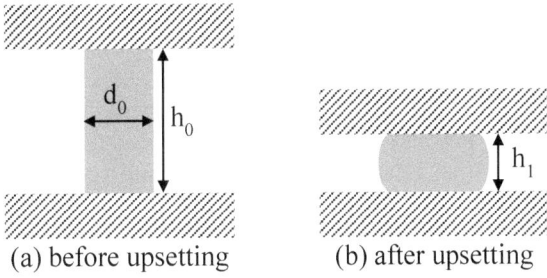

(a) before upsetting　　　　(b) after upsetting

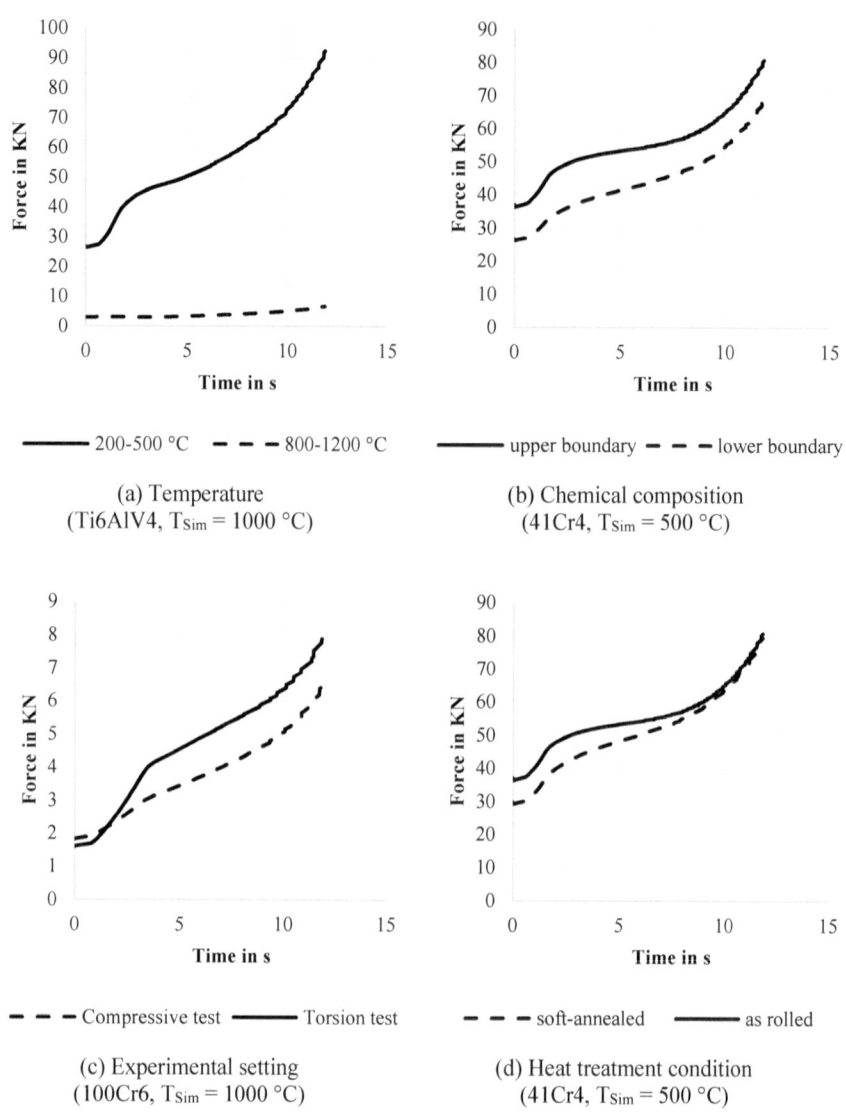

Fig. 3 FEM results for a compressive test using flow curves with different validity ranges regarding **a** temperature, **b** chemical composition, **c** experimental setting and **d** heat treatment condition

The experiment's setting also influences the simulation results (see Fig. 3c): by adding flow curves conducted through compressive and torsion tests, the resulting force differs clearly. Obviously, simulating a compressive test, the flow curve should have been recorded within a compression test. In addition, the results are affected by different heat treatment conditions, which can be seen in Fig. 3d.

The presented examples in Fig. 3 clearly highlight the crucial role of aligning the flow curve with process parameters, such as temperature, chemical composition, experimental setting, and heat treatment conditions. It is evident that validated materials datasets obtained from practical material investigations that are consistent with the parameters and the main stress state in the forming process can substantially impact the simulation results and enhance their accuracy.

3 Recrystallization Model to Calculate Grain Size Evolution

One topic many FEM users are not aware of is the inclusion of microstructural or recrystallization models for the calculation of grain size distribution and grain growth. This is accomplished by calculating the static and dynamic recrystallization processes. The results of the recrystallized fractions and the grain size distribution are among others essential for the calculation of the mechanical properties.

There are numerous microstructural models and approaches, which have been recently summarized in [3]. As the authors describe, there are on the one hand empirical plasticity models, which do not show a high accuracy between process and simulation results and require a lot of experimental data. On the other hand, there are physical plasticity and microstructural evolution models as well as computational approaches which show good results with high agreement between simulation and industrial results and are easy to implement into FEM simulation. Also, Hallberg [4] provides a comprehensive review of different approaches to modeling recrystallization in metals and discusses the challenges in developing accurate recrystallization models, such as the need for experimental data for model validation and the difficulty of incorporating complex physical processes. Overall, the importance of accurately predicting recrystallization for improving the performance of materials is highlighted. Especially for industrial applications, microstructural or recrystallization models should not be very time-consuming to implement and use in the FEM simulation.

Well, the recrystallization model described in this contribution is a physically based microstructural model which is designed for hot forming processes. The semi-empirical model is based on Sellars and Whiteman [5, 6] with modifications of Lehnert/Cuong [7]. A more detailed insight into the assumptions is shown in [8].

For now, the focus will be on the application of the recrystallization model. Biba et al. [8] validated the accuracy of the recrystallization model in QForm UK by comparing the predicted with experimentally measured grain sizes for an Inconel 718 (see Fig. 4). Thus, 3-step manufacturing process (nosing, flattening, and closed-die forging) for a structural component was simulated to predict the grain size. The authors conclude that this approach can provide useful insights into the recrystallization behavior and help optimize forming processes to reach a better product quality.

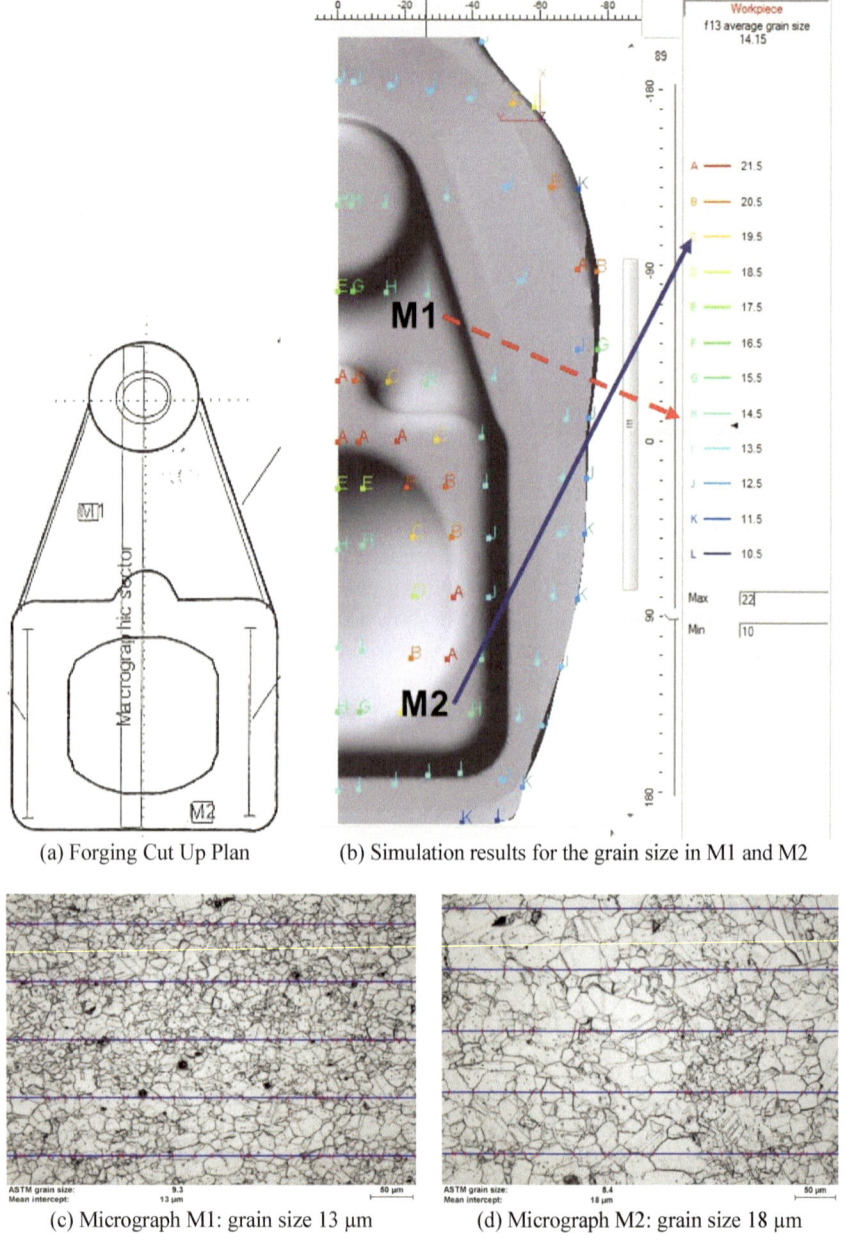

(a) Forging Cut Up Plan   (b) Simulation results for the grain size in M1 and M2

(c) Micrograph M1: grain size 13 μm   (d) Micrograph M2: grain size 18 μm

Fig. 4 Comparison of the grain size evolution for a real forming process and FEM simulation according to [8]

For the same nickel-based alloy, the accuracy of the recrystallization model was demonstrated in [9] within the optimization of the manufacturing process of a turbine disk, which is characterized by four cylindrical upsetting operations and a final shaping in the closed-die with subsequent trimming to the contour of the finished part.

Also, in [10] an entire open-die forging process was simulated using FEM simulation to optimize the process sequence. The FEM results were validated by comparing them with metallographically determined grain sizes and thermographic images. Additionally, the microstructure model was utilized for temperature control optimization in a bar mill, as well as for bar tempering.

4 Calculation of Phase Transformation

CCT diagrams are well-known for the prediction of phase transformation. Sometimes the required material dataset is unavailable. In this context, neuronal networks can be used for the calculation of phase transformation based on an extensive data evaluation of CCT diagrams. This so-called transformation model from the materials database MatILDa® is used to describe the transformation behavior and is applicable to a selected range of analyzes of a steel grade or limited steel group and can be described as a function of chemical analysis, cooling rate, and austenitizing temperature.

The simulation of the transformation behavior of a low-alloyed NiCrMo-forged steel (BS S154) using the QForm UK software is shown in [11]. As first, the available experimental data was analyzed, including dilatometric measurements and metallography. Based on this data, a material model was developed using a combination of empirical and physical approaches, which could predict the transformation behavior of the steel during heat treatment. The model was validated and tested for a forged fork from the aerospace industry: the tensile strength from experiments in the range of 925 and 990 N/mm^2 showed good agreement compared to the mean value of 950 N/mm^2 from FEM simulation. Thus, the developed transformation model showed a high level of prediction quality.

Furthermore, this neuronal network calculating phase transformation is utilized in the temperature and microstructure simulation to realistically determine the distribution of forming intensity in a KOCKS-3-roll RSB® block [12].

5 Summary

Material data and models can be used for the simulation of any forming and heat treatment process. Several examples involving real forming processes and process chains were given illustrating the diverse application fields of material and FEM simulation. The shown results highlight the benefits of precise material data and models in FEM simulation projects. In future, accurate and validated material datasets selected for the specific metallic alloy should be established as standard in FEM simulation.

Finally, the authors provide a summary of factors—decisive for a high standard in simulation—that users must consider in order to achieve a high level of agreement between simulation and experimental results. These factors include:

- using material data and models for your specific alloy,
- verifying the validity ranges and recording conditions of material data sets to ensure they are appropriate for the specific forming process (such as temperature, true strain, strain rate, or stress state),
- integrating all aspects of material simulation to achieve realistic simulation results and
- expanding the functions and models according to process parameters with appropriate materials expertise.

References

1. Borowikow A, Bambach MD, Wehage D (2021) Einfluss von Fließkurven auf die Berechnung des Kraft-und Arbeitsbedarfs bei der Simulation von Warmumformprozessen. massivUMFORMUNG, pp 24–29
2. Borowikow A, Schafstall H, Blei H, Wehage D, Borowikow M (2004) Integrierte Gefügemodellierung bei der FEM-Simulation mit Hilfe der Werkstoffdatenbank "MatILDa®". Kompetenzzentrum Neue Materialien Bayreuth
3. Jo SY, Hong S, Han HN, Lee MG (2023) Modeling and simulation of steel rolling with microstructure evolution: an overview. Steel Res Int 94:1–21
4. Hallberg H (2011) Approaches to modeling of recrystallization. Metals 1(1):16–48
5. Sellars CM, Whiteman JA (1976) Controlled rolling processing of HSLA-steels. Proc Prod Technol Conf York
6. Sellars CM, Whiteman JA (1979) Recrystallization and grain growth in hot rolling. Met Sci 13:187–194
7. Cuong ND (1991) Mathematische Modellierung und Simulierung der Gefügebildungsvorgänge beim Warmwalzen in Kalibern, vorzugsweise beim Walzen von Stabstahl und Draht. Dissertation TU Bergakademie Freiberg
8. Biba N, Borowikow A, Wehage D (2011) Simulation of recrystallisation and grain size evolution in hot metal forming. In: American institute of physics AIP conference proceedings, vol 1353, no 1, pp 127–132
9. Biba N, Borowikow A, Wehage D (2015) Möglichkeiten und Grenzen der simulationsbasierten Prozesskettenoptimierung. Dargestellt am Beispiel eines Schmiedeerzeugnisses aus einer Nickel-Basislegierung. Internationale Konferenz "Neuere Entwicklungen in der Massivumformung". Fellbach bei Stuttgart
10. Borowikow A, Wehage D, Blei H (2007) Modell zur Gefüge-und Eigenschaftsberechnung für online und offline Anwendungen, XXVI. Verformungskundliches Kolloquium, Planneralm, AT, pp 123–137
11. Doktorowski A, Biba N, Borowikow A, Wehage D (2011) Simulation of the transformation behaviour of low-alloyed NiCrMo-forged steels-from data analysis to material model
12. Kruse M, Schuck M, Borowikow A (2012) Innovations in simulation of microstructure developments. Mater Sci Forum 706–709:2170–2175

Optimization of the Heat Transfer Simulation Time During 3D Printing of PLA Material

Kandy Benié, Abel Cherouat, Thierry Barrière, and Vincent Placet

Abstract This work focused on optimizing the simulation time for heat transfer during the Fused Deposition Modeling process, which is necessary for a recent tensile property optimization study mentioned in the literature. The approach for optimizing the simulation time involved conducting a comparative analysis of various mesh sizes and simulation step times, assessing their influence on the resulting temperature profiles. The heat transfer simulation of the printed material was done with COMSOL Multiphysics FEA and the normal and extremely fine meshes as well as the simulation step times of $\Delta t = 0.01$ and 0.05 s were considered. The results showed that all the combinations resulting from these simulation parameters were equivalent in terms of both temperature profile results and the results of the tensile property optimization study, but that the simulation time was minimized by using the normal mesh with $\Delta t = 0.05$ s with a simulation time ten times shorter than in the case of the extremely fine mesh with $\Delta t = 0.01$ s.

Keywords 3D Printing · Heat Transfer · Simulation Time

1 Introduction

Manufacturing a part can be done subtractively, formatively, or additively. Additive manufacturing, more commonly known as 3D printing in a non-technical context, is the most recent of these methods and is increasingly used in major industries. Fused deposition modeling (FDM) is the most widely used additive manufacturing process for rapid prototyping. It involves passing solid polymer material through a heated

K. Benié (✉) · A. Cherouat
Laboratory of Automatic Generation of Meshing and Advanced Methods (UR-GAMMA3),
Université de Technologie de Troyes, 12 Rue Marie Curie, 10300 Troyes, France
e-mail: kandy.benie@utt.fr

K. Benié · T. Barrière · V. Placet
Department of Applied Mechanics (DMA), Université Bourgogne Franche-Comté, FEMTO-ST
Institute, CNRS/UFC/ENSMM/UTBM, 24 Chemin de L'épitaphe, 25000 Besançon, France

J. Kusiak et al. (eds.), *Numerical Methods in Industrial Forming Processes*, Lecture Notes in Mechanical Engineering, https://doi.org/10.1007/978-3-031-58006-2_6

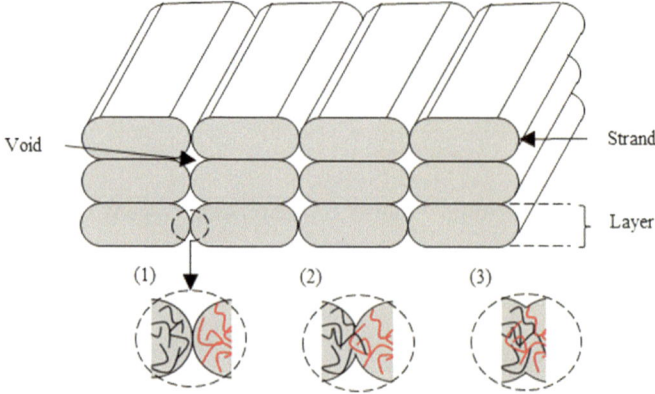

Fig. 1 Bond formation process through sintering: (1) surface contact; (2) bond formation and molecular diffusion at the interface; (3) bond and molecular diffusion growth

cavity where the material is melted and extruded through a nozzle. The extruded material is then deposited on the printing plate according to the desired geometry to build the final 3D part layer by layer [1]. This operating principle means that the part produced by 3D printing is composed of several layers that are made up of partially bonded strands (Fig. 1). This partial bonding between the strands makes the mechanical properties of the 3D parts depend on the mechanical properties of the printed material, the void between the strands and the strand-to-strand bond strength [2].

This makes the mechanical properties of parts made by 3D printing inferior to those of conventional processes such as injection and thermoforming process for plastic parts [3]. However, many authors have shown that the printing parameters used affect the mechanical properties of the printed parts [4]. However, due to the interdependence of these parameters, it is difficult to precisely identify their influence on the mechanical properties, and thus to choose the optimal printing parameters. To overcome this problem, Benié et al. [5] have proposed a study to identify the printing parameters allowing to maximize the mechanical properties of the printed parts without having to carry out costly and time-consuming experiments.

The aim of this work was to define an optimization parameter linked to the printing parameters, the value of which would be an indicator of the mechanical performance of the printed part. To achieve this, the main physical phenomena involved in printing, namely diffusion, coalescence, and crystallization, were studied. Contrary to the existing literature, where these phenomena were studied independently of one another (e.g. study of diffusion by [6], study of coalescence by [2], and study of crystallization by [7]), the study proposed by Benié et al. consisted in coupling all these phenomena in order to derive a numerical value named DCC (Diffusion, Coalescence and Crystallization), whose maximization would lead to the best tensile properties.

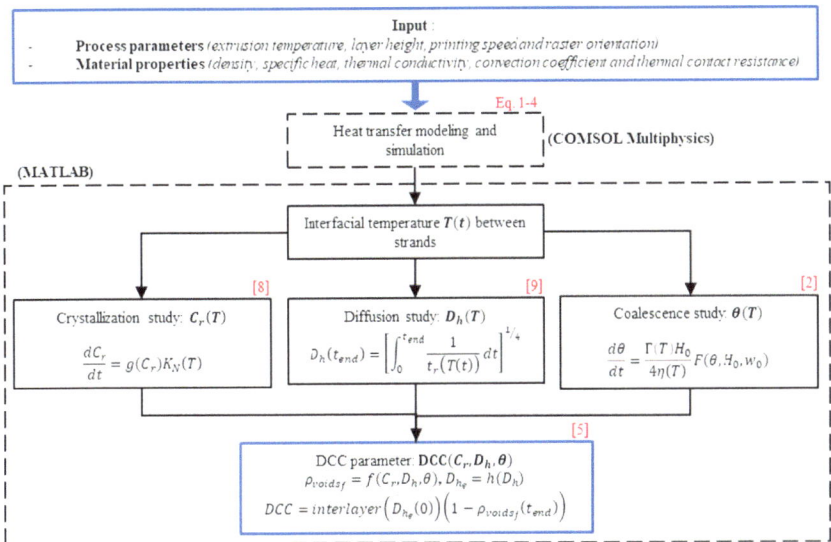

Fig. 2 Procedure for calculating the DCC parameter

At the end of the study, it has been shown that the DCC values were excellent indicators of the mechanical performance of printed parts. However, a problem arises when determining this DCC parameter.

Indeed, the physical phenomena under consideration are temperature-dependent. Thus, to study them, as Fig. 2 shows, it is essential to know the temperature profile during 3D printing. Since the temperature profile was determined by numerical simulation, to identify the optimum DCC values, it is necessary to simulate the heat transfer for all the combinations studied. In [5] for example, sixteen-parameter combinations were studied with simulation times between 2 and 8 h per combination, which is considerably high and makes the DCC study lose its time saving dimension.

In this paper, different mesh sizes and different simulation step times were studied in order to optimize the simulation time and to make the DCC study more interesting. Although the approach proposed for this paper was basic, it achieved the desired objectives.

2 Modeling and Numerical Simulation

2.1 Modeling

In this paper, the model used was the same as that used by Benié et al. [5]. It is a 2D model based on the successive activation of the strands for each layer of the printing part. Indeed, starting from a 2D geometry of the part to be printed, the strands of the part are activated one after the other in order to allow the heat exchange between the adjacent strands. To do this, the time required to go from one strand to another is determined before the numerical simulation.

During 3D printing, many authors have shown that the deposition of new strands tends to change the temperature in the previously printed strands. However, for the printing of intralayer strands, this temperature modification can be neglected contrary to the case of interlayer strands where the influence of the deposition of new strands is more pronounced. Indeed, the contact surface between the interlayer strands is very large compared to that of the intralayer strands. This leads to a higher heat exchange for the interlayer strands than for the intralayer strands. The work of Sun et al. [10] illustrates this well, since the temperature profile obtained experimentally when printing a 38 × 38 mm × 30-layer part shows that the printing of interlayer strands leads to temperature peaks, whereas the printing of intralayer strands shows no peaks and was not easily identifiable on the temperature profile. From this point of view, only two intralayer cords and all layers have been considered in the model as shown in Fig. 3a.

Partial Differential Heat Equation: the heat conduction equation in the polymer domain was described by the partial differential equation of the transient conduction Eq. 1 where ρ is the density, C_p the specific heat and k the thermal conductivity (the heat source term for the crystallization is supposed negligible).

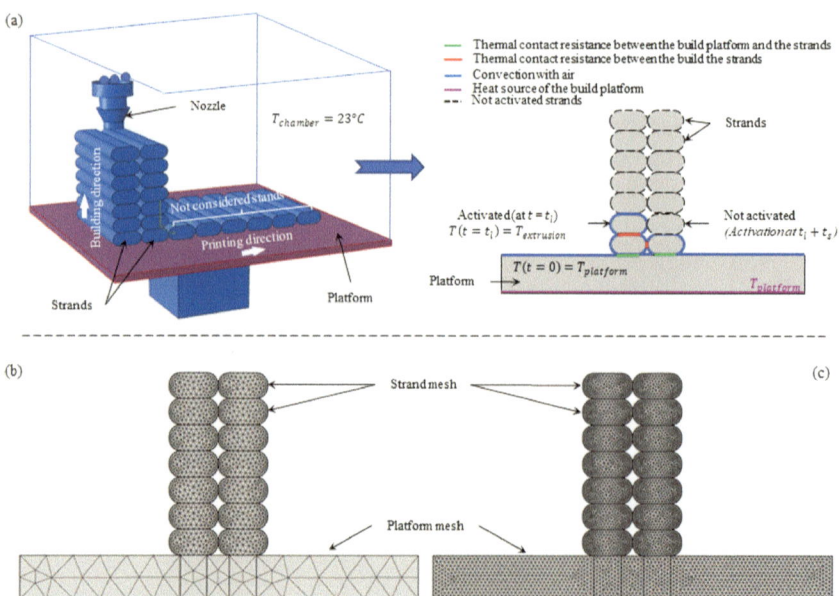

Fig. 3. **a** Initial and boundary conditions of the heat transfer during printing; **b** normal mesh (M1) and **c** extremely fine mesh (M2)

$$\rho C_p(\dot{T}) = \vec{\nabla} \cdot \left(k \vec{\nabla} T \right) \tag{1}$$

The Boundary and Initial Conditions (Fig. 3a):

- The exposure of the external surface of the polymer to the air leads to its cooling by convection (the radiative effects are neglected) which is described by Eq. 2 as:

$$-k \vec{\nabla} T \cdot \vec{n} = h(T - T_{\text{chamber}}) \tag{2}$$

where $\vec{n}$ is the outward normal vector of domain and h the convection heat transfer coefficient and T_{chamber} is the temperature inside the chamber.
- The heat conduction between each domain in contact (strand-strand and build platform-stand) was given by Eq. 3 using the thermal contact resistances (*TCR*):

$$\begin{cases} -k_{\text{strand}} \vec{\nabla} T \cdot \vec{n}_{\text{strand}} = \frac{1}{TCR_{\text{strand}}}(T^+ - T^-) & Strand - Strand \\ -k_{\text{platform}} \vec{\nabla} T \cdot \vec{n}_{\text{platform}} = \frac{1}{TCR_{\text{platform}}}(T^+ - T^-) & Platform - Strand \end{cases} \tag{3}$$

where T^+ and T^- are the temperatures on both sides of the interface. Low values of *TCR* were considered and $TCR_{\text{strand}} = TCR_{\text{platform}}$ since their influences were negligible on the simulation results
- At initial time, the temperature in each domain was supposed to be known and uniform as (Eq. 4):

$$\begin{cases} T(t = 0) = T_{\text{extrusion}} & On\ the\ new\ strand \\ T(t = 0) = T_{\text{platform}} & On\ the\ build,\ platform \end{cases} \tag{4}$$

Furthermore, it was assumed that the heat transfer was performed exclusively by thermal contact and convection, that the crystallization had no influence on the heat transfer and that the thermal properties of PLA were fixed during the heat transfer (Table 1).

Table 1 Thermal properties of material

Material	ρ (kg/m^3)	C_p (J/kg K)	k (W/m K)	h (W/m^2 K)	TCR (m^2 K/W)
Strands: PLA	1250 [11]	1179 [11]	0.28 [11]	5 [12]	10^{-5} [13]
Plate: PEI	1270 [14]	2000 [15]	0.2 [15]	5 [12]	10^{-5} [13]

3 Numerical Simulation of 3D Printing

To develop the finite element equations, the partial differential equations Eq. 1 must be restated in an integral form called the weak form. A weak form of the differential equations is equivalent to the governing equation and boundary conditions, i.e. the strong form. To solve the partial differential by the finite element method, we used the weighted residue method in the Galerkin formulation in which, we multiplied by an arbitrary temperature T^* and integrate over the polymer domain S as Eq. 5:

$$W\left(T, T^*\right) = \iint_S T^*\left(\rho C_p(\dot{T}) - \vec{\nabla}.\left(k\,\vec{\nabla}\,T\right)\right) dS \forall T^* \tag{5}$$

The 2D numerical simulations were performed with COMSOL Multiphysics 6.0 using a time dependent study of the heat transfer in solids interface. The boundary conditions were implemented from the COMSOL integrated nodes of convection and thermal contact and the resolution method used was the implicit Backward Differentiation Formula (BDF) method.

The implicit method is to write the time derivative of the temperature as Eq. 6 where Δt is the step time. Starting from the initial condition T(t = 0), the nodal temperature is estimated at each instant by increments of successive time Δt. This problem could be solved with direct integration methods over time (Euler method explicit or implicit, semi-explicit methods, the Crank–Nicholson method, etc.).

$$\dot{T} = \frac{T_{t+\Delta t} - T_t}{\Delta t} \tag{6}$$

Two predefined meshes in COMSOL were studied: the normal mesh M1 (212 triangular linear finite elements for each strand and the global model has 6925 dof) and the extremely fine mesh M2 (556 triangular linear finite elements for each strand and the global model had 41,155 dof) as shown in Fig. 3b and c. Also, two different simulation step times were studied: $\Delta t = 0.01$ and $\Delta t = 0.05$ s.

For each of these cases, four combinations of printing parameters were considered. They are presented in Table 2. At the end of simulations, the temperatures at the interface of the strands were used from probes positioned at each thermal contact and measuring the average temperature of the interface to calculate the DCC parameter. Thus, the DCC values obtained were used to identify any correlation between the different simulation parameters. The simulations were performed on an Intel(R) Core(TM) i5-4590 CPU (Fig. 4).

Table 2 Specimen nomenclature and printing parameters used for the study

Specimen	Raster orientation (°) (see Fig.)	Extrusion temperature (°C)	Printing speed (mm/s)	Layer height (mm)	Platform temperature (°C)
A_L	0	200	50	0.4	40
BL	0	200	50	0.6	40
AT	90	200	50	0.4	40
BT	90	200	50	0.6	40

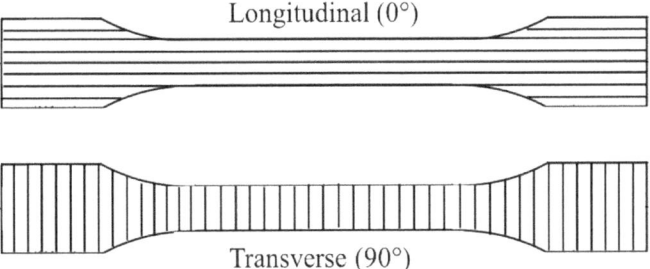

Longitudinal (0°)

Transverse (90°)

Fig. 4 Orientation of the strands in the specimens according to the raster orientation

4 Results and Discussions

To be able to validate the heat transfer simulation, a numerical simulation is conducted for a multi-layer wall manufactured by the FDM process. Note that the simulation is applied on ABS as material to be able to match the condition of existing experimental results in the literature. Figure 5 represents the comparison of the results of numerical simulation and the results of an experimental study derived from the literature [13]. The agreement between the simulation and experimental results validates the heat transfer modeling approach conducted in this work.

First of all, no convergence problems were encountered during the simulations. The results of the numerical simulation of the heat transfer presented in paragraph 2.b are shown in Table 3 for the specimen B_T and showed that in a given strand the temperature was not uniform and that no heat transfer took place with non-activated strands until their activation.

The temperature iso-values are given in Table 4 for the B_T combination and show that regardless of the simulation parameters used, the final temperatures are almost identical. As well, to be more precise, the final values of the interfacial temperatures between the first two strands were collected for all combinations of printing parameters in order to identify the impact of the simulation parameters on the temperature. The results are presented in Table 5 and show that only the mesh size has an impact on the temperature although it is minimal. Indeed, for the same mesh size, the final interfacial temperature hardly varied when using a simulation step time Δt of 0.01

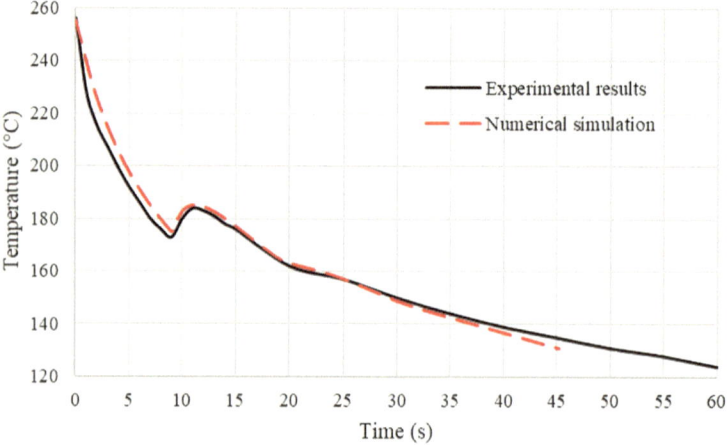

Fig. 5 Validation of the heat transfer model with experimental results derived from literature on a multi-layer wall geometry [13]

or 0.05 s. On the other hand, for a fixed simulation step time, moving from one mesh size to another led to a slight variation of the temperature of the order of 0.1 °C maximum. This is negligible. Therefore, the interfacial temperature was only slightly affected by the simulation parameters used.

For the different mesh sizes and simulation step times studied, the DCC parameter was determined for four combinations of printing parameters. The resulting DCC values were used to compare the simulation methods.

The results presented in Table 6 show that there was an excellent correlation between the DCC values obtained for different mesh sizes. Thus, using the extremely fine mesh or the normal mesh led to the same conclusions regarding the mechanical performance of the printed parts. Indeed, in both cases, the A_L combination was the best (highest DCC value) and the B_T combination was the worst (lowest DCC value).

Moreover, the correlation between the DCC values obtained in all cases was at least 0.9998 with respect to the most accurate case (extremely fine mesh and $\Delta t = 0.01$ s). This showed that all the simulation parameters studied here could be used for the DCC study and the interpretations of the final DCC results would be the same. In this sense, it is preferable to use simulation parameters that minimize the simulation time.

The lowest simulation time was obtained with the normal mesh and $\Delta t = 0.05$ s where the simulation lasted 26 min and 12 s for A_L combination against 5 h 4 min and 5 s in the case of the longest simulation time with extremely fine mesh and $\Delta t = 0.01$ s. This means that the simulation time was divided by 11 between these two cases. The same was true for all combinations of printing parameters where there was a factor of at least 10 between the simulation times of these two cases without impacting the interpretation of the DCC values.

Table 3 Iso-values of temperature during the printing of B_T specimen with M1 mesh and $\Delta t = 0.05$ s

$t = 0.2$ s	$t = 0.25$ s	$t = 186$ s
Layer 1/strand 1 printing	Layer 1/strand 2 printing	Layer 5/strand 2 printing

$t = 279$ s	$t = 280$ s	$t = 325$ s
Layer 7/strand 2 printing	Temperature during cooling	Final temperature

Tensile tests were carried out to determine the Young's moduli, maximum stresses, and fracture strains of the different printing parameter combinations considered in this work. These tensile properties were then used to study their correlation with the DCC values. Thus, considering Table 7 presenting the values of the tensile properties for the four combinations of printing parameters and the correlation between these properties and the DCC values obtained for different simulation parameters, it appears that the correlations remained almost equivalent whatever the simulation parameters used for each tensile property. This means that for all simulation parameters, the DCC values remained excellent indicators of mechanical performance.

Therefore, among the simulation parameters studied in this work, the normal mesh with $\Delta t = 0.05$ s is the best combination to optimize the simulation time in order to guarantee the time saving dimension of the DCC parameter besides being a good indicator of mechanical performances.

Table 4 Iso-value of temperature at the end of the simulation for B_T specimen

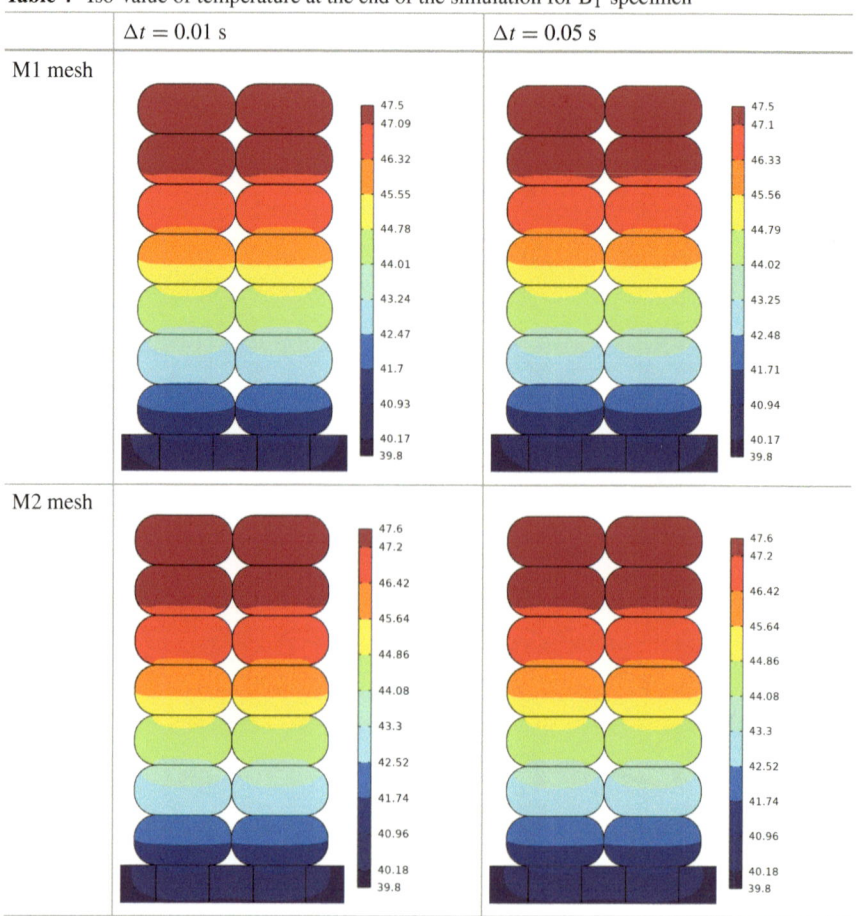

Table 5 Final interfacial temperatures for different simulation parameters

Specimen	M1 mesh		M2 mesh	
	$\Delta t = 0.01$ s	$\Delta t = 0.05$ s	$\Delta t = 0.01$ s	$\Delta t = 0.05$ s
A_L	43.26 °C	43.26 °C	43.3 °C	43.3 °C
B_L	47.88 °C	47.88 °C	47.98 °C	47.99 °C
A_T	42.41 °C	42.41 °C	42.45 °C	42.45 °C
B_T	46.55 °C	46.55 °C	46.64 °C	46.64 °C

Table 6 Simulation times and DCC results obtained for different simulation parameters

Specimen	M1 mesh		M2 mesh	
	$\Delta t = 0.01$ s	$\Delta t = 0.05$ s	$\Delta t = 0.01$ s	$\Delta t = 0.05$ s
Simulation time				
A_L	2 h 42 min 26 s	26 min 12 s	5 h 04 min 05 s	1 h 46 min 03
B_L	1 h 06 min 56 s	13 min 08 s	2 h 16 min 40 s	25 min 36 s
A_T	3 h 25 min 55 s	26 min 48 s	5 h 15 min 18 s	52 min 51 s
B_T	1 h 29 min 48 s	15 min 16 s	2 h 38 min 24 s	27 min 07 s
DCC values				
A_L	0.9472	0.9486	0.9478	0.9490
B_L	0.9301	0.9318	0.9310	0.9324
A_T	0.9168	0.9179	0.9170	0.9181
B_T	0.9071	0.9083	0.9074	0.9085
r^2				
	0.9998	0.9999	1	0.9999

Table 7 Tensile properties of specimens and correlation between these properties and the DCC values obtained from different simulation parameters

	Specimen	Young's modulus (GPa)	Tensile strength (MPa)	Fracture strain (%)
	A_L	3.19 ± 0.10	45.1 ± 1.8	1.57 ± 0.05
	B_L	3.11 ± 0.08	31.5 ± 3.4	1.03 ± 0.13
	A_T	3.01 ± 0.03	25.5 ± 0.9	0.92 ± 0.05
	B_T	2.94 ± 0.11	22.8 ± 0.3	0.83 ± 0.04
	r^2 with DCC values			
M1 mesh	$\Delta t = 0.01$ s	0.9958	0.9583	0.9001
	$\Delta t = 0.05$ s	0.9931	0.9552	0.8943
M2 mesh	$\Delta t = 0.01$ s	0.9933	0.9549	0.8937
	$\Delta t = 0.05$ s	0.9941	0.9526	0.8898

5 Conclusion

In this work, a study on the optimization of the simulation time of heat transfer during 3D printing has been carried out in order to guarantee time saving in the identification of the printing parameters maximizing the mechanical properties of the printed parts through the DCC study introduced in [5].

To perform this study, a 2D model based on stepwise strand activation was used and the numerical simulation was performed under COMSOL Multiphysics 6.0 as in [5]. The optimization study was subsequently performed by varying the mesh size and simulation step time between two levels each: the mesh size between the normal and the extremely fine mesh integrated in COMSOL and the step time Δt between 0.01 and 0.05 s. For each of these simulation parameters, the temperature profiles from the numerical simulations for four combinations of printing parameters were used to determine the DCC values. From these DCC values obtained for all simulation parameters, correlation studies were performed to identify the influence of the simulation parameters on the DCC study.

During the simulations, no convergence problems were encountered and the results showed that for each combination of printing parameters studied, the final temperature values varied very little from one combination of simulation parameters to another. Also, it was shown that not only was there a very good correlation between the DCC values obtained from all the simulation parameter combinations with a coefficient of determination r^2 of at least 0.9998, but also that in all the simulation cases, the DCC parameter remained a good indicator of the mechanical performance of the printed parts. Therefore, any combination of simulation parameters studied could be used to perform the DCC study.

The major differentiation between the different simulation cases being the simulation time and in order to minimize it, in this work, the use of the normal mesh with $\Delta t = 0.05$ s was the best combination since the simulation time of this combination was the smallest and was at least ten times shorter than the longest simulation time for the combination using the extremely fine mesh with $\Delta t = 0.01$ s.

Thus, although the approach proposed in this article was basic, it was possible to give the DCC parameter its full value, which is to quickly determine the printing parameters that maximize the mechanical properties of the printed parts without performing material-intensive experiments.

References

1. Brenken B, Barocio E, Favaloro A, Kunc V, Pipes RN (2018) Fused filament fabrication of fiber-reinforced polymers: a review. Addit Manuf 21:1–16
2. Garzon-Hernandez S, Garcia-Gonzalez D, Jérusalem A, Arias A (2020) Design of FDM 3D printed polymers: an experimental-modelling methodology for the prediction of mechanical properties. Mater Des 188:108414
3. Le Duigou A, Correa D, Ueda M, Matsuzaki R, Castro M (2020) A review of 3D and 4D printing of natural fibre biocomposites. Mater Des 194:108911
4. Dey A, Yodo N (2019) A systematic survey of FDM process parameter optimization and their influence on part characteristics. J Manuf Mater Process 3:64
5. Benié K, Barrière T, Placet V, Cherouat A (2023) Introducing a new optimization parameter based on diffusion, coalescence and crystallization to maximize the tensile properties of additive manufacturing parts. Addit Manuf 69:103538
6. Costa SF, Duarte JF, Covas JA (2017) Estimation of filament and adhesion development in fused deposition techniques. J Mater Process Technol 245:167–179
7. Balani SB (2019) Additive manufacturing of the high-performance thermoplastics: experimental study and numerical simulation of the fused filament fabrication. Ph.D. thesis
8. Patel RM, Spruiell JE (19991) Crystallization kinetics during polymer processing—analysis of available approaches for process modeling. Polym Eng Sci 31:730–738
9. Yang F, Pitchumani R (2002) Healing of thermoplastic polymers at an interface under nonisothermal conditions. Macromolecules 35:3123–3224

10. Sun Q, Rizvi GM, Bellehumeur CT, Gu P (2008) Effect of processing conditions on the bonding quality of FDM polymer filaments. Rapid Prototyp J 14:72–80
11. Noel H, Sadou A, Glouannec P, Feller J-B, Antar Z (2009) Étude expérimentale des propriétés thermo-optiques de matériaux polymères composites pour la réalisation d'absorbeurs solaires. In: Congrès Français de Thermique, SFT 2009. Vannes, pp 26–29
12. Aressy M (2013) Etude et modélisation de la cristallisation du Polylactide (PLA) en vue de l'optimisation du procédé de rotomoulage. Ph.D. thesis
13. Lepoivre A, Boyard N, Levy A, Sobotka V (2020) Heat transfer and adhesion study for the FFF additive manufacturing process. Procedia Manuf 47:948–955
14. Li M, Jiang Z, Hu B, Zhai W (2020) Fused deposition modeling of hierarchical porous polyetherimide assisted by an in-situ CO_2 foaming technology. Compos Sci Technol 200:108454
15. PEI online supplier: thermal properties of PEI. https://www.azom.com/article.aspx?ArticleID=1883. Llast accessed 22 May 2022

Optimization of Process Variables in Warm Extrusion of a Mg–Al–Zn Alloy by Numerical Simulation

Reeturaj Tamuly, D. Ravi Kumar, and S. Aravindan

Abstract Magnesium alloys are extensively used in various fields such as automobile, defense, and aerospace due to their high specific strength and good damping properties. Extrusion is one of the important metal forming processes that is used to process the cast alloy into various useful shapes at high production rates. It is also known to improve the workability and strength of the material. Due to these advantages, interest on the extrusion of magnesium alloys has significantly increased in the recent past. But due to poor formability at room temperature, extrusion of Mg alloys in the warm forming temperature range is an important process in the fabrication of these alloys. Optimization of process variables is critical for the minimization of extrusion load and prevention of defects. In this work, numerical simulations of the extrusion process of a Mg–Al–Zn (AZ31) alloy are carried out to predict the optimum combination of process parameters, namely, extrusion ratio, extrusion speed, and extrusion temperature. Hansel Spittle model was used to define the flow stress curve in the simulation. The effect of the process parameters on extrusion load was studied, and the optimum die angle required to minimize the load required for extrusion was predicted.

Keywords Mg alloy · Extrusion · Warm forming · Finite element simulation

1 Introduction

The applications of magnesium alloys have gradually increased due to their high specific strength and good damping properties [1, 2]. These alloys are processed into various useful shapes by forming processes such as forging and extrusion. Extrusion is used to produce tubes, bars, rods, and other solid/hollow profiles at high production rates. The extrusion of magnesium alloys is difficult due to their HCP structure

R. Tamuly (✉) · D. R. Kumar · S. Aravindan
Department of Mechanical Engineering, Indian Institute of Technology Delhi, New Delhi 110016, India
e-mail: reeturaj.tamuly@mech.iitd.ac.in

J. Kusiak et al. (eds.), *Numerical Methods in Industrial Forming Processes*, Lecture Notes in Mechanical Engineering, https://doi.org/10.1007/978-3-031-58006-2_7

which has a limited number of independent slip systems for plastic deformation and the additional gliding planes slide only at elevated temperatures of more than 225 °C [3] and hence warm extrusion of Mg alloys is preferred. In warm extrusion, the die geometry and the process parameters like extrusion temperature, extrusion ratio, and extrusion speed need to be optimized to minimize the extrusion force. Optimization of the process parameters has been a challenge for many researchers [4–6]. In process optimization, simulation of extrusion using the finite element method helps reduce the problems encountered during the extrusion in practice. The thermo-mechanical simulation of Mg alloys is very complex and hence information is scarce. Chandrasekaran and Shyan John [5] studied two-dimensional finite element (FE) simulation of AZ31 alloy which is one of the most commonly used and commercially available Mg alloys. Lapovok et al. [7] used FE simulation to construct an extrusion limit diagram for AZ31 alloy at different extrusion ratios. Hsiang et al. [8] studied the influence of the processing parameters on the extrusion of Mg alloy tubes at elevated temperatures. Hu et al. [9] studied the effect of die geometry on the extrusion of wrought Mg.

Most of the previous works were carried out to develop the optimized die design and the process parameters but very few attempts were made to study the implementation of genetic algorithm (GA) for optimization of process parameters for the Mg alloy extrusion process. In this work, a full factorial technique was used to design the set of simulations for warm extrusion of AZ31 alloy using QForm software, followed by a study of the significance of the process parameters on extrusion load by using ANOVA and the process parameters have been optimized for obtaining minimum peak load using both GA and Minitab Response optimizer tool.

2 Methodology

2.1 Material Selection

The typical composition of AZ31 alloy is given in Table 1.

The physical properties of the Mg alloy are given in Table 2.

H13 steel is used as the material for the ram and die and its typical composition is given in Table 3. H13 steel is selected as it has a high resistance to thermal cracks (fatigue) due to its excellent toughness in hot conditions.

Table 1 Chemical composition of AZ31 alloy

Elements	Mg	Al	Zn	Mn	Cu	Fe
wt%	94.8–96.6	2.5–3.5	0.6–1.4	0.15	0.05	0.005

Table 2 Physical properties of AZ31 alloy

Property	Value
Density (kg/m^3)	1740
Thermal conductivity (W/mk)	96
Specific heat (J/kgK)	960

Table 3 Chemical composition of H13 steel

Element	Fe	C	Si	Mn	P	S	Cr	Mo	V
wt%	89.3–91.8	0.35–0.45	0.9–1.2	0.25–0.55	0.03	0.03	4.5–5.5	1.2–1.7	0.85–1.15

3 Simulation of Warm Extrusion

In the present work, the simulation of the extrusion process of AZ31 alloy was carried out by QForm FE software. The design parameters used in extrusion simulation are given in Table 4.

The tool drive conditions were assigned in such a way that the ram could move in the same direction as that of the billet during deformation and the die-container assembly remains fixed. The stopping conditions are implemented in such a way that the simulation stops until the minimum distance between the ram and the die is zero. A hydraulic press with a maximum capacity of 100MN was selected. The material model and friction model used in the simulations are discussed in the following sub-sections. A schematic of the direct extrusion process is shown in Fig. 1.

Material Model. The flow stress values were obtained from the QForm database for a strain range of 0.002 to 3 at strain rates of 0.01, 0.1, 1, 3, and 20 s^{-1} and temperatures of 300, 350, 400, 450 °C. Hansel Spittel model was used for flow stress dependence on the strain, strain rate, and temperature as shown in Eq. (1).

$$\sigma = A e^{m1T} T^{m9} \varepsilon^{m2} e^{m4/\varepsilon} (1 + \varepsilon)^{m5T} e^{m7\varepsilon} \dot{\varepsilon}^{m3} \dot{\varepsilon}^{m8T} \tag{1}$$

Table 4 Design parameters used in extrusion simulation

Design parameters	Dimensions
Billet diameter	25 mm
Billet length	40 mm
Extrusion ratio	10:1
Exit diameter of billet	7.9 mm
Container diameter	25.1 mm
Container length	60 mm
Collar corner radius, R_1	0.8 mm
Die land radius, R_2	0.5 mm
Die relief taper, a	5°

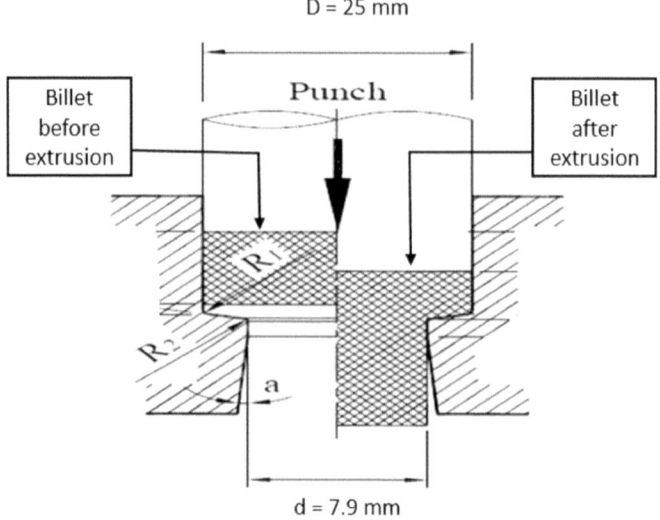

D = 25 mm

d = 7.9 mm

Fig. 1 Schematic representation of forward extrusion process of AZ31 alloy

where σ = Stress in MPa, ε = Strain, $\dot{\varepsilon}$ = Strain rate in s^{-1}, T = temperature in °C at which the test was done. The material constants A, m_1, m_2, m_3, m_4, m_5, m_7, m_8, and m_9 were calculated using the linear regression method in Minitab software and the values are presented in Table 5. The constants were then used to define the material model in extrusion simulations.

Considering the material to be isotropic, the Von Mises yield criterion was used to define the yielding behavior of the material.

Friction model. Siebel's law was used to define friction at the tool workpiece interface. According to the Siebel friction model, shear stresses τ (in MPa) on the contacting surface of a workpiece depend only on the friction factor m, and flow stress σ_0 (in MPa) of the workpiece material as given below.

$$\tau = m.\frac{\sigma 0}{\sqrt{3}} \tag{2}$$

The friction factor m was considered as 0.4 which indicates dry frictional conditions with well-cleaned surfaces [10].

Meshing. The workpiece and the tools were meshed using triangular elements. For the workpiece, mesh adaptation parameter which is the ratio of maximum element

Table 5 Material constants of Hansel–Spittel model

A (MPa)	m_1	m_2	m_3	m_4	m_5	m_7	m_8	m_9
4.567e12	0.00909	−0.1299	−0.0894	−0.00226	0.002508	−0.3909	0.000406	−4.83

Fig. 2 Simulation result showing variation of element size of the workpiece at a ram displacement of 10 mm

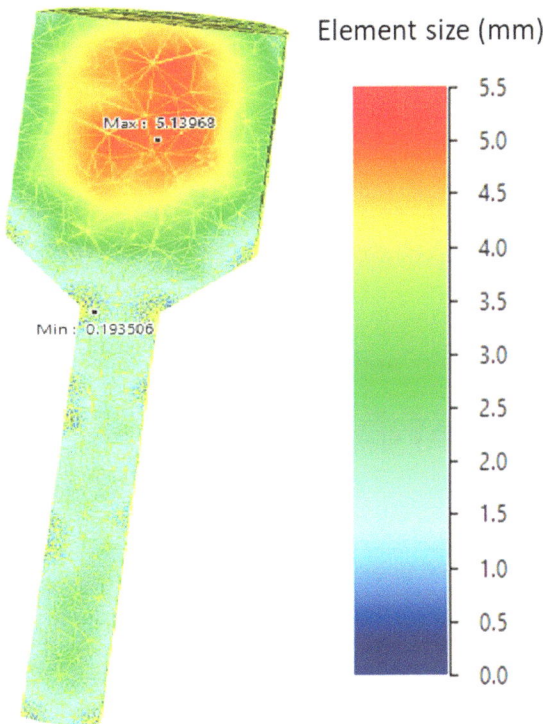

size to the size of any element in the mesh was defined. The maximum element size is the normalized size that the solver determines based on the dimensions and shape of a solid. The minimum and the maximum adaptation parameters were considered as 1 and 15, respectively. In a simulation with extrusion temperature of 400 °C, extrusion speed of 0.5 mm/s and semi-die angle of 45°, the maximum and the minimum element size before the simulation were 5.68 mm and 0.14 mm, respectively whereas after the simulation, they were 2.52 mm and 0.13 mm, respectively. The variation of element size of the workpiece during the simulation at a ram displacement of 10 mm is shown in Fig. 2.

4 Variation of Warm Extrusion Process Parameters

Usually, in an extrusion process, process parameters like extrusion speed, working temperature, and die angle are of major concern. In the present work on the warm extrusion of AZ31 alloy, the effect of these three process parameters on extrusion load was studied through numerical simulations. The variation of parameters in the simulation is shown in Table 6

Table 6 Levels of process parameters

Symbol	Process parameters	Level 1	Level 2	Level 3	Level 4	Level 5	Level 6
A	Extrusion temperature (°C)	300	350	400			
B	Semi-die angle (°)	15	22.5	30	45	60	75
C	Extrusion speed (mm/s)	0.5	1	2			

The process parameters were varied with full factorial design in the simulations.

5   Results and Discussion

5.1   *FEM Simulations*

The results obtained from the simulation of direct extrusion of AZ31 alloy in a warm temperature range are described in this section. Figure 3 shows the arrangement of tools (die, container, and ram) and the billet in the FE model. The stress and strain distributions of the workpiece at a ram displacement of 10 mm are shown in Fig. 4a and b respectively. The peak load required for extrusion in each case was predicted from the load v/s displacement graph as shown in Fig. 5. The minimum load obtained from simulations was 1.23×10^5 N at an extrusion temperature of 400 °C, a semi-die angle of 45°, and an extrusion speed of 0.5 mm/s.

Further analyses were done using ANOVA and GA to identify the significance of each of the extrusion parameters and optimize the process parameters to minimize the extrusion force.

Fig. 3 Simulation of the extrusion process indicating the arrangement of tools (die container and ram) and the billet

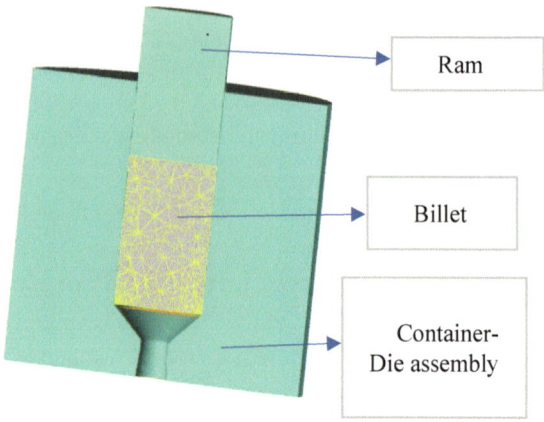

Ram

Billet

Container-
Die assembly

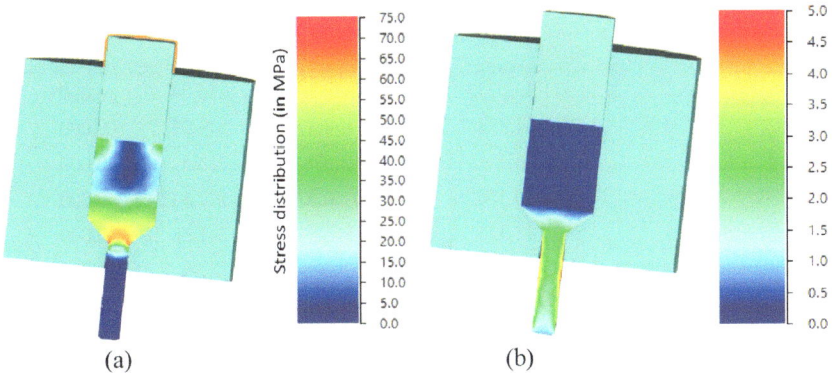

Fig. 4 Simulation results showing **a** stress distribution and **b** strain distribution at a ram displacement of 10 mm

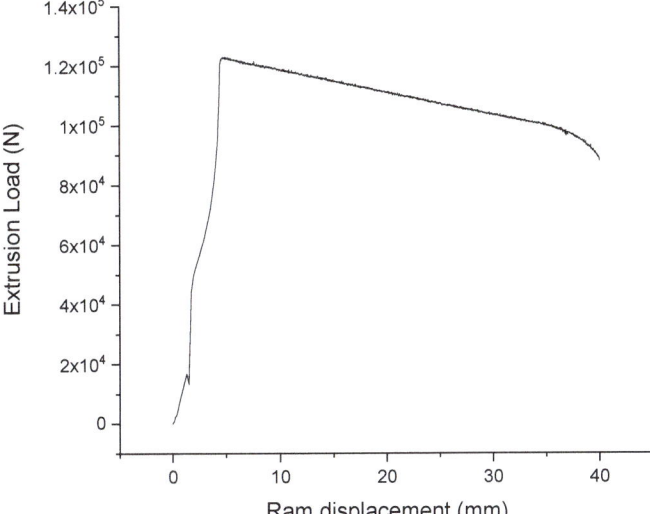

Fig. 5 Variation of extrusion load with displacement

6 Developing Empirical Relationships

To incorporate the effect of the three process variables (extrusion temperature, extrusion speed, and semi-die angles) and their interaction on the peak extrusion load, the polynomial function can be written as

$$Y = a0 + a1(A) + a2(B) + a3(C) + a11(A2) \\ + a22(B2) + a33(C2) + a12(AB) + a23(BC) + a13(AC) \qquad (3)$$

Table 7 Analysis of variance

Source	DF	Adj SS	Adj MS	F-value	P-value
Regression	7	44,571,303,733	6,367,329,105	670.24	0.000
A	1	2,374,364,878	2,374,364,878	249.93	0.000
B	1	201,683,678	201,683,678	21.23	0.000
A^2	1	1,686,650,370	1,686,650,370	177.54	0.000
B^2	1	2,293,145,336	2,293,145,336	241.38	0.000
C^2	1	131,951,208	131,951,208	13.89	0.001
AB	1	168,284,211	168,284,211	17.71	0.000
AC	1	223,020,643	223,020,643	23.48	0.000
Error	46	437,002,193	9,500,048		
Total	53	45,008,305,926			

where Y is the extrusion load (N), A is the extrusion temperature (°C), B is the semi-die angle (°), and C is the extrusion speed (mm/s). A^2, B^2, and C^2 are the quadratic effects of the process parameters. AB, AC, and BC are the interaction effects between the process parameters. a_0, a_1, a_2, a_3, a_{11}, a_{22}, a_{33}, a_{12}, a_{23}, a_{13} are the coefficients. In the above equation, all the factors may not affect the response significantly. To identify the relative significance of both individual and interaction factors, the ANOVA analysis was used.

Table 7 shows the sum of squares (SS), mean of square (MS), mean square factor to the square of error mean ratio (F), degree of freedom (DF), and the P value to check the fitness of the established model. The model presents $F_{0.05,7,46}^{\text{standard}} = 2.216$, for extrusion load, where $F^{\text{regression}} > F_{0.05,7,46}^{\text{standard}}$ which confirms that the model is adequate.

7 Regression Equation

A confidence level of 95% was considered for framing the regression model so that the significance of each of the factors on the response is high, and the corresponding P value should be less than or equal to 0.05. From Table 7, it is observed that the P value of the model is less than or equal to 0.001, which indicates the fitness of the model is satisfactory. From the F-value in Table 7, it is observed that the extrusion temperature has remarkable linear effects on the extrusion load followed by the semi-die angle. Moreover, the quadratic level of all three parameters has a significant influence on the extrusion load. The Pareto chart clearly shows the contribution of each of the parameters in Fig. 6. The chart shows the absolute values of the standardized effects from the largest effect to the smallest effect. The chart also plots a reference line to indicate the effects which are statistically significant across the reference line which is 2.01. It is observed that the extrusion temperature has the highest contribution to the

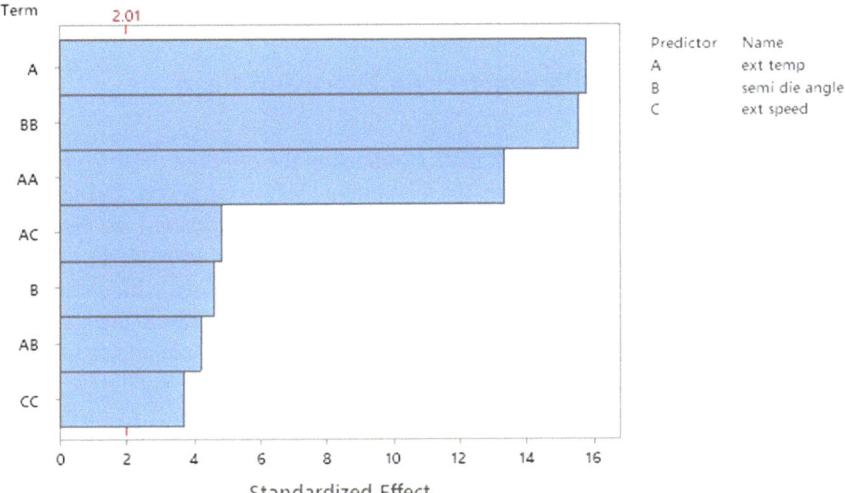

Fig. 6 Pareto chart indicating the contribution of each of the process parameters (linear, quadratic, and interaction effects)

extrusion load, followed by the quadratic effect of the semi-die angle and extrusion temperature and then the interaction effects of all three process parameters.

The developed empirical relationship is shown in Eq. 4.

$$Y = 975786 - 3959A - 937B + 4.742A2 + 18.93B2$$
$$- 5267C2 - 2.049AB + 50.5AC \tag{4}$$

8 Checking the Adequacy of the Relationships

The R^2 value indicates the goodness of fit for a model as shown in Table 8. Here, the value of the adjusted R^2 is 98.88% which is very high and hence indicates a high significance of the model. The predicted R^2 is also in good agreement with the adjusted R^2. Moreover, the model prediction ability is commonly measured by the value of predicted R^2 which is 98.58% in this case. This indicates that the model provides a good prediction and can be used for experiments in further studies.

Table 8 R^2 values

S	R^2	R^2 (adj)	R^2 (pred)
3082.21	99.03%	98.88%	98.58%

9 Effect of Process Parameters on Extrusion Load

Figure 7 shows the interaction plot matrix indicating the effect of all three process parameters at various levels of extrusion temperature, extrusion speed, and semi-die angle on peak extrusion load. It is observed that the peak extrusion load decreases with an increase in extrusion temperature which is due to a decrease in the flow stress of the material. Moreover, the peak extrusion load increases with an increase in extrusion speed which is due to an increase in the flow stress of the material. It is also observed that the peak extrusion load is higher at low semi-die angles which is due to high frictional forces at the billet and tool interface and higher at high semi-die angles due to the formation of the dead metal zone at the end of the extrusion process. However, there is an optimum die angle where the extrusion load is minimum.

Figure 8a and b show the dead metal zone in the die region for extrusion dies with a semi-die angle of 60° and 75° respectively. It is observed that a higher volume of dead metal in the die region increases the load at the end of the extrusion process due to high die angles. Moreover, extrusion defects like piping defect are likely to occur at higher die angles as shown in Fig. 8b. Hence, in the simulations, it was observed that the extrusion load is higher for dies with higher semi-die angles which is due to a high volume of dead metal at the end of the extrusion process.

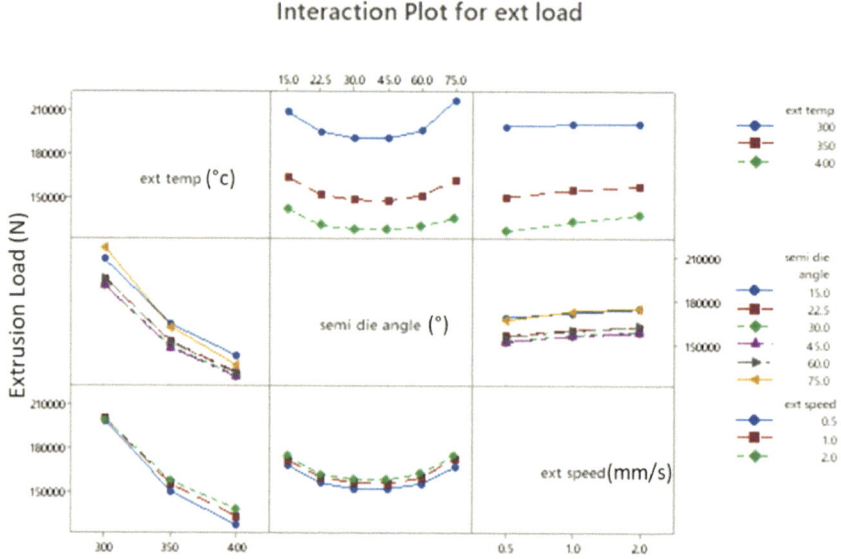

Fig. 7 Interaction plot matrix indicating the effect of extrusion parameters on peak extrusion load

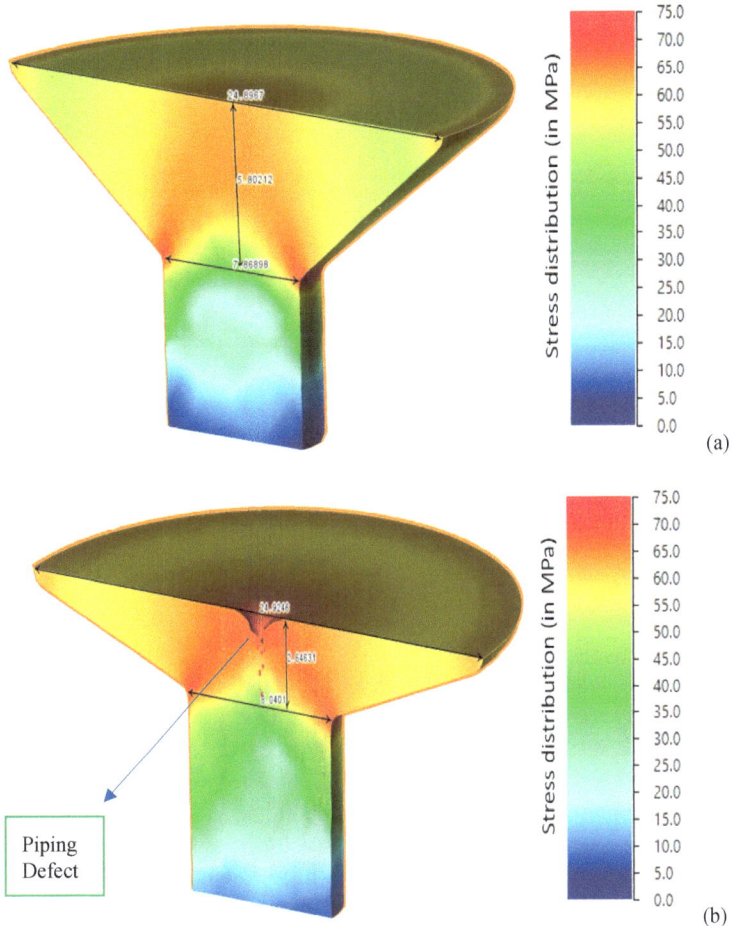

Fig. 8 Sectional view of the dead metal zone in the die region for **a** extrusion die with 60° semi-die angle **b** extrusion die with 75° semi-die angle

10 Optimization of Process Parameters by Genetic Algorithm and Minitab Response Optimizer

To achieve minimum extrusion load, the extrusion parameters were optimized based on the results obtained from simulations by GA and Minitab response optimizer tool.

MATLAB17a program using a "genetic algorithm toolbox" was used to optimize the process parameters. For minimizing the extrusion peak load, Eq. 4 was used as the fitness function. The boundary values used for process parameters are as follows: extrusion temperature between 300 and 400 °C, semi-die angle between 15° and 75°, and extrusion speed between 0.5 and 1 mm/s. The genetic operators, namely,

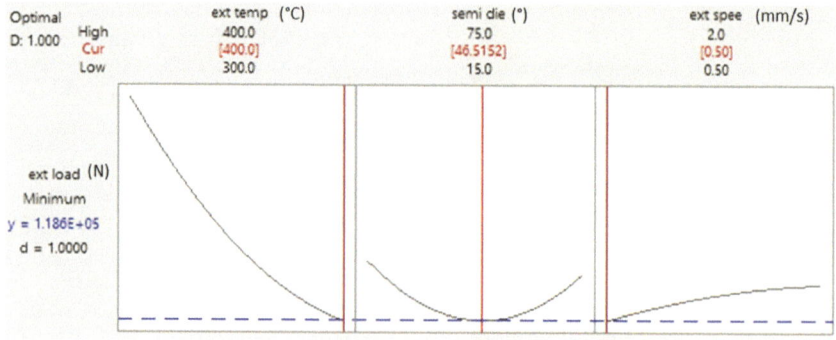

Fig. 9 Results obtained from the MiniTab response optimizer

mutation and crossover were used with a probability of 0.2 and 0.6 respectively. Similar conditions were considered by Jurkovic et al. for the optimization of extrusion parameters [11]. After 98 iterations/generations and for a population size of 50, the optimized values for minimizing the extrusion peak load are observed to be an extrusion temperature of 398.67 °C, semi-die angle of 46.186° and extrusion speed of 0.504 mm/s for a peak extrusion load of 1.129×10^5 N.

Minitab response optimizer was also used to compare the optimized results obtained by GA as shown in Fig. 9. Here, the optimum process parameters at which the peak extrusion load is minimum are an extrusion temperature of 400 °C, semi-die angle of 46.515°, and an extrusion speed of 0.5 mm/s for a peak extrusion load of 1.186×10^5 N. It is observed that the optimum process parameter values predicted by GA are also close to that predicted by the Minitab response optimizer.

From the optimum solutions obtained from both the Minitab response optimizer and GA, it can be concluded that the optimum process parameters for warm extrusion of this alloy for industrial applications are an extrusion temperature of 400 °C, a semi-die angle of 45°, and an extrusion speed of 0.5 mm/s. The utility of FE simulations to identify the optimal process variables for warm extrusion has been demonstrated in this work.

11 Conclusion

The warm extrusion of AZ31 alloy was studied using FE simulations by QForm software, and the simulation results were analyzed using Minitab statistical software package. The process parameters (extrusion temperature, semi-die angle, and extrusion speed) were varied at various levels using full factorial design. ANOVA analysis was done to study the significance of each of the process parameters on the extrusion load. Both GA and Minitab response optimizer tools were employed to obtain the optimum process parameters at which the peak extrusion load is minimum. The summary of the work done is discussed below.

- ANOVA analysis indicates that the linear effect of extrusion temperature has the highest contribution to the extrusion load, followed by the quadratic effect of semi-die angle and extrusion temperature. Although the interaction effect of each of the process parameters is relatively less, their contribution cannot be ignored.
- GA optimization tool predicts that the optimum process parameters are an extrusion temperature of 398.67 °C, extrusion speed of 0.504 mm/sec, and semi-die angle of 46.186° for a minimum peak extrusion load of 1.129×10^5 N.
- The Minitab response optimizer tool predicts that the optimum process parameters are an extrusion temperature of 400 °C, extrusion speed of 0.5 mm/sec, and semi-die angle of 46.515° for a minimum peak extrusion load of 1.186×10^5 N. The optimum values predicted by the response optimizer tool are close to those predicted by the GA optimization tool.
- From the optimum solutions obtained from both the Minitab response optimizer and GA, it can be concluded that the optimum process parameters for warm extrusion of this alloy for industrial applications are an extrusion temperature of 400 °C, a semi-die angle of 45°, and an extrusion speed of 0.5 mm/s.

12 Future Scope

The present paper is focused only on the optimization of process parameters for minimization of peak load during warm extrusion of AZ31 alloy using FE simulations and optimization tools. Since the extrusion ratio considered in the simulations is very small, which is 10:1, the AZ31 alloy may not reach the fracture limit during the process. The prediction of failure using a damage model in the simulations will be carried out in future. Also, the simulation results need to be validated through experiments and then the efficacy of the optimization techniques can be established accurately.

Acknowledgements The author would like to acknowledge the Ministry of Human Resource Development (MHRD), Government of India for receiving the assistantship during this research work.

References

1. Jurkovic Z, Jurkovic M, Buljan S (2006) Optimization of extrusion force prediction model using different techniques. J Achievements Mater Manuf Eng 17:353–356
2. Ouyang YX, Sui GX, Yang R, Zhuang GS, Ouyang YX, Sui GX, Yang R, Zhuang GS (2007) Preparation and mechanical properties of magnesium salt whisker/ABS composites preparation and mechanical properties of magnesium salt whisker/ABS composites. Mater Manuf Proocess 6914
3. Hu LJ, Peng YH, Li DY, Zhang SR, Hu LJ, Peng YH, Li DY, Zhang (2010) Influence of dynamic recrystallization on tensile properties of AZ31B magnesium alloy sheet influence of dynamic

recrystallization on rensile properties of AZ31B magnesium alloy sheet. Mater Manuf Process 6914

4. Siegert K, Jäger S, Vulcan M, Wizemann C (2005) Forming tubes, extrusions and sheet metal from magnesium AZ31. Mater Sci Forum 489:499–508

5. Chandrasekaran M, John YMS (2004) Effect of materials and temperature on the forward extrusion of magnesium alloys. Mater Sci Eng A 381:308–319

6. Li L, Zhou J, Duszczyk J (2006) Determination of a constitutive relationship for AZ31B magnesium alloy and validation through comparison between simulated and real extrusion. J Mater Process Technol 172:372–380

7. Lapovok RY, Barnett MR, Davies CHJ (2004) Construction of extrusion limit diagram for AZ31 magnesium alloy by FE simulation. J Mater Process Technol 146:408–414

8. Hsiang SH, Lin YW (2007) Investigation of the influence of process parameters on hot extrusion of magnesium alloy tubes. J Mater Process Technol 193:292–299

9. Hu HE, Wang XY, Deng L (2014) Comparative study of hot-processing maps for 6061 aluminium alloy constructed from power constitutive equation and hyperbolic sine constitutive equation. Mater Sci Technol 30:1321–1327

10. Ayer Ö (2019) A forming load analysis for extrusion process of AZ31 magnesium. Trans Nonferrous Metals Soc China 29:741–753

11. Jurković Z, Brezočnik M, Grizelj B, Mandić V (n.d.) Optimization of extrusion process by genetic algorithms and conventional techniques. Tehnički vjesnik 27–33

Deformation of Elastically Inhomogeneous Solid Modelled with Diffuse Interface Approach

Aarne Pohjonen

Abstract Diffuse interfaces are useful tools for simulating evolution of material phases, and they are fundamentally important part of the phase field method. Since, in solid state, different phases can have different elastic constants, it is important to be able to incorporate them into the numerical simulations. In current study, the theory of deformation of elastically inhomogeneous solid with diffuse interface approach and application of relatively simple numerical method for solving the problem are described. Deformation of a solid containing an elastical inhomogeneity is simulated for two cases for compression in two-dimensional plane strain condition: a harder inclusion and a softer inclusion.

Keywords Elasticity · Numerical methods · Elastic inhomogeneity · Strain · Strain distribution · Stress · Stress concentrator

1 Introduction

The elastic problem of material containing inhomogeneities is almost as old as the continuum elasticity theory itself: the foundations of the continuum elasticity theory were laid in the first part of the nineteenth century, and by the end of the century it had become obvious that it is necessary to consider the effect of inhomogeneities in the material when analysing the actual performance of a solid structure [1].

Although there exist some important analytic solutions to certain problems involving elastic inhomogeneities, most notably the Eshelby inclusion problem, there exist serious mathematical difficulties which hinder the development of general analytic solutions [2].

The problems associated with the analytical theory can be circumvented by the usage of numerical methods, which have become useful standard tools for analysis. Although the usage of numerical methods has their own difficulties associated with

A. Pohjonen (✉)
Materials and Mechanical Engineering Department, University of Oulu, Oulu, Finland
e-mail: Aarne.Pohjonen@iki.fi

© The Rightsholder, under exclusive licence to [Springer Nature Switzerland AG], part of Springer Nature 2024
J. Kusiak et al. (eds.), *Numerical Methods in Industrial Forming Processes*, Lecture Notes in Mechanical Engineering, https://doi.org/10.1007/978-3-031-58006-2_8

accuracy, stability and performance, they provide reasonably straightforward general application to many different types of problems.

A useful approximation for simulating a system containing elastical and structural inhomogenieity is provided by the so-called diffuse interface approach, where the phase boundary is modelled as smooth but rapid transition between phases [2–5]. This approach allows for straightforward calculation of the phase boundary evolution, merging and disappearance of the field regions, and it was used in the previous study to investigate phase region evolution [6].

In the current article, the aim is to investigate an elastic inhomogeneity located inside of strained surrounding material using the diffuse phase boundary approach. The current study complements the work presented in [6] by considering numerical experiments with phases that have different elastic constants.

2 Theory

Current approach is based on the formalism presented in [3] with the extension for different elastic constants for different phases described in [6]. For completeness, the mathematical description is derived from first principles of linear elasticity, including the description for transformation strain (i.e. stress-free strain of a phase, eigenstrain).

Assume that there exist two phases, α and γ, in a solid material. The function $\phi(\mathbf{r})$ describes the fraction of α phase at material position $\mathbf{r}(t)$. When $\phi = 0$, the local transformation strain (eigenstrain) is 0. The formation of the α phase causes the local structure of the material to change, so that when $\phi = 1$, the local stress-free strain (the eigenstrain) is ϵ_{kl}^{00}.

The initial material position of unstrained γ phase is described by the coordinates $\mathbf{w} = (w_1, w_2, w_3)$ and the current material position is $\mathbf{r} = (r_1, r_2, r_3)$. The displacement of the material is $\mathbf{u} = \mathbf{r} - \mathbf{w}$. The total strain of the material is $\epsilon_{kl}^{tot} = \frac{1}{2}\left(\frac{\partial u_k}{\partial w_l} + \frac{\partial u_l}{\partial w_k}\right)$. The elastic strain field of the γ phase is $\epsilon_{kl}^{e\gamma} = \epsilon_{kl}^{tot}$. On the other hand, the elastic strain field of the alpha phase takes in to account the stress-free transformation strain so that $\epsilon_{kl}^{e\alpha} = \epsilon_{kl}^{tot} - \epsilon_{kl}^{00}$.

To describe the strain and stress partitioning within the diffuse interface, a phase fraction weighed mixing model is adopted. The calculated quantities are volume fraction weighted averages, where a grid point value represents the local average. The local weighed average is obtained by replacing the volume integral with the product of the average and the local volume represented by the gridpoint. It is assumed that the local volume V_L can be separated in the volumes occupied by the α and γ phases, denoted by V_L^α and V_L^γ, respectively. In the case of strain $\int_{V_L} \epsilon_{ij} = \int_{V_L^\alpha} \epsilon_{ij}^\alpha + \int_{V_L^\gamma} \epsilon_{ij}^\gamma \Rightarrow V_L\langle\epsilon_{ij}\rangle = V_L^\alpha\langle\epsilon_{ij}^\alpha\rangle + V_L^\gamma\langle\epsilon_{ij}^\gamma\rangle$, where $\langle\epsilon_{ij}\rangle$ is average strain in volume V_L, $\langle\epsilon_{ij}^\alpha\rangle$ and $\langle\epsilon_{ij}^\gamma\rangle$ are the average local stresses in the α and γ regions V_L^α and V_L^γ, respectively. Dividing both sides with V_L yields $\langle\epsilon_{ij}\rangle = \frac{V_L^\alpha}{V_L}\langle\epsilon_{ij}^\alpha\rangle + \frac{V_L^\gamma}{V_L}\langle\epsilon_{ij}^\gamma\rangle$. Since the local volume fraction of α phase is $\phi = \frac{V_L^\alpha}{V_L}$ and only two phases are present, the

Fig. 1 Schematic showing the boundary conditions and the α and γ regions. The arrows schematically indicate the (exaggerated) movement of the material at the interface relative to the position in the absence of external strain

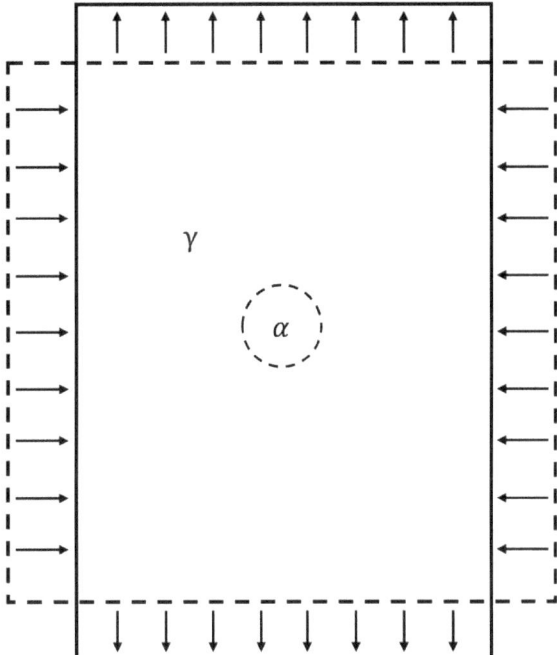

3.1.1 Hard Inclusion, $E_\alpha > E_\gamma$

As a first case, we consider the case where the inclusion, represented by the α region, is harder than the surrounding γ region under the vertical compression. The ratio $E_\alpha = 2E_\gamma$ was used in the numerical calculation.

The calculated von Mises stress of the gamma phase in a system containing a hard inclusion is shown in Fig. 2. It can be seen in a and b that in the γ region the stress is concentrated near the interface, but within the diffuse interface the value of σ_{VM}^γ drops. The reason for this is that in the diffuse interface the harder α constituent shields the γ region, and this effect becomes more pronounced when the fraction of α increases within the diffuse interface region. The von Mises stress of the α region is highest near the boundaries, as shown in c and d. Comparing b and c it can be seen that the von Mises stress of the α phase σ_{VM}^γ is higher everywhere inside of the inclusion than the corresponding far-field value of the γ region.

3.1.2 Soft Inclusion, $E_\alpha < E_\gamma$

As a second case, we consider the case where the inclusion, represented by the α region, is softer than the surrounding γ region under the vertical compression. In this case the value $E_\alpha = E_\gamma/2$ was used.

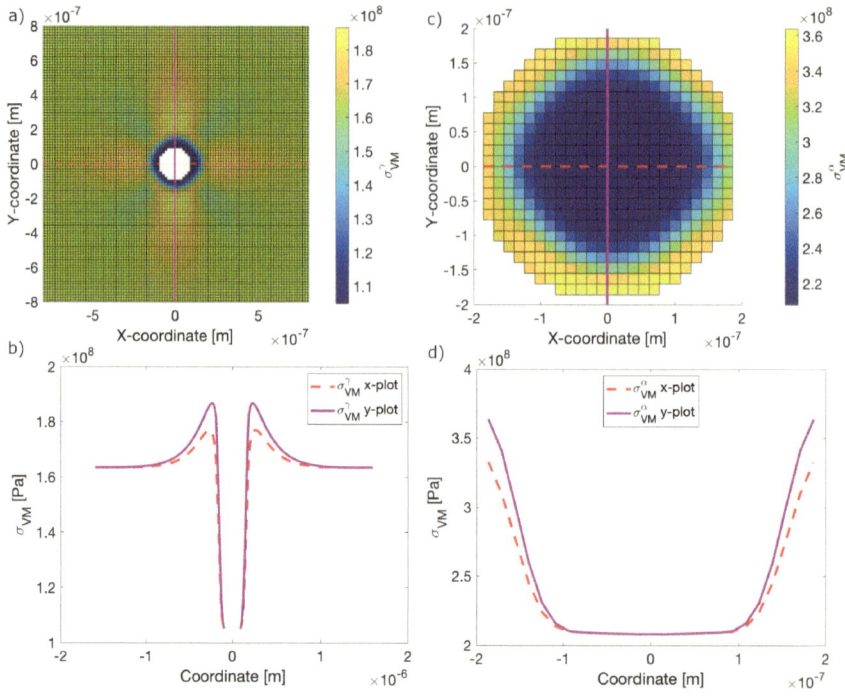

Fig. 2 Hard inclusion ($E_\alpha > E_\gamma$). **a** The von Mises stress of the γ region σ_{VM}^γ, **b** Plotted values along the lines shown in **a**, **c** the von Mises stress σ_{VM}^α of the α region, **d** plotted values along the lines shown in **c**

The calculated von Mises stress of the system containing the soft inclusion is shown in Fig. 3. It can be seen that the result is opposite of the one observed for the hard inclusion. In this case, the softer alpha region gives into the surrounding stress when analysed along the horizontal and vertical lines passing through the centre of the inclusion.

4 Conclusions

The description for the deformation of elastically inhomogeneous material with diffuse interface approach was derived from the principles of linear elasticity. Two numerical experiments were conducted for a two-dimensional solid containing the inhomogeneity under horizontal compression in plane strain condition: a softer inhomogeneity and a harder inhomogeneity. The numerical experiments of the current study complement the earlier results, where transformation strains were considered [6].

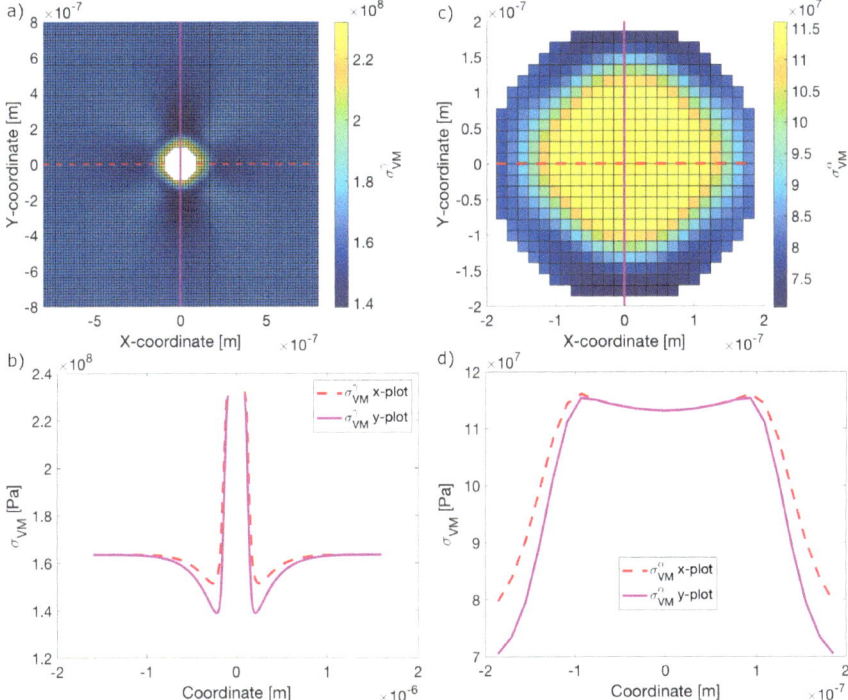

Fig. 3 Soft inclusion ($E_\alpha < E_\gamma$). **a** The von Mises stress of the γ region σ_{VM}^γ, **b** Plotted values along the lines shown in **a**, **c** the von Mises stress σ_{VM}^α of the α region, **d** plotted values along the lines shown in **c**

Acknowledgements Jane ja Aatos Erkon säätiö (JAES) and Tiina ja Antti Herlinin säätiö (TAHS) for their financial supports on Advanced Steels for Green Planet project are gratefully acknowledged. Professor J. Kõmi is acknowledged for funding acquisition.

References

1. Tokovyy Yuriy, Ma Chien-Ching (2019) Elastic analysis of inhomogeneous solids: history and development in brief. J Mech 35(5):613–626
2. Wang YU, Jin YM, Khachaturyan AG (2002) Phase field microelasticity theory and modeling of elastically and structurally inhomogeneous solid. J Appl Phys 92(3):1351–1360
3. Malik A, Yeddu HK, Amberg G, Borgenstam A, Ågren J (2012) Three dimensional elasto-plastic phase field simulation of martensitic transformation in polycrystal. Mater Sci Eng: A 556:221–232
4. Yamanaka A, Takaki T, Tomita Y (2008) Elastoplastic phase-field simulation of self-and plastic accommodations in cubic $\rightarrow$ tetragonal martensitic transformation. Mater Sci Eng: A 491(1–2):378–384

5. Leo PH, Lowengrub JS, Jou H-J (1998) A diffuse interface model for microstructural evolution in elastically stressed solids. Acta Mater 46(6):2113–2130
6. Pohjonen A (2023) Full field model describing phase front propagation, transformation strains, chemical partitioning, and diffusion in solid-solid phase transformations. Adv Theory Simul, 2200771
7. Landau LD, Lifšic EM, Lifshitz EM, Kosevich AM, Pitaevskii LP (1986) Theory of elasticity: volume 7, vol 7. Elsevier
8. Bouville Mathieu, Ahluwalia Rajeev (2006) Interplay between diffusive and displacive phase transformations: time-temperature-transformation diagrams and microstructures. Phys Rev Lett 97(5):055701
9. Pohjonen A (2021) Solving partial differential equations in deformed grids by estimating local average gradients with planes. In: Journal of physics: conference series, vol 2090. IOP Publishing, p 012069
10. Online document. Matlab documentation. https://se.mathworks.com/help/matlab/. Accessed 21 Mar 2023
11. Batra RC (2006) Elements of continuum mechanics. Aiaa

Effect of Work Roll Surface Warming on Hot Strip Temperature Development in Industrial Scale Virtual Rolling Model

Joonas Ilmola, Joni Paananen, and Jari Larkiola

Abstract Digitalization is taking place in the steel industry and novel virtual rolling models will be implemented into industrial systems. Virtual rolling models are mainly assembled of individual models for modeling the rolling process considering mechanical, thermal and metallurgical phenomena. Advanced virtual rolling models even predict mechanical properties based on modeled metallurgical state. These models for certain phenomena do not usually consider the constantly changing process parameters which affect the temperature of rolled strip in the industrial process. In the hot rolling industry, temperature development is of particular importance. To calculate temperature development accurately throughout the rolling process, multiple thermal boundary conditions must be considered. In this investigation, we used the FE-model to study how parameters of rolling pass, work roll coolers and full length transfer strip affect warming of a work roll surface. These results were utilized in prediction of work roll surface temperature in contact between the work roll and the strip over full length strip rolling in virtual rolling model to model development of strip temperature in full scale industrial hot strip rolling process.

Keywords FEM · Thermal conductance · Work roll warming · Virtual rolling model

1 Introduction

The steel industry is taking a huge leap towards green transition and fossil free steels during the next decade. This requires significant investments in new technologies and processes. Modeling plays a substantial role in this field. Specifically virtual rolling models have an important part in this transition. Authors have developed a virtual rolling model to obtain calculated thermal, mechanical and metallurgical state of the

J. Ilmola (✉) · J. Paananen · J. Larkiola
Materials and Mechanical Engineering, Centre for Advanced Steels Research, University of Oulu, P.O. Box 4200, 90014 Oulu, Finland
e-mail: joonas.ilmola@oulu.fi

J. Kusiak et al. (eds.), *Numerical Methods in Industrial Forming Processes*, Lecture Notes in Mechanical Engineering, https://doi.org/10.1007/978-3-031-58006-2_9

steel strip throughout the rolling process. The model uses detailed line layout which is required for correct positioning of boundary conditions, e.g. descalers, rolling stands, coil box, water cooling units, etc. Process and product data is needed to define parameters like alloying and dimensions of the slab, pass schedule, water flows among other predefined process parameters. The calculation begins when a slab is discharged from the walking beam furnace, see Fig. 1. Next the slab is transferred to descaling and after this to roughing process. The following process steps are coil boxing, finishing rolling and accelerated water cooling. The slab position on the production line is controlled according to process data, and all boundary conditions are affecting at a correct position in layout. Figure 1 presents the temperature development throughout the rolling process for the top and the bottom surfaces and center thickness of the strip. On an x-axis is the line layout. The thickness reduction of the strip over the process is also depicted in Fig. 1 with black dotted line.

In this investigation we have focused on one feature of our virtual rolling model, a simplified sub-model to define development of surface temperature of the work roll (WR) in hot strip rolling process and how it affects the temperature development of the strip in virtual rolling model.

The most crucial parameter of the virtual rolling model is temperature. Temperature of the strip is coupled with all phenomena occurring in hot strip rolling and miscalculation of temperature weakens reliability of calculated metallurgical and mechanical phenomena. The virtual rolling model considers continuously changing conditions of manufacturing process by reading the process data. However, there are also parameters that are not measured in the process data, e.g. surface temperature of the WR. Surface of the WR is subjected to sequential heating and cooling cycles

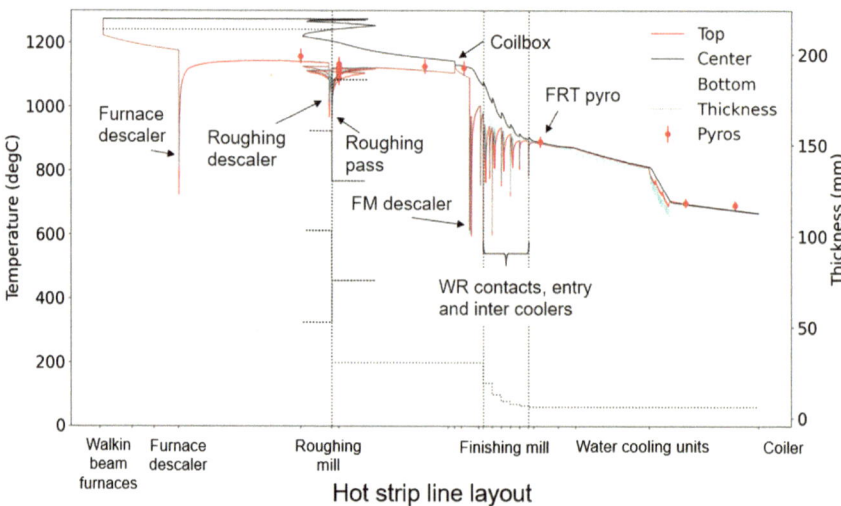

Fig. 1 Calculated temperature development of top and bottom surfaces and center thickness of the strip in virtual rolling model. Calculated temperature curves are compared to measured pyrometer values from industrial process

with the net result of positive heat influx [1]. In industrial process, this may lead to significant thermal expansion of the work rolls affecting the roll bite geometry and further the flatness of the strip and wearing conditions [2, 3]. The developed virtual rolling model is 1-dimensional and only changes in thermal conductance between the strip and WR are considered. However, alternating conditions in thermal contact conductance along the produced strip must be calculated in virtual rolling model. Several studies on the topic focus on heating and cooling cycles during one WR revolution [1, 4, 5], while virtual rolling model must calculate how the WR surface temperature evolves over the full length strip rolling as studied in [6]. The number of the WRs in virtual rolling model is 14 and surface temperature must be updated for all of those at every time step which set limitations for computing time of this individual sub-model. The virtual rolling model must be fast to be used as an online calculation tool in industrial systems and thus the simplified model for calculation of WR surface temperature is developed. The model is based on results of detailed Finite Element (FE)-model presented in this paper.

2   Configuration and Results of FE-Model

FE-model for six stands finishing mill is presented by authors in previously published articles [7, 8]. The modeling principles, physics and mechanics of this model are based on the same articles. The model utilizes non-linear facilities of a coupled temperature-displacement solver of the software package Abaqus™ Explicit. Finely discretized geometry is meshed with plane strain CPE4RT thermally coupled quadrilateral elements with enhanced hourglass control. Eight elements were used over the strip thickness and element length in rolling direction was 5 mm in the strip and on the contact surface of the WR. Exceptionally, 2 mm element length in rolling and 4 mm in vertical direction was used for the simulation 2 in Table 1 to ensure proper contact between the strip and the WR. FE-model contains automatic roll gap control which calculate the contact pressure and friction shear stress distributions in roll bite to predict roll force. Setup calculations and roll gap control are introduced by authors in [9]. The FE-model of the finishing mill is modified as a single stand model to simulate heat transfer during 75 m long transfer strip rolling on the first rolling stand of the finishing mill. Single stand FE-model is shown in Fig. 2.

2.1   Thermal Boundary Conditions of WR

Surface of the WR goes through sequential heating and cooling cycles during rolling pass. To obtain reliable surface temperature development, all thermal boundary conditions affecting the WR must be considered. The main heat transfer mechanisms in this case are heat transfer by conduction from surface of the WR to inner of the WR, thermal contact conductance between the strip and the WR, heat radiation

Table 1 DOE for WR temperature evolution in single pass FE-simulation

Simulation	Transfer strip initial temperature (°C)	WR circumferential velocity (m/s)	Reduction (%)
Sim. 1	800	5	12
Sim. 2	900	5	3
Sim. 3	900	1	12
Sim. 4	900	5	12
Sim. 5	900	10	12
Sim. 6	900	5	48
Sim. 7	1000	5	12

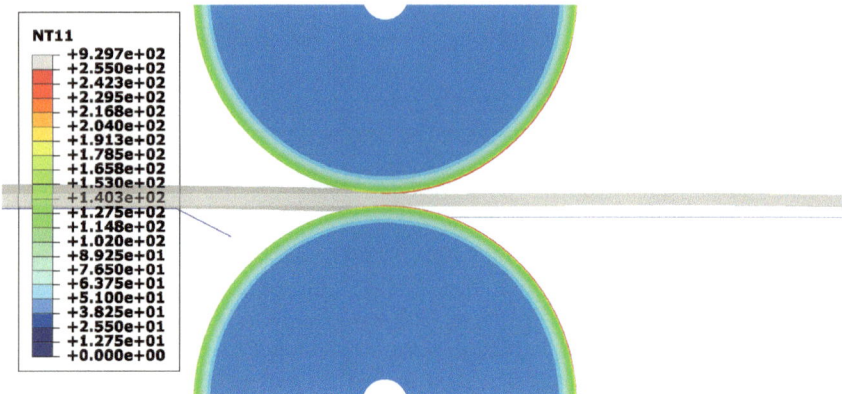

Fig. 2 FE-model of the single stand transfer bar rolling

and convection to ambient and between the strip and WR when there is no mechanical contact. And the most efficient cooling mechanism is water cooling nozzles.

The heat transfer efficiency of water cooling highly depends on the temperature of steel surface, see Fig. 3 (cooling water is assumed to be near room temperature, 20 °C). At high surface temperature of steel, a steam film (Leidenfrost effect) forms and isolating steam layer between the surface of the steel and liquid water weakens the heat transfer [10]. Because effective heat transfer coefficient (EHTC) is significantly dependent on surface temperature of WR, it must be considered in the FE-model. EHTC on the surface of WR due to water cooling spray is defined by Eq. 1, used by Martin in [10] and originally experimentally determined in [11].

$$\omega_c = 3.16 \cdot 10^9 \zeta_0 \zeta_1 [(u - u_0) - \zeta_2(u - u_3)]^{-2.455} \dot{\theta}^{0.616}, \qquad (1)$$

where ζ_0, ζ_1 and ζ_2 are defined in the following Eqs. 2, 3 and 4:

$$\zeta_0 = \begin{cases} 0 \ u < u_0 \\ \frac{u-u_0}{u_1-u_0} \ u \in [u_0, u_1], \\ 1 \ u > u_1 \end{cases} \tag{2}$$

$$\zeta_1 = \left(1 - \frac{1}{1 + \exp\left(\frac{u-u_2}{40}\right)}\right), \tag{3}$$

$$\zeta_2 = \left(1 - \frac{1}{1 + \exp\left(\frac{u-u_3}{10}\right)}\right) \tag{4}$$

with $u_0 = 273$ K, $u_1 = 425$ K, $u_2 = 573$ K, $u_3 = 973$ K, u is the node temperature of the WR surface and $\dot{\theta}$ is the water flux (l/(min·m²)). It must be noticed that different water nozzle geometries and used water pressure may affect the EHTC on different systems and thus parameters of equations may need readjustment. EHTC is implemented into FE-model using VDFLUX-subroutine. In the VDFLUX, user must determine heat flux q on the picked surface and thus EHTC must be multiplied with temperature difference between cooling water and WR surface, $q = \omega_c \cdot (T_{WR} - T_{\text{water}})$. The cooling water temperature was set to 20 °C. Water flux $\dot{\theta} = 1000 l/(\text{min} \cdot \text{m}^2)$ in Fig. 3.

Schematic Fig. 4 shows assembly of cooling nozzles used in FE-model. Figure 4 shows simulated temperature history of WR surface node during one revolution. Numbers in the schematic roll Fig. 4 show the thermal boundary conditions of WR surface and the same numbering in right side chart shows the temperature history corresponding to the boundary conditions. WR and the strip are assumed to be initially isothermal, and temperatures are predefined as 45 and 900 °C at the beginning of simulation, respectively. In the first phase (Fig. 4), the contact with hot strip heats the

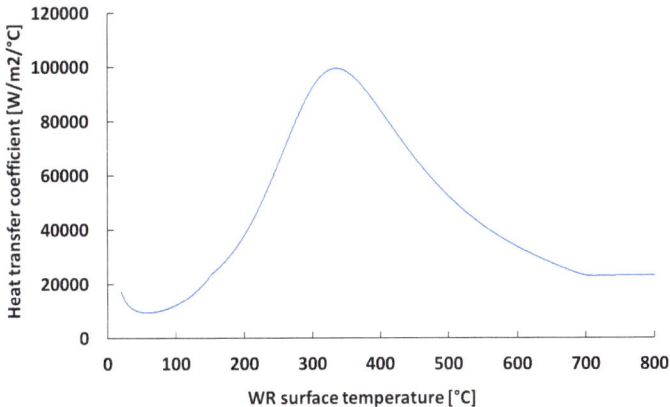

Fig. 3 Effective heat transfer coefficient values for forced convection as function of WR surface temperature

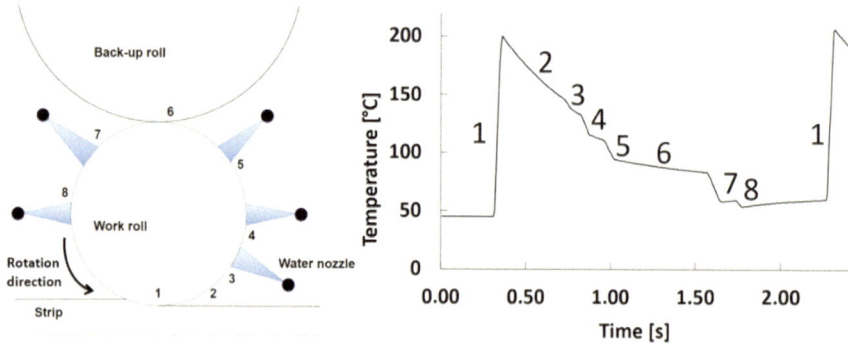

Fig. 4 Schematic figure of WR thermal boundary conditions and modeled temperature history of WR surface during one revolution (Sim. 3, Table 1)

WR surface at 198.6 °C. Magnitudes of convection, radiation and thermal conductance are applied as experimentally determined coefficients in [12]. The second phase contains heat conduction below the WR surface and heat radiation from strip to WR surface. Conductance in WR transfers heat efficiently from surface, simultaneously cooling the WR surface. Heat radiation has a minor effect compared to conductance. Next three phases consist of WR water cooling nozzles. The water impingement areas of the WR cooling nozzles are determined based on areas limited by cartesian coordinates due to revolving WR. The cooling efficiency of these units depends highly on water flux and pressure as well as temperature gradient between cooling water and WR surface. In addition, the heat conduction inside the WR decreases during the rolling period, and the WR surface achieves balanced heating and cooling cycles depending on pass schedule (Fig. 5). During phase six heat conduction cools the WR surface and WR is in contact with back-up roll (BUR). Heat conductance between WR and BUR can be assumed to be insignificant [1]. Thus, the contact is ignored due to relatively short contact length which would have required very fine discretization on contact surfaces, increasing computing time significantly. The last two phases of revolution include two cooling nozzles before next contact with strip surface. Surface temperature of the WR increases after phase eight due to thermal conductance from inside the WR to the surface.

2.2 FEM Simulations and Results

Heat transfer phenomena in single pass transfer strip rolling were simulated in 7 different cases. The entry thickness of the transfer strip was set to 39.59 mm. Simulations were designed using three different reductions, circumferential velocities of WR and initial temperatures of transfer strips, see Table 1.

FE-simulations presented in Table 1 are shown in Figs. 5 and 6. Figure 5 shows the effect of rolling speed on the heating and cooling cycles. Only circumferential

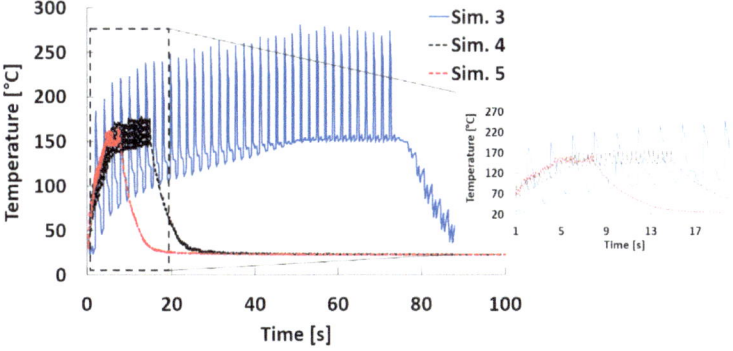

Fig. 5 WR surface temperatures of simulations 3, 4 and 5 with different circumferential velocities of WR

velocity of the WR was deviated between simulations 3, 4 and 5. Simulation 3 was run with the slowest rolling speed and long contact times with the strip and water cooling sprays cause high temperature variations on the WR surface. Simulation 1 requires the longest time to reach state of equilibrium between sequential revolutions. In equilibrium state, the lowest temperature during a single revolution is 148 °C after phase six in Fig. 4 and highest temperature 278 °C right after strip contact. When rolling speed increases in simulations 4 and 5, the state of equilibrium is reached faster as the same amount of revolution is achieved in a shorter time. Temperature differences during a single revolution are also significantly smaller. However, the lowest equilibrium temperature 148 °C after every revolution is approximately the same for all three simulations 3, 4 and 5. So the rolling speed has an effect only on time when the equilibrium state is achieved. It is also worth noting that idling revolutions after the rolling pass cools the surface of WR relatively fast regardless of rolling speed.

The effect of different reductions and transfer strip surface temperatures on warming of WR surface is depicted in Fig. 5a and b. With a small 3% reduction

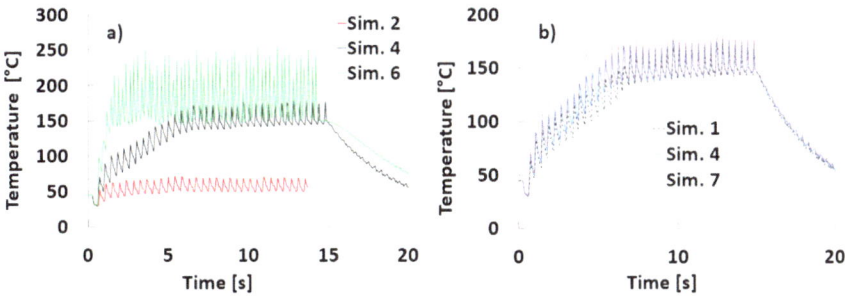

Fig. 6 WR surface temperatures of simulations **a** 2, 4 and 6 with different thickness reductions and **b** 1, 4 and 7 with different transfer bar temperatures

in simulation 2, the WR warming is almost imperceptible and equilibrium temperature is around 50 °C. This results from small reduction and very short contact length with rolled strip when there is less time for heat transfer and because the surface of the WR remains relatively cold after strip contact the cooling boundary conditions can keep equilibrium temperature low. The difference is clear compared to the high 48% reduction in simulation 6 where equilibrium is found near 150 °C which is the same with all other simulations except simulation no. 2. Also, simulations 1, 4 and 7 achieve the state of equilibrium at around 150 °C.

Simulations 3 and 6 have the highest temperature variations during a single revolution of the WR at equilibrium state. This is explained by slow circumferential WR velocity 1 m/s which leads to long contact time with all boundary conditions in simulation 3 and high thickness reduction 48% which emphasizes the heat transfer in strip contact due to long contact time in simulation 6. In any case, these two simulations achieve the same equilibrium state temperature of 150 °C with other simulations excluding simulation 2. This address that the higher surface temperature of the WR intensifies the cooling effect of water according to Eq. 1 and nearly the same equilibrium temperature of 150 °C between the heating strip contact cycle and cooling cycles is achieved with different rolling parameters.

3 Work Roll Warming Model for the Fast Virtual Rolling Model

To utilize FEM results in the fast virtual rolling simulations, a model based on the FEM results was developed. The FEM results were approximated so that the time t_2, required to reach 'equilibrium'. Equilibrium is assumed to be reached when the temperature at the end of contact is T_{high} and cools to T_{low} during one revolution before the beginning of next contact. The t_1 is the time required to reach 90% of these temperatures. These four parameters were collected from FEM results, see Fig. 7. It was assumed that the heating of work roll would depend on the following parameters: strip temperature, contact length, contact time and cooling time. A Python library called scikit-learn was used to create linear regression predictor objects that read those four parameters and outputs t_1, t_2, T_{high} and T_{low}. The input and output parameters for the linear regression predictor object are listed in Table 2. T_{strip} is the temperature of the strip before the roll contact, $t_{contact}$ is the time that a single point on work roll surface is in contact with strip during one revolution and $t_{cooling}$ is the time that the point on the surface of WR cools before it hits the surface of the strip again. R in Table 2 is the thickness reduction. Values are obtained from simulations presented in Table 1. Once all these parameters are known, the work roll surface temperature can be approximated so that it is $T_{beginning}$ (Eq. 5) in the beginning of a single contact and rises linearly during the contact so that it is T_{end} (Eq. 6) at the end of that contact.

$$T_{\text{beginning}} = \begin{cases} T_0 + t/t_1(0.9T_{\text{low}} - T_0), t \leq t_1 \\ T_{\text{low}}(0.9 + 0.1(t - t_1)(t_2 - t_1)), t_1 < t < t_2 \\ T_{\text{low}}, t \geq t_2 \end{cases} \qquad (5)$$

$$T_{\text{end}} = \begin{cases} T_0 + t/t_1(0.9T_{\text{high}} - T_0), t \leq t_1 \\ T_{\text{high}}(0.9 + 0.1(t - t_1)/(t_2 - t_1)), t_1 < t < t_2 \\ T_{\text{high}}, t \geq t_2 \end{cases} \qquad (6)$$

where t is the time since the head of the strip entered the roll bite and T_0 is the initial temperature of the WR surface.

After the rolling pass is finished, the work roll surface temperature drops following Newton's law of cooling as in Eq. 7.

$$T(t) = T_0 + (T_{\text{low}} - T_0)\exp(kt_c) \qquad (7)$$

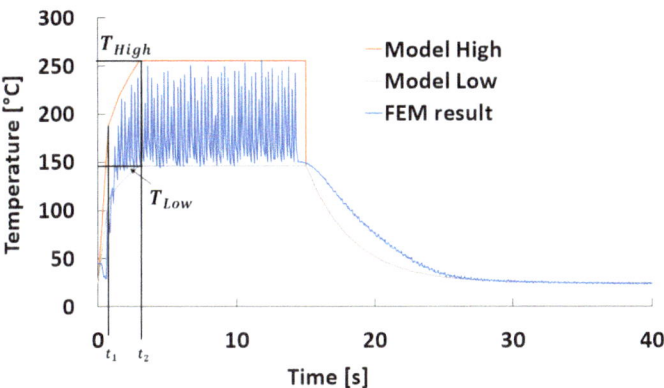

Fig. 7 Result of FE-model utilized to derivate simplified equations for predicting WR surface temperature in virtual rolling model

Table 2 The input and output parameters for the linear regression predictor object

T_{strip}	t_{contact}	t_{cooling}	R	t_1	t_2	T_{low}	T_{high}
800	11.5	425.9	0.12	6	10	142	173
900	11.5	425.9	0.12	5	10	142	178
1000	11.5	425.9	0.12	3	11	143	182
900	57.6	2129.6	0.12	15	46	147	280
900	5.8	213.0	0.12	5	6	145	165
900	5.8	431.7	0.03	2	5	48	70
900	23.1	414.3	0.48	1	3	146	255

where t_c is the time since the finishing of the last pass, and k is a coefficient that was fitted to be −0.3.

4   Results and Discussion

Developed Eqs. 5–7 for predicting WR surface temperature in roughing and finishing mill processes are implemented in the virtual rolling model. The WR warming model is utilized to calculate development of strip temperature in virtual rolling model. The same sub-model for predicting WR warming in virtual rolling model was implemented for all WRs in the model. The virtual rolling model considers rolling pass mechanics and all thermal boundary conditions on the production line as well as metallurgical phenomena. The focus in the following results is only on temperature model of the virtual rolling model. Figure 8 presents the influence of WR surface temperature on the temperature development of bottom surface of the strip. Figures 9 and 10 show two different rolling cases: coil boxed and passed-through which means that strip is finishing rolled directly after roughing passes without coil boxing in hot strip production. Calculated temperature history for strip head and tail in both cases is shown. Calculation points are located 10% lengthwise from both ends of the strip so that thermal conductance in lengthwise direction does not have effect on temperature in calculation point (1D-model). The model layout mimics hot strip mill of SSAB Raahe. In the calculated cases all features of virtual rolling mill have been used and modeled cases compared to measured process data. Calculated and measured finishing rolling temperature (FRT) are depicted in Figs. 8, 9 and 10. FRT is the bottom surface temperature of the strip after the last finishing mill rolling pass. This temperature is critical for planning suitable water cooling practices and thus accurate temperature prediction for the FRT is required in virtual rolling mill.

Two simulations throughout the virtual rolling model were run and compared in Fig. 8. In the other simulation constant WR surface temperature of 20 °C was used and another one utilized a developed WR warming model in thermal contact between the WR and the strip. Both simulations were calculated for the tail part of the strip when WR surface is clearly heated due to long lasting strip contact. Results in Fig. 8 are depicted on the finishing mill area after coil boxing process in virtual rolling model. Both calculated temperature curves of bottom strip surface and the measured FRT are presented in Fig. 8. The difference between temperature paths shows how significant difference in temperature development of strip surface occurs without considering the WR warming in virtual rolling model. On the x-axis is used normalized process time starting from discharge from walking beam furnace and ending to coiling.

The simulated strip in Fig. 9 is produced utilizing coil boxing process. It can be seen how the tail of the strip spent a long time coiled in the coil box. There is a clear difference in finishing mill rolling starting temperature (RST) between strip head and tail. The RST is used in Figs. 9 and 10 to point out the bottom surface temperature of the strip before finishing mill descaling process. There is no measured value for

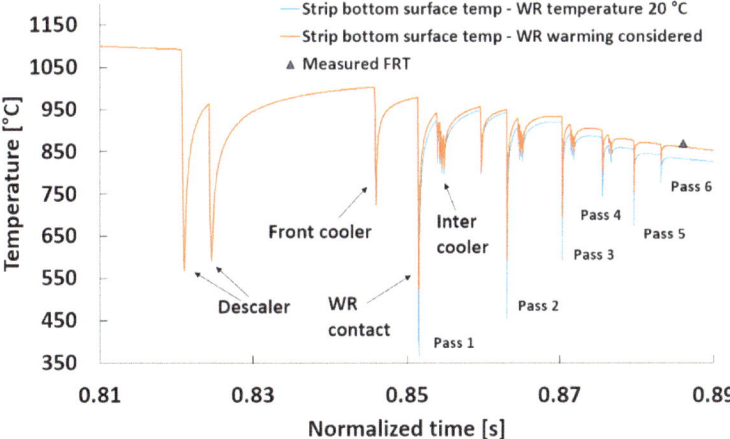

Fig. 8 Effect of WR surface warming on temperature development of bottom surface of the strip. Simulation with constant WR surface temperature (20 °C) and simulation with WR warming model implemented in virtual rolling model are presented

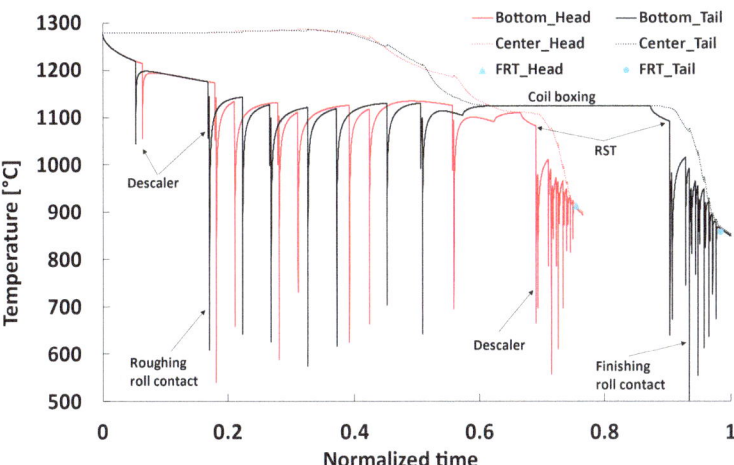

Fig. 9 Calculated temperature history of strip head and tail (bottom surface and center of the strip) in coil boxed strip production

RST from process data. According to calculated temperature paths in Fig. 9, the finishing mill process of the strip head starts at lower RST than rolling of the tail. And because the FRT of the head of the strip is also higher than the FRT of the tail which finishing rolling process is completed much later, there must have been adjustments of finishing rolling parameters between rolling the head and the tail of the strip. This is explained by deceleration of tandem mill from 4.7 to 3.7 m/s (6th stand rolling speed) between the strip head and tail. This influences the time at which

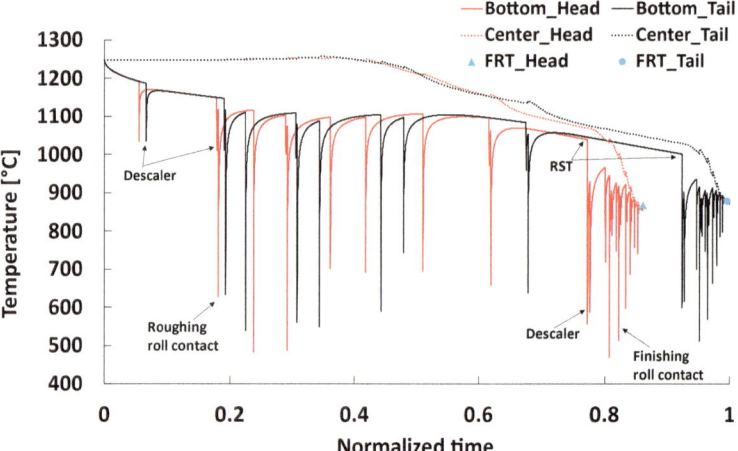

Fig. 10 Calculated temperature history of strip head and tail (bottom surface and center of the strip) in passed-through strip production

boundary conditions affect the virtual rolling mill due to which the tail of the strip loses more thermal energy than head of the strip. All heat transfer mechanisms last for a longer time and less adiabatic heat is produced in the strip. Measured FRT values confirm the results.

The other case in Fig. 10 is produced without coil boxing process. Unlike in the coil boxed case the finishing mill has accelerated from 4.2 to 5.1 m/s between the strip head and tail, leading to slightly higher FRT at strip tail. Measured FRT values agree with the calculated temperatures. In both cases, the new model for WR surface warming was used and according to results it is suitable for predicting WR surface temperatures in virtual rolling model. Ignoring WR surface temperature variations during the rolling period inevitably leads to significant temperature errors in virtual rolling model.

Results prove how essential continuously changing boundary conditions are for the virtual rolling model which calculates the entire hot rolling process. To simulate temperature development throughout the hot strip production line, all thermal boundary conditions must be considered. Simulation results of virtual rolling model were compared to the measured FRT because there are generally several minutes of rolling process past at this point and it is one of the most critical temperatures in the process regarding the final properties of the strip. So, the model must be able to predict FRT reliably. According to simulation results, the WR warming model predicts reliably WR surface temperature throughout the rolling process. The WR warming model is only one parameter which is changing constantly in industrial process. For that reason, the presented results contained all features of virtual rolling model. These are virtual rolling automation, temperature model, calculation module for metallurgical phenomena and thermo-mechanical module for rolling passes. The virtual rolling automation mimics the industrial automation system controlling the strip position, velocity and changes of direction in roughing. The temperature module acknowledges all thermal phenomena and boundary conditions and delivers them to other modules during the simulation. The metallurgical module calculates recrystallization, grain growth and phase transformations which are further utilized in thermo-mechanical rolling module which in turn calculate phenomena in roll bite predicting roll force and WR flattening based on contact pressure and friction shear stress between the strip and WR.

As an afterthought the thickness reduction parameter for simulation 2 was poorly chosen because such a small reduction is not generally used in hot strip rolling process. On further investigation 6-stands FE-model of finishing mill can be used to study temperature development of hot strip and WR surfaces in multi-stand rolling process. Also multiple strips could be simulated sequentially to investigate evolution of equilibrium temperature over rolling periods.

5 Conclusion

Work roll heating and cooling cycles were modeled with detailed FE-model to find out what kind of temperature variations work roll surface goes through during the rolling pass. It was also unknown that the work roll surface would reach equilibrium state at some point of rolling full length strip. Thermal cycles and equilibrium state were established in results of FE-model and a simplified version for virtual rolling model was derived from the FE-results. In validation results the temperature development of strip in virtual rolling model was modeled. The correct FRT value was obtained for hot strip rolling cases which means that the temperature development throughout the hot strip rolling process was modeled correctly. Calculated FRT after finishing mill corresponded with the measured FRT value from industrial process. The validity of the temperature calculation between strip head and tail was compared to the measured FRT with good correlation.

It must be considered that every cooling system for WR surface cooling is unique and water flows, impingements areas and water pressure may differ significantly. Thus, parameters used for effective heat transfer equations in WR water cooling in this study may vary noticeably in validation to other cooling systems. This may result in a different equilibrium temperature on WR surface in steady state rolling.

The virtual rolling model also considers the most essential thermal boundary conditions and includes virtual rolling automation which both are required to calculate the temperature evolution of the whole hot strip rolling process reliably.

Acknowledgements This research has been carried out in co-operation with SSAB Europe and funding from Business Finland FOSSA-project. The authors also acknowledge the valuable technical input and knowhow to the research by M.Sc. J. Paavola and M.Sc Oskari Seppälä from the research group of Material and Mechanical Engineering at the University of Oulu.

References

1. Guerrero MP, Flores CR, Pérez A, Colás R (1999) Modelling heat transfer in hot rolling work rolls. J Mater Process Technol 94:52–59
2. Fapiano DJ (1982) Crown, feather and flatness in hot rolling. Iron Steel Eng59: 29–35
3. Sheppard T, Roberts JM (1973) Shape control and correction in strip and sheet. Int Metall Rev 18:1–18
4. Lin ZC, Chen CC (1995) Three-dimensional heat-transfer and thermal-expansion analysis of the work roll during rolling. J Mater Process Tech 49:125–147
5. Johnson RE, Keanini RG (1998) An asymptotic model of work roll heat transfer in strip rolling. Int J Heat Mass Transf 41:871–879
6. Serajzadeh S, Taheri AK (2004) Modelling work-roll temperature variations in hot strip rolling. Int J Model Simul 24:42–50
7. Ilmola J, Seppälä O, Leinonen O, Pohjonen A, Larkiola J, Jokisaari J and Putaansuu E (2018) Multiphysical FE-analysis of a front-end bending phenomenon in a hot strip mill. AIP Conf Proc 1960
8. Ilmola J, Pohjonen A, Seppälä O, Leinonen O, Larkiola J, Jokisaari J, Putaansuu E and Lehtikangas P (2018) Coupled multiscale and multiphysical analysis of hot steel strip mill and microstructure formation during water cooling. Procedia Manufacturing
9. Ilmola J, Seppälä O, Pohjonen A, Larkiola J (2022) Virtual rolling automation and setup calculations for six stands FEM finishing mill. IOP Conf Ser Mater Sci Eng 1270: 012060
10. Martin DC (2011) Selected heat conduction problems in thermomechanical treatment of steel, University of Oulu
11. Hodgson P, Browne KM, Collison D, Pham T, Gibbs R (1991) A mathematical model to simulate the thermomechanical processing of steelitle. Quenching carburizing. Int Fed Heat Treat Surf Eng 139–59
12. Leinonen O, Ilmola J, Seppälä O, Pohjonen A (2018) Experimental determination of heat transfer coefficients in roll bite and air cooling for computer simulations of 1100 MPa carbon steel rolling

Soft Sensor Model of Phase Transformation During Flow Forming of Metastable Austenitic Steel AISI 304L

Julian Rozo Vasquez[ID], **Lukas Kersting**[ID]**, Bahman Arian,**
Werner Homberg[ID]**, Ansgar Trächtler, and Frank Walther**[ID]

Abstract This paper deals with the modeling of a soft sensor for detecting α'-martensite evolution from the micromagnetic signals that are measured during the reverse flow forming of metastable AISI 304L austenitic steel. This model can be prospectively used inside a closed-loop property-controlled flow forming process. To achieve this, optimization by means of a non-linear regression of experimental data was carried out. To collect the experimental data, specimens were produced by flow forming seamless tubes at room temperature. Using a combination of production parameters (like the infeed depth and feed rate), specimens with different α'-martensite contents and wall-thickness reductions were produced. An equation to compute α'-martensite from both specific production-process parameters and micromagnetic Barkhausen noise (MBN) measurements was obtained using numerical methods. In this process, the behavior of the quantity of interest (namely, the α'-martensite content) was mathematically evaluated with respect to non-destructive MBN data and the feed rate that was used to produce the components. A combination of exponential and potential functions was defined as the ansatz functions of the model. The obtained model was validated online and offline during the real flow forming of workpieces, obtaining average deviations of up to 7% α'-martensite with respect to the model. The implementation of the soft sensor model for property-controlled production represents an important milestone for producing high-added-value components on the basis of a well-understood process-microstructure-property relationship.

J. R. Vasquez (✉) · F. Walther
Chair of Materials Test Engineering (WPT), TU Dortmund University, Baroper Str. 303, 44227 Dortmund, Germany
e-mail: julian.rozo@tu-dortmund.de

L. Kersting · A. Trächtler
Fraunhofer Institute for Mechatronic Systems Design (IEM), Zukunftsmeile 1, 33102 Paderborn, Germany

B. Arian · W. Homberg
Forming and Machining Technology (LUF), Paderborn University, Pohlweg 53, 33098 Paderborn, Germany

© The Rightsholder, under exclusive licence to [Springer Nature Switzerland AG], part of Springer Nature 2024
J. Kusiak et al. (eds.), *Numerical Methods in Industrial Forming Processes*, Lecture Notes in Mechanical Engineering, https://doi.org/10.1007/978-3-031-58006-2_10

117

Keywords Material behavior modeling · Optimization · Non-linear regression · Soft sensor · Micromagnetic testing · Phase transformation · Flow forming · Closed-loop property control

1 Introduction

The production of components by flow forming holds a remarkable importance in a wide range of industrial fields due to its manufacturing flexibility, its resource optimization, and the possibility of producing components with not only a desired geometry but also specific mechanical properties [1]. The use of austenitic stainless steel as a raw material in combination with advanced manufacturing techniques makes the production of high-performance products possible. During the metal-forming of metastable austenitic steel, plastic deformation not only changes the geometrical characteristics but also the microstructure. In particular, a strain-induced phase transformation from metastable austenite to α'-martensite may occur; this modifies the magnetic and mechanical properties [2, 3].

Modeling in metal forming processes plays a crucial role in the evolution, optimization, and description of the process limits. Analytic and simulative modeling approaches include methods like slab, slip-line field, upper bound, finite difference, and finite elements [4, 5]. Some investigations focus on material behavior and the formability of sheet metal [6, 7]. Other studies have described models of phase transformation in metal-forming processes [8–10]. Regarding the modeling of α'-martensite phase transformation during plastic deformation, some general mathematical models can be found in the literature [3, 11, 12]; among these, the most representative is the Olson-Cohen model [13].

The primary goal of the overall research project is to develop a property-controlled flow-forming process. The implementation of online closed-loop control within metal forming processes enables the application-oriented and efficient production of high-quality components [14]. The control systems require the development of novel soft sensors to monitor the evolution of the properties during the plastic deformation. By definition, a soft sensor allows for the transformation of measured secondary variables into more useful process knowledge by means of a model [15]. A soft sensor integrates a hardware part (which, in this case, is a micromagnetic sensor device) and a software part (or soft sensor model) to transform the measured signals into numerical values of the variable(s) of interest. In this work, a soft sensor model for detecting α'-martensite evolution from measurable micromagnetic properties in combination with process parameters was developed. To achieve this, optimization by means of a non-linear regression of experimental data was carried out. This kind of optimization has been widely used in many branches of science in such applications as economics, operations research, network analysis, and the optimal design of mechanical or electrical systems [16]. In a material-behavior model during metal forming, this represents a novel approach.

2 Materials and Methods

2.1 Specimen Production by Means of Reverse Flow Forming

The specimens were produced using a PLB 400 single roller spinning machine that was manufactured by Leifeld Metal Spinning GmbH (Ahlen, Germany). The raw material for the specimens consisted of seamless tubes that were made of metastable AISI 304L austenitic steel (X2CrNi18-9, 1.4307) with an outer diameter of 80 mm. The specimens were produced at room temperature with a constant rotational speed of $n = 30$ rpm. The flow-forming operation was performed using a single roller arrangement in order to reduce the wall thickness of the tubes (Fig. 1a). To produce specimens that combined different geometrical features and properties, the following process parameters were used: infeed depths (the positions of the roller tool on the y-axis) of 1, 2, and 3 mm, and feed rates (the speeds of the roller tool in the axial direction [x-axis]) of 0.1, 0.2, 0.3, 0.4, and 0.5 mm/s (Fig. 1b). To ensure the reliability of the results, each flow-forming experiment was repeated three times.

Figure 2a illustrates the dimensions of the produced specimens. The initial wall thickness (w_{IC}) was reduced according to the three different infeed depths that were used to produce the resulting wall-thickness reduction (Δw). The final appearance of each of the specimens is shown in Fig. 2b.

Starting from the initial condition (IC), which corresponded to the metastable austenite, the forming process generated the illustrated forming zone (FZ). For each

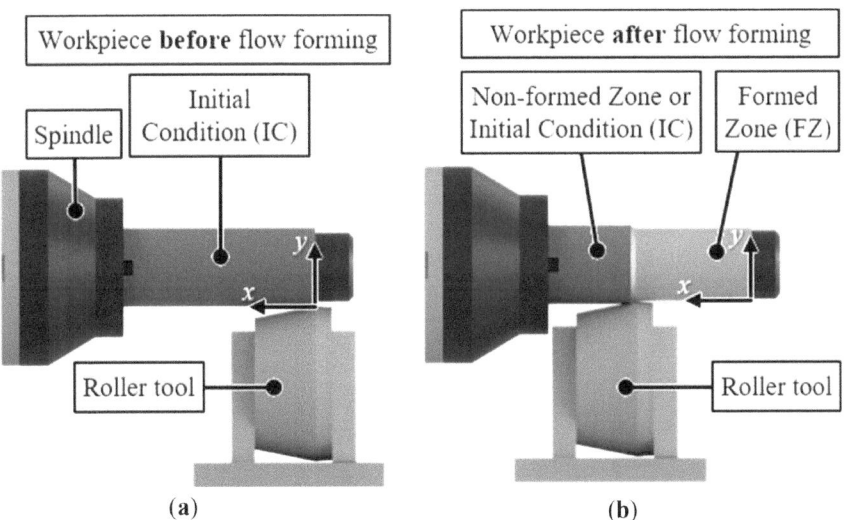

Fig. 1 Setup for production of specimens: **a** before reverse flow forming; **b** after reverse flow forming

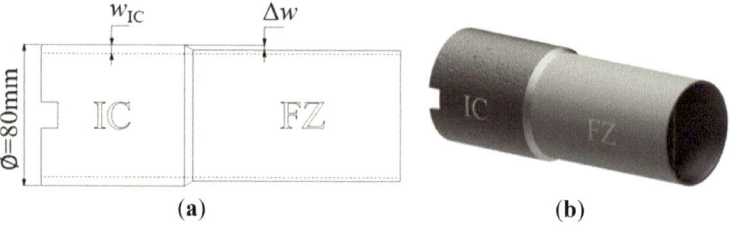

Fig. 2 Workpiece geometry: **a** dimensions; **b** final appearance

specimen, the forming zone corresponded to a plastic deformation state with a defined amount of strain-induced α'-martensite depending on the process parameters.

2.2 Procurement of Experimental Data

The experimental data that were used in the modeling of the soft sensor were obtained by means of characterizing the evolution of geometrical features and properties like wall-thickness reduction (Δw), α'-martensite amount, roughness, and the micromagnetic parameters of the workpieces. For this investigation, a total of 45 specimens were produced: three infeed depths × five feed rates × three repetitions.

Figure 3 illustrates the points that were measured to obtain the experimental data. Each specimen was measured around four equally spaced axes (A–D). Along each axis, six points were recorded for the forming zone (FZ) and one for the initial condition (IC). In summary, 28 measurements were carried out on each specimen, which adds up to a total of 1260 measurements for each characterization technique.

The conducted characterization techniques for acquiring the experimental data are described as follows:

Wall-thickness Reduction, Average Roughness, and α'-Martensite Measurements. The wall-thickness reduction (Δw) was measured offline on the produced specimens by using a D4R50 mechanical gage (Kroeplin GmbH, Schluechtern, Germany). The arithmetic mean roughness (R_a) of the formed specimens was measured with a MarSurf M300 tactile device (Mahr GmbH, Goettingen, Germany).

Fig. 3 Points measured for acquisition of experimental data

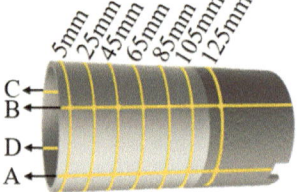

The α'-martensite content measurements were used as reference values in order to characterize the phase transformation during the plastic deformation of the material. The data was recorded by means of an FMP30 Feritscope (Helmut Fischer GmbH, Sindelfingen, Germany). This device works under the magneto-inductive principle to measure the ferritic phase content, which is correlated with the strain-induced α'-martensite (according to [17]). These measurements could not be used for closed-loop property control, as the online acquisition and real-time transmission of the data to a control system was not possible. As a result, the use of micromagnetic testing was a convenient alternative as a soft sensor for property control.

Micromagnetic Measurements. By means of the magnetic Barkhausen noise (MBN) analysis, different material features such as lattice defects, dislocation density, phase changes, grain sizes, and residual stresses could be detected. In previous publications, the suitability of MBN testing has been demonstrated for detecting the amounts of strain-induced α'-martensite phases—especially by means of the maximum amplitudes of the MBN profiles (M_{max}) [18–20]. MBN testing has been used by many authors for detecting ferromagnetic α'-martensite phase amounts [21–23]. In this work, the micromagnetic measurements were performed using the 3MA-II system (Fraunhofer IZFP, Saarbruecken, Germany), with the SN18201sensor in contact with the outer surfaces of the specimens. The measurements were carried out by emulating the real conditions for the online measurements. This meant protecting the sensor head against wear with two layers of Kapton bands and the appropriate testing parameters for the online application. The micromagnetic parameters for the measurements were as follows: a 200 Hz magnetization frequency, an 80 A/cm magnetization amplitude, a 15° magnetic phase offset, a 20 dB gain, and without bandpass filters on. This allowed us to perform an integral measurement of the entire wall thickness.

3 Results and Discussion

3.1 Wall Thickness, Roughness, α'-Martensite Content, and MBN Signals

As mentioned in Sect. 2.2, four different characterization methods were used on the 45 specimens in order to obtain the experimental data for the soft sensor model. A summary of the quantitative results of the measurements is presented in Table 1.

Table 1 Quantitative results for characterizations of wall-thickness reduction (Δw), α'-martensite content (α'), and maximum amplitude of MBN (M_{max}) and their respective standard deviations for flow formed workpieces

Feed rate f (mm/ s)	Infeed depth y (mm)	Δw(mm)	Std. dev. Δw (mm)	α' (%)	Std. dev.α' (%)	M_{max} (V)	Std. dev. M_{max} (V)	Roughness R_a (μm)	Std. dev. R_a (μm)
0.1	0	–	–	0.87	0.48	0.1130	0.0006	4.33	0.82
	1	0.191	0.035	45.95	2.75	0.8247	0.0893		
	2	0.473	0.026	52.16	3.83	1.2537	0.1536		
	3	0.774	0.017	70.94	0.52	1.8408	0.0690		
0.2	0	–	–	0.55	0.14	0.1130	0.0008	4.74	2.04
	1	0.160	0.015	24.13	0.50	0.7737	0.0634		
	2	0.373	0.008	38.39	4.68	1.2145	0.1177		
	3	0.551	0.019	43.55	5.01	1.4560	0.1719		
0.3	0	–	–	0.59	0.35	0.1127	0.0003	6.02	1.63
	1	0.135	0.013	22.56	2.95	1.0260	0.0842		
	2	0.332	0.014	30.94	3.21	1.2764	0.1781		
	3	0.462	0.007	32.10	3.87	1.2970	0.1223		
0.4	0	–	–	0.94	0.44	0.1129	0.0005	8.83	0.79
	1	0.142	0.005	15.61	0.26	1.0795	0.3467		
	2	0.303	0.011	27.25	1.78	1.1409	0.1841		
	3	0.404	0.019	32.70	1.79	1.5013	0.1259		
0.5	0	–	–	0.67	0.42	0.1119	0.0004	11.83	0.93
	1	0.134	0.025	14.66	1.42	0.9132	0.1528		
	2	0.279	0.010	21.00	2.39	0.9857	0.1559		
	3	0.370	0.017	23.59	3.28	1.1088	0.2230		

3.2 Correlation Between α'-Martensite, M_{max}, and Feed Rate

Figure 4a illustrates the evolution of the α'-martensite content with respect to the wall-thickness reduction (Δw). Keeping constant production parameters like the feed rate and varying infeed depth, more α'-martensite was present on the microstructure of the flow formed workpieces due to the increased plastic deformation with higher Δw. The lower feed rates of the roller tool on the x-axis during the flow forming process (see Fig. 1) entailed the higher local deformation of the areas that were in contact with the roller tool; this led to increased α'-martensite formation. In this way, the strain-induced deformation promoted the transformation of the austenite into α'-martensite. These aspects have been discussed in previous publications (e.g., [18, 19]).

Evaluating the roughness on produced specimens is crucial for applying non-destructive testing by means of MBN measurements. As has been demonstrated in earlier papers, higher roughness values increase the numbers of detected MBN

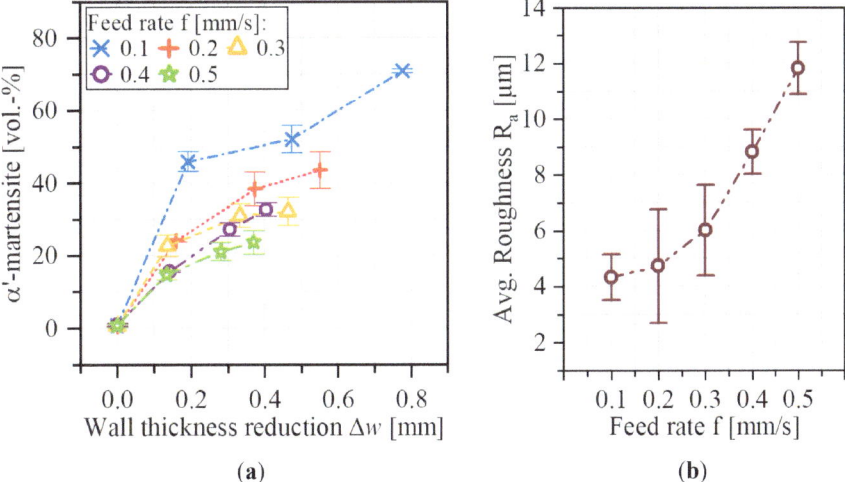

Fig. 4 **a** Correlation between α'-martensite and wall-thickness reduction for flow formed specimens at different feed rates; **b** influence of feed rate on average roughness of specimens

signals [24, 25]. Since roughness cannot be measured online during the production process, their effects can be considered by means of other variables. The feed rate of the roller tool during the flow forming of the specimens has a great influence on the roughness of flow-formed components. From our numerical results (Table 1, Fig. 4b), higher feed rates increased the average roughness (R_a). At lower feed rates, the distance between the microgrooves that were produced during the flow-forming process was smaller; this entailed lower roughness values. For this reason, the feed rate was chosen as an important production parameter for the soft sensor modeling.

Figure 5 illustrates the correlation between the α'-martensite content and the MBN measurements (specifically, M_{max}); the data is plotted for feed rates between 0.1 and 0.5 mm/s. The results showed a good correlation for feed rates between 0.1 and 0.3 mm/s; however, the separation between the measured data was not clear enough for feed rates of 0.3 through 0.5 mm/s. In Fig. 6, a 3D plot of the experimental data allows one to interpret the results in a clearer way.

3.3 Non-linear Regression by Means of Data-Fitting Optimization

Definition of Ansatz Function. The Ansatz function must be defined as an initial assumption for solving the problem. In this case, the starting equation must describe the behavior of the quantity of interest (namely, the α'-martensite) with respect to secondary measurable quantities like maximum MBN amplitude (M_{max}) and process

Fig. 5 Correlation between α'-martensite and maximum MBN amplitude M_{max} for flow-formed specimens at different feed rates

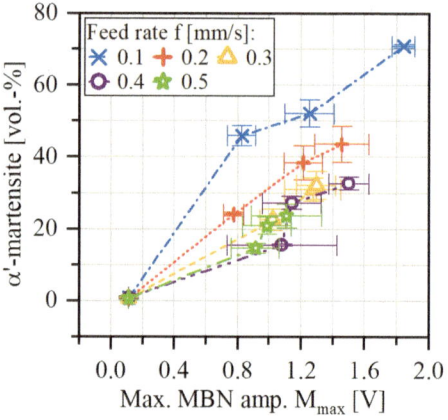

Fig. 6 Three-dimensional plot of correlations among α'-martensite, maximum MBN amplitude M_{max}, and feed rate for flow-formed specimens

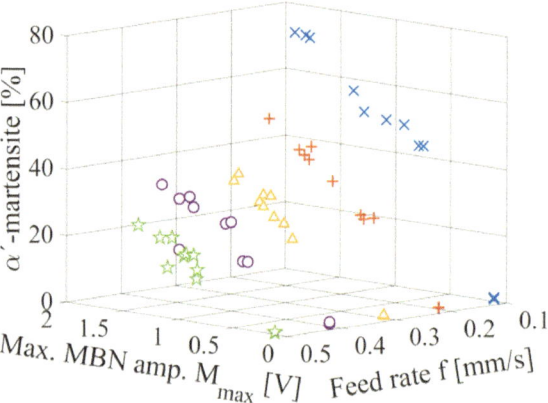

parameters like the feed rate. As previously discussed, the selected feed rate for the production of specimens had a remarkable influence on the roughness of our specimens; this also had an impact on the measured M_{max} signals.

In Fig. 7a, a non-linear regression of the experimental data of the α'-martensite with respect to M_{max} was carried out using the Simple Fit tool from OriginPro (2021b). A good fitting of the points was obtained by means of an exponential function. Similarly, a potential function performed the best description of the behavior of the α'-martensite with the feed rate. The product of both functions was expected to generate the combined effect of the variables. The final Ansatz function is expressed in Eq. 1.

$$\alpha\prime = \beta_1 \left(1 - \beta_2^{M_{max}}\right)\left(f^{\beta_3}\right) \tag{1}$$

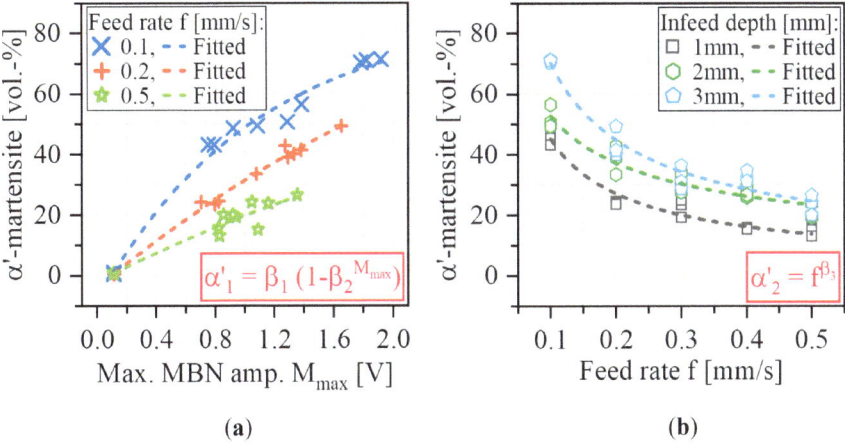

Fig. 7 Non-linear regression based on experimental data for α'-martensite content depending on **a** maximum MBN amplitude M_{max} and **b** feed rate f

The problem is now to find the numerical values of the β_n coefficients that better fit the Ansatz equation to the experimental data.

Derivatives of Ansatz Function. To calculate the numerical values of the β_n coefficients in Eq. 1, it is necessary to compute the partial derivatives of the Ansatz function with respect to each coefficient. The partial derivatives are detailed in Eqs. 2 through 4.

$$\partial\alpha/\partial\beta_1 = \left(1 - \beta_2{}^{M_{max}}\right)\left(f^{\beta_3}\right) \tag{2}$$

$$\partial\alpha/\partial\beta_2 = -\beta_1(M_{max})\left(\beta_2{}^{M_{max}-1}\right)\left(f^{\beta_3}\right) \tag{3}$$

$$\partial\alpha/\partial\beta_3 = \beta_1\left(1 - \beta_2{}^{M_{max}}\right)\left(f^{\beta_3}\right)(ln(f)) \tag{4}$$

Optimization by Means of Objective Function. The definition of an objective function or cost function is crucial for solving the optimization problem. The problem is formulated in terms of a real function of several real variables. The goal of the optimization is to find a set of arguments that provide a minimal function value [16]. For this optimization, the convergence is defined by means of Newton's method, where f is the objective function, and $\widehat{\alpha'}$ is the minimizer. In the implementation, the objective function (Eq. 5) and its derivatives (Eq. 6) are used.

$$f_{n+1} = f_n + 1/2\left(\alpha'_n - \widehat{\alpha'}\right)^2 \tag{5}$$

$$df_{n+1} = df_n + \left(\alpha'_n - \widehat{\alpha'}\right)d\alpha'_n \tag{6}$$

MATLAB Implementation. The implementation in MATLAB for the solution of the problem involves the inclusion and plotting of the experimental data, the formulation of the Ansatz function (and its derivatives), and the convergence function. To compute the unknown β_n coefficients, the initial values were obtained by means of the non-linear regressions that were discussed previously in this section (Fig. 7). The *fmincon solver* was used to perform the optimization and find the quantitative values of the coefficients.

Surface and Equation of Soft Sensor Model. The numerical results of the unknown β_n parameters that were calculated by means of the implementation in MATLAB were $\beta_1 = 37.9134$, $\beta_2 = 0.6495$, and $\beta_3 = -0.534$. Including the numerical values of the coefficients in the Ansatz equation, the soft sensor model can be expressed by means of Eq. 7.

$$\alpha' = 37.9134\left(1 - 0.6495^{M_{max}}\right)\left(f^{-0.5309}\right) \tag{7}$$

Figure 8 shows a 3D plot of Eq. 7. The experimental data is plotted for a qualitative comparison to the soft sensor model.

Quantitative Error Analysis of Soft Sensor Model with Respect to Experimental Data. The deviation of the α'-martensite that was calculated with the model and the measured experimental data is plotted in Fig. 9. The quantitative values of the maximal deviation between the calculated and experimental data of the α'-martensite

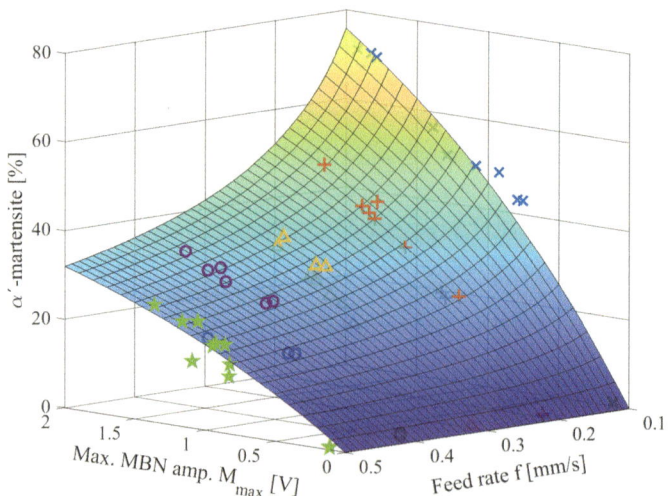

Fig. 8 Surface of soft sensor model resulting from data-fitting optimization by means of non-linear regression of experimental data (experimental data plotted with symbols)

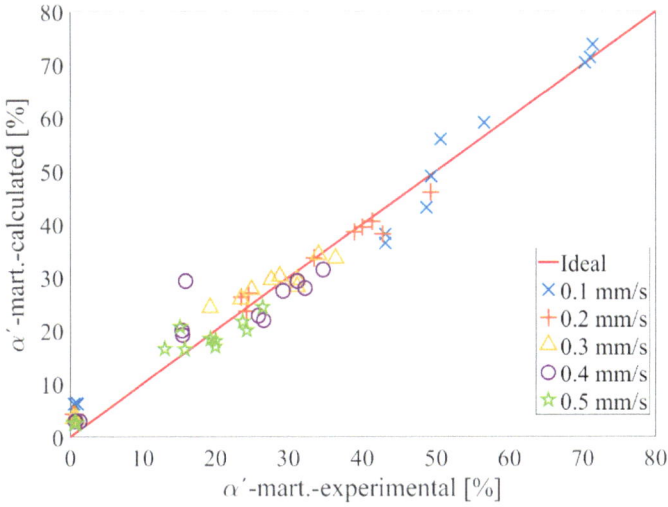

Fig. 9 Deviation plot of α'-martensite calculated by model vs. experimental data

are summarized in Table 2. The highest deviation was 13.4 α'-%; however, most of the deviation values were below 6%.

The root means square error (RMSE) of the soft sensor model was calculated for the group of data on each value of the feed rate according to Eq. 8; these results are presented in Table 2. The calculated errors were below 5α'-%. It can be concluded that the α'-martensite amount that was calculated with the soft sensor model was accurate with respect to the experimental data. A new challenge for applying the soft sensor model consisted of comparing the online measurements that were recorded under real production conditions with the offline α'-martensite measurements that were taken on the produced specimens. This is the topic of the next section.

$$RMSE = \sqrt{\sum_{i=1}^{N} \left(\alpha'_{calculated} - \alpha'_{experimental} \right)^2 / N} \qquad (8)$$

Table 2 Deviation and root mean square error (RMSE) of soft sensor model with respect to experimental data for different feed rates

Feed rate f (mm/s)	0.1	0.2	0.3	0.4	0.5
α'-deviation [α'-%]	6.73	4.61	5.02	13.40	5.50
RMSE [α'-%]	4.36	2.70	2.69	4.94	2.80

3.4 Online Validation of Soft Sensor Model

Since the goal of the soft sensor is to apply it in property-controlled flow-forming processes, validation under real production conditions is necessary. Figure 10 illustrates the setup for the online measurement of α'-martensite by means of the soft sensor model. For the experiments, the physical 3MA-II sensor was coupled with the soft sensor model, which was implemented in the machine control via a TCP/IP approach that was developed by the authors (see [26]). The micromagnetic soft sensor parameters were set to be equal to the characterization that was described in Sect. 2.2. The online measurements were recorded during the formation of the AISI 304L tubes with a constant infeed depth of 2 mm and feed rates of 0.1, 0.3, and 0.5 mm/s (Tests 1, 2, and 3, respectively). The data was recorded along the x-position. The flow forming process required a liquid coolant to refrigerate the workpieces and keep them at a constant temperature; this coolant had no remarkable influence on the measured signals.

The results of the online measurements of the α'-martensite amount by means of the soft sensor are plotted in Fig. 11 versus the x-position (see Fig. 10—right). The highest α'-martensite amounts were detected in Test 1, which was performed with the lowest feed rate of the roller tool. The lowest values of α'-martensite corresponded to Test 3 due to the high feed rate that was used during the production of the components. In all of the performed tests, the infeed depth of the roller tool remained constant (at 2 mm of the outer surface of the tubes). The results of the online measurements under the real production conditions agreed with the measurements that were performed on the specimens that were produced to generate the experimental data that was used to develop the soft sensor model (see Sect. 3.2).

The produced specimens were measured offline with the FMP30 Feritscope to determine the reference values of the amounts of α'-martensite on each test. To

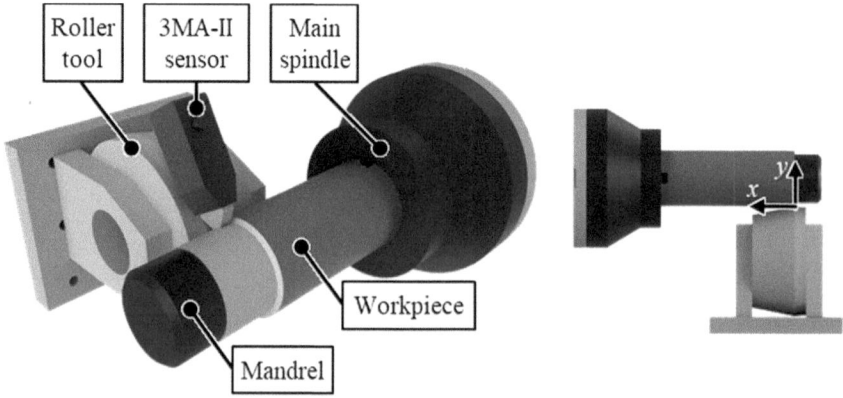

Fig. 10 Setup for online measurements of α'-martensite by means of soft sensor model using 3MA-II micromagnetic sensor during flow forming of workpiece with infeed depth of 2 mm and feed rates of 0.1, 0.3, and 0.5 mm/s

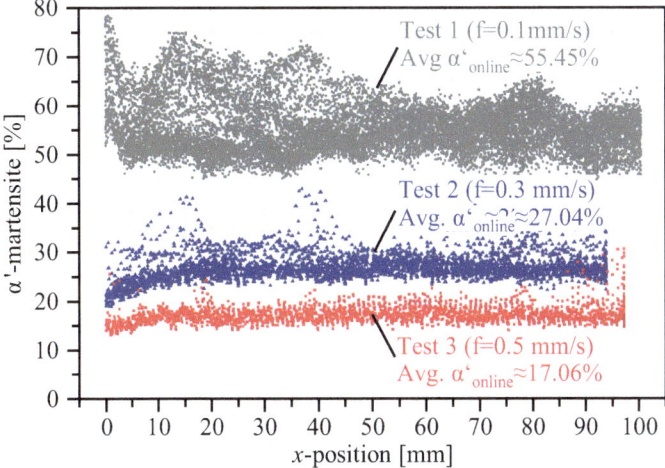

Fig. 11 Online measurements of α'-martensite by means of soft sensor model during flow forming of workpiece with infeed depth of 2 mm and feed rates of 0.1, 0.3, and 0.5 mm/s

increase the resolution of the α'-martensite measurements, the measured points along the *x*-position and around the A-D axes were increased with respect to the measurements that were performed in Sect. 2.2 (see Fig. 3). The results of the offline measurements (plotted in Fig. 12) corresponded well with the online measurements that used the soft sensor model.

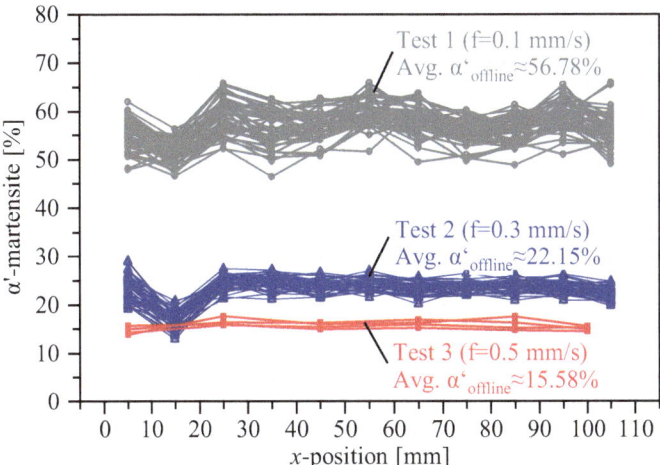

Fig. 12 Offline measurements of α'-martensite by means of Feritscope after flow forming of workpiece with infeed depth of 2 mm and feed rates of 0.1, 0.3, and 0.5 mm/s

Table 3 Average α'-martensite and deviation measured online with soft sensor model compared to offline reference measurements with Feritscope for different feed rates

	Feed rate f (mm/s)	Avg. α'-martensite online (α'-%)	Avg. α'-martensit offline (α'-%)	Deviation-α' (α'-%)
Test 1	0.1	55.45	56.78	1.3
Test 2	0.3	27.04	22.15	4.89
Test 3	0.5	17.06	15.58	1.48

Qualitatively, the measured α'-martensite amount for each test could be well-differentiated. The results of Test 1 showed the greatest scattering in both the online and offline measurements; in both cases, the data was between 45 and 65 α'-%. However, peaks of up to 80 α'-% were recorded between 0 and 40 mm of the x-position in the online measurements. These could be the effects of surface roughness areas that were considered to be disturbances. In Test 2, the drop that was measured offline at 15 mm in the x-position was not detected in the online measurements. In this region, the lower α'-martensite amount could not be detected on the online measurements due to the large measuring spot of the micromagnetic sensor that was used in the soft sensor (which produced delays in the detected signals). The online and offline signals that were measured in Test 3 corresponded very well. Some peaks were detected in the online measurement from 70 mm in the x-position; these can be evaluated as measurement noises since they were not representative.

A quantitative analysis was carried out with the average values of the α'-martensite amount that was calculated for each test of the online and offline measurements; these results are reported in Figs. 11 and 12 and summarized in Table 3. The maximal deviation of the α'-martensite amount could be found in Test 2 due to the non-detected drop in the online measurement at 15 mm of the x-position. These deviations were acceptable within the window operation of the soft sensor since the repeatability of the measurements that were made with the Feritscope reached up to 5 α'-%.

3.5 Future Perspective for Application of Soft Sensor in Closed-Loop Property Control

The proposed soft sensor can be prospectively used inside a closed-loop property control of the α'-martensite volume fraction during the flow-forming process (as shown in Fig. 13). As discussed, M_{max} is measured online during the manufacturing of the specimen and transformed to the α'-martensite volume fraction by the soft sensor model. Subsequently, the control variable α'-martensite volume fraction is compared to the desired α'-martensite value by the controller. The correction signal from the controller output directly manipulates process parameters such as the feed rate and infeed depth or the tool trajectory of the flow-forming machine during

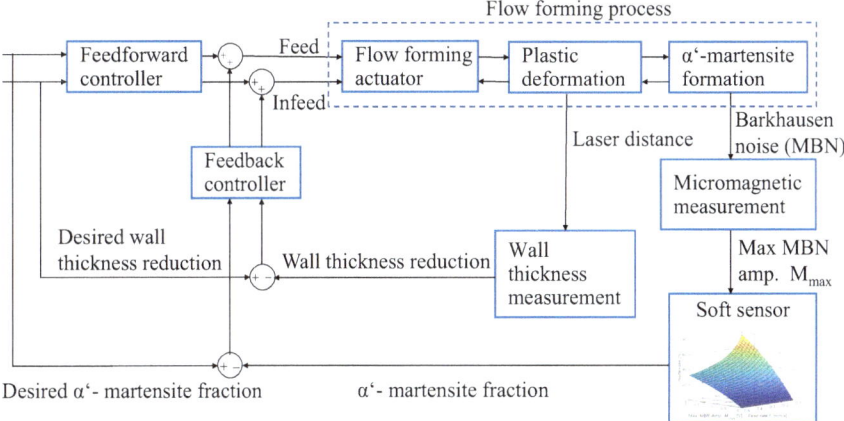

Fig. 13 Soft sensor-based multi-variable control concept of closed-loop property control during flow forming of metastable austenitic stainless steels (*note* for simplicity, feed-dependence of soft sensor model is not drawn)

the manufacture of the workpiece. Thus, the α'-martensite fraction of the manufactured forming part is set to the desired value. The wall-thickness closed-loop control [26] complements the soft sensor-based control of the α'-martensite to concurrently ensure the desired geometry of the part. Overall, the closed-loop property control is, therefore, a multi-variable control. For the design of such a control, additional numerical model-based approaches can be used in terms of control-oriented real-time simulation models (cf. [26, 27] for further information).

4 Conclusions

To achieve the property-controlled production of flow formed components, a soft sensor model was proposed and validated by means of two strategies. The soft sensor aims to compute the variable of interest for the property control (namely, the α'-martensite content), which is not directly measurable within the process. Secondary measurable variables like the maximum amplitude of the magnetic Barkhausen noise were successfully correlated with the variable of interest. The influence of process parameters such as the feed rate were included in the soft sensor model, as they play a crucial role in the forming process, the phase transformation of the material, and the online measurements. The model was obtained through optimization by means of a non-linear regression of experimental data. The validation of the soft sensor model first took place with respect to the experimental data that was used on the modeling and then online under real flow forming-process conditions. The influence on the measurements of different factors like the liquid coolant was evaluated.

The developed soft sensor model requires additional measurements that combine different production parameters and strategies in order to complete the validation for use on a property-controlled production. The operation of the sensor under real conditions demands an investigation of the compensations of any disturbances of the real process. There is also an intent to look for an alternative instrumentation for detecting α'-martensite with lower sensitivities to disturbances (such as the roughness of a specimen). In this sense, sensors and devices working under eddy-current principles are promising.

Acknowledgements The authors would like to thank the German Research Foundation (Deutsche Forschungsgemeinschaft, DFG) for their support of the depicted research within the priority program SPP 2183 "Property-Controlled Forming Processes," through project No. 424335026 "Property Control During Spinning of Metastable Austenites." Special thanks to Dr. Cesar Polindara for the excellent discussion about the optimization model and his cooperation in implementing the model in MATLAB.

References

1. Marini D, Cunningham D, Xirouchakis P, Corney J (2016) Flow forming: a review of research methodologies, prediction models and their applications. Int J Mech Eng 7:285–315
2. Sohrabi MJ, Naghizadeh M, Mirzadeh H (2020) Deformation-induced martensite in austenitic stainless steels: a review. Archiv Civ Mech Eng. https://doi.org/10.1007/s43452-020-00130-1
3. Angel T (1954) Formation of martensite in austenitic stainless steels: effects of deformation, temperature, and composition. J Iron Steel Inst 5:165–174
4. Dixit US (2020) Modeling of metal forming: a review. In: Mechanics of materials in modern manufacturing methods and processing techniques. Elsevier, pp 1–30
5. Volk W, Groche P, Brosius A, Ghiotti A, Kinsey BL, Liewald M, Madej L, Min J, Yanagimoto J (2019) Models and modelling for process limits in metal forming. CIRP Ann. https://doi.org/10.1016/j.cirp.2019.05.007
6. Bruschi S, Altan T, Banabic D, Bariani PF, Brosius A, Cao J, Ghiotti A, Khraisheh M, Merklein M, Tekkaya AE (2014) Testing and modelling of material behaviour and formability in sheet metal forming. CIRP Ann. https://doi.org/10.1016/j.cirp.2014.05.005
7. Dixit US, Joshi SN, Davim JP (2011) Incorporation of material behavior in modeling of metal forming and machining processes: a review. Mater Des. https://doi.org/10.1016/j.matdes.2011.03.049
8. Pietrzyk M, Kuziak R (2012) Modelling phase transformations in steel. In: Microstructure evolution in metal forming processes. Elsevier, pp 145–179
9. Ostwald R, Bartel T, Menzel A (2012) Modelling and simulation of phase-transformations and plasticity in steel. In: 1st international conference on thermo-mechanically graded materials, pp 93–98
10. Behrens B-A, Olle P (2007) Consideration of phase transformations in numerical simulation of press hardening. Steel Res Int 784–790
11. Mangonon PL, Thomas G (1970) The martensite phases in 304 stainless steel. Metall Trans. https://doi.org/10.1007/BF02642003
12. Smaga M, Walther F, Eifler D (2006) Investigation and modelling of the plasticity-induced martensite formation in metastable austenites. Int J Mater Res. https://doi.org/10.3139/146.101396
13. Olson GB, Cohen M (1975) Kinetics of strain-induced martensitic nucleation. Metall Trans A 6A:791–795

14. Polyblank JA, Allwood JM, Duncan SR (2014) Closed-loop control of product properties in metal forming: a review and prospectus. J Mater Process Technol. https://doi.org/10.1016/j.jmatprotec.2014.04.014
15. Torgashov A, Snegirev O, Yang F (2022) Methyl sec-butyl ether content estimation in MTBE products via clustering-based adaptive nonlinear soft sensors. In: 14th international symposium on process systems engineering, vol. 49. Computer Aided Chemical Engineering, Elsevier, pp 1387–1392
16. Madsen K, Nielsen HB (2010) Introduction to optimization and data fitting. Informatics and Mathematical Modelling, Technical University of Denmark, DTU (2010)
17. Talonen J, Aspegren P, Hänninen H (2004) Comparison of different methods for measuring strain induced α-martensite content in austenitic steels. Mater Sci Technol. https://doi.org/10.1179/026708304X4367
18. Rozo Vasquez J, Arian B, Riepold M, Homberg W, Trächtler A, Walther F (2020) Microstructural investigation on phase transformation during flow forming of the metastable austenite AISI 304. In: Neidel (ed) 16–18th September (2020)—Fortschritte in der Metallographie, pp 75–81
19. Arian B, Homberg W, Vasquez JR, Walther F, Riepold M, Trächtler A (2021) Forming of metastable austenitic stainless steel tubes with axially graded martensite content by flow-forming. ESAFORM. https://doi.org/10.25518/esaform21.2759
20. Baak N, Hajavifard R, Lücker L, Rozo Vasquez J, Strodick S, Teschke M, Walther F (2021) Micromagnetic approaches for microstructure analysis and capability assessment. Mater Char. https://doi.org/10.1016/j.matchar.2021.111189
21. Astudillo MRN, Nicolás MN, Ruzzante J, Gómez MP, Ferrari GC, Padovese LR, Pumarega MIL (2015) Correlation between martensitic phase transformation and magnetic Barkhausen noise of AISI 304 steel. Procedia Mater Sci. https://doi.org/10.1016/j.mspro.2015.05.014
22. Neslušan M, Minárik P, Čilliková M, Kolařík K, Rubešová K (2019) Barkhausen noise emission in tool steel X210Cr12 after semi-solid processing. Mater Charact. https://doi.org/10.1016/j.matchar.2019.109891
23. Haušild P, Kolařík K, Karlík M (2013) Characterization of strain-induced martensitic transformation in A301 stainless steel by Barkhausen noise measurement. Mater Des. https://doi.org/10.1016/j.matdes.2012.08.058
24. Rozo Vasquez J, Arian B, Riepold M, Homberg W, Trächtler A, Walther F (2021) Magnetic Barkhausen noise analysis for microstructural effects separation during flow forming of metastable austenite 304L. In: Proceedings of the 11th international workshop NDT in progress 2021, pp 1–9
25. Knyazeva M, Rozo Vasquez J, Gondecki L, Weibring M, Pöhl F, Kipp M, Tenberge P, Theisen W, Walther F, Biermann D (2018) Micro-magnetic and microstructural characterization of wear progress on case-hardened 16MnCr5 gear wheels. Materials. https://doi.org/10.3390/ma11112290
26. Kersting L, Arian B, Vasquez JR, Trächtler A, Homberg W, Walther F (2022) Innovative online measurement and modelling approach for property-controlled flow forming processes. Key Eng Mater. https://doi.org/10.4028/p-yp2hj3
27. Kersting L, Arian B, Rozo Vasquez J, Trächtler A, Homberg W, Walther F (2023) Echtzeit-fähige Modellierung eines innovativen Drückwalzprozesses für die eigenschaftsgeregelte Bauteilfertigung. At—Automatisierungstechnik. https://doi.org/10.1515/auto-2022-0106

Modeling the Morphology Evolution of Recrystallized Grains of Aluminum Alloys Under Oriented Growth and Anisotropic Dispersoid Pinning

Ruxue Liu, Zhiwu Zhang, Baocheng Zhou, Guowei Zhou, and Dayong Li

Abstract Static recrystallization takes place during the annealing of deformed metals, and the resulting fine or coarse, equiaxed or lath-like recrystallized grains together with various textures can have a significant effect on the material's performance. In this work, oriented growth and anisotropic dispersoid pinning effects on boundary migration are taken into account in a cellular automaton model to simulate the morphology evolution of recrystallized grain in aluminum alloy. Diverse grain morphology can be predicted when varying the settings of annealing condition, and the underlying physical mechanism is discussed based on the simulated results.

Keywords Oriented growth · Dispersoid pinning · Cellular automaton

1 Introduction

The static recrystallization (SRX) that occurs during annealing cold or hot worked metals causes dramatical change of microstructure. Fine or coarse, equiaxed or lath-like recrystallized grains together with various textures can be generated, which have a significant impact on the material's performance [1]. The morphology and texture of recrystallized grains are determined by various factors such as deformed microstructure, second phase particles, and annealing conditions. For Al–Mn alloys, for example, studies [2, 3] have shown that high temperature annealing after cold rolling produces fine equiaxed recrystallized grains with *Cube* as the main texture

R. Liu · Z. Zhang · D. Li (✉)
School of Mechanical Engineering, Shanghai Jiao Tong University, Shanghai 200240, China
e-mail: dyli@sjtu.edu.cn

B. Zhou
Chinalco Materials Application Research Institute Co., Ltd., Beijing 102209, China

G. Zhou
School of Naval Architecture, Ocean and Civil Engineering, Shanghai Jiao Tong University, Shanghai 200240, China

J. Kusiak et al. (eds.), *Numerical Methods in Industrial Forming Processes*, Lecture Notes in Mechanical Engineering, https://doi.org/10.1007/978-3-031-58006-2_11

component. However, during low temperature or non-isothermal annealing, dynamic precipitation brings a strong pinning effect on the boundary migration of nuclei, resulting in coarse lath-like recrystallized grains, with high strength of P and ND-rotated Cube textures, which is closely related to the high mobility of the special grain boundary ($\Sigma 7, 40° <111>$) and the uneven distribution of the dispersed particles [4]. Many numerical approaches such as cellular automaton [5, 6] (CA), Monte Carlo [7], and phase field [8, 9] models have been developed to simulate the microstructural evolution over the thermo-mechanical processes like SRX. However, there is no model so far to consider the anisotropy of recrystallized grain growth, which may come from the fibrous morphology of deformed grains, the orientation dependence of grain boundary mobility and the inhomogeneous distribution of dispersed particles.

The CA model has gained popularity because of its length scale calibrations and high efficiency. The core algorithms of the CA technique mainly include discrete spatial and temporal evolution of complex systems by applying local or global probabilistic or deterministic transformation rules on the location of a lattice. In this work, oriented growth and anisotropic dispersoid pinning effects on boundary migration are incorporated into a CA model to simulate the morphology evolution of recrystallized grain in cold-rolled aluminum alloy. Diverse grain shape and textures can be predicted when varying the parameters of annealing condition, and the underlying physical mechanism is discussed based on the simulated results.

2   Model Setup

2.1   Cellular Automaton Framework

A cellular automaton with periodic borders and a cubic grid is employed in this work. State variables such as Euler angles, grain number, and dislocation density are assigned to each cell. The crystallographic misorientation between each cell and its neighbors is calculated in every step, and the grain boundary is regarded as existing when the misorientation angle is larger than a critical value. Whether the cell at grain boundary is transformed to the target neighbors is determined by the transition state variable:

$$f_{\text{tran}} = \sum_{i=1}^{N_{\text{nbrs}}} G_i \frac{v_i \Delta t}{l_0} \tag{1}$$

where N_{nbrs} is the total number of all von Neumann neighbor cells (6 in 3D space), v_i is the velocity at which the neighbor i encroaches the central cell of interest, i.e., migration rate, Δt is the time increment, l_0 is the cell size, G_i is a logical factor that is assigned 1 only if the maximum v_i of all neighbors is greater than 0 and the cell of neighbor i belongs to the same grain as that with the maximum v_i. Otherwise, G_i is 0. The value of f_{tran} of all cells will be saved and followed in each step. As f_{tran}

reaches 1, cell state variables like orientation, grain number, and dislocation density are switched, with the neighbor having the highest v_i at present being the target. The f_{tran} will be reset to 0 after the switch.

A time increment that adjusts dynamically with the maximal migration speed is used in this model:

$$\Delta t = \frac{l_0}{v_{max}} \tag{2}$$

where v_{max} is the highest migration velocity among all boundaries, which is not constant throughout the simulation process.

2.2 Boundary Migration Velocity

The migration rate v of grain boundary can be described by the following equation in accordance with Burke and Turnbull's theory [10]:

$$v = MP \tag{3}$$

where M and P are the boundary mobility and net driving pressure, respectively. M is a parameter related to boundary misorientation angle θ and temperature T, which can be expressed using the equation as below [11, 12]:

$$M = M_0 \exp\left(-\frac{Q_b}{RT}\right)\left(1 - \exp\left(-5\left(\frac{\theta}{\theta_m}\right)^4\right)\right) \tag{4}$$

where M_0 is the pre-factor, Q_b is the activation energy for boundary diffusion, R is the general gas constant, θ_m is the defined critical misorientation angle for high angle boundary (15° is assumed here).

Equation (4) depicts the classical expression of grain boundary mobility, which only considers the relevance of misorientation angle, i.e., grain boundary mobility increases with misorientation angle from 0 to θ_m, and reaches saturation at θ_m. A new orientation dependent factor q_s is introduced in this model, and a higher value of 5 is assigned for the Σ7 (40° <111>) boundary and 1 for the others. Hence, the pre-factor M_0 in Eq. (4) is replaced by $q_s M_0$. Note that a boundary segment with rotation angle of 30~46° and rotation axis of ~15° to <111> (crystallographic direction) is regarded as Σ7 boundary in this work.

The driving pressure acting on the boundary can be calculated as follows:

$$P = P_s + P_c + P_z \tag{5}$$

where P_s is the stored energy difference of grains on both sides of the boundary and can be formulated as

$$P_s = 0.5\mu b^2 \Delta\rho \qquad (6)$$

where μ is the temperature dependent shear modulus, b is the length of Burgers vector, $\Delta\rho$ is the difference of dislocation density on both sides of grain boundary.

P_c refers to the driving force of curvature, which can be positive or negative, depending on the decrease or increase of local interface energy at grain boundaries. The Kremeyer model [13], given by Eqs. (7) and (8), is employed in this study because of its simplicity:

$$P_c = \gamma_s \kappa \qquad (7)$$

$$\kappa = \frac{A}{l_0} \frac{kink - N_i}{N + 1} \qquad (8)$$

where κ is the curvature obtained by Kremeyer approach, A is the shape factor (0.394 in 3D space), $kink$ is the number of cells required to create a flat grain boundary ($kink$ = 75 in 3D space), N is the total number of CA cells in a region containing two layers adjacent to the cell of interest ($N = 124$ in 3D space), and N_i is the number of cells belonging to the same grain with the central cell of interest in this region. γ_s is the boundary energy, and can be calculated using the Read-Shockley equation [14]:

$$\gamma_s = \begin{cases} \gamma_m, \theta \geq \theta_m \\ \frac{\gamma_m \theta}{\theta_m} \left(1 - \ln \frac{\theta}{\theta_m}\right), \theta \leq \theta_m \end{cases} \qquad (9)$$

where γ_m is the energy of high angle boundary. A lower energy for 40° <111> boundary of aluminum by 5~10% compared with other "random" high angle boundaries has been reported [15]. As a result, a lower pinning effect by particles is expected since the pinning pressure is proportional to the boundary energy. Hence, in this study, a lower boundary energy $\gamma_m' = 0.9\gamma_m$ is adopted on these special boundaries.

The last term P_z in Eq. (5) is the Zener drag pressure on boundary provided by the fine dispersive particles, and can be classically expressed as [16]:

$$P_z = -\frac{3f_p\gamma_s}{2r_p} \qquad (10)$$

where f_p is the volume fraction of dispersive particles, r_p is the average particle radius. Equation (10) gives a simplified but practical way to calculate the pinning effect of particles evenly distributed on the boundary. However, the particle distribution after plastic deformation, such as rolling, will be not uniform anymore and banded along the rolling direction. In addition, the axial ratio of particles is usually ignored since the orientation of the ellipsoidal axis prior to rolling is assumed to be random. However, following rolling deformation, the particles' long axis is frequently getting parallel to the rolling direction. Both the anisotropy of the particle distribution and the orientation of the ellipsoid axis can result in a greater pinning resistance in the normal direction (ND) direction and less in the rolling direction (RD). Therefore, an anisotropic pinning factor w_i is introduced into Eq. (10):

$$P_{z-i} = -\frac{3f_p\gamma_s}{2r_p w_i} \qquad (11)$$

where $w_i = \exp(\beta e_p \varepsilon_i)$, β is a constant and here is 0.5, e_p is the average axial ratio of particles, ε is the strain and i represents the three directions of RD, TD (transverse direction), and ND. Therefore, an anisotropic pinning force is imposed, and it becomes more pronounced with the increase of rolling strain and average axial ratio of particles.

2.3 Microstructure Initialization for Annealing

In order to simulate the recrystallized grain growth behavior during annealing, a reasonable cold-rolled microstructure must be initialized first, which mainly includes plate-like (or fibrous for 2D space) grain morphology and typical rolling textures such as *Copper*, *S*, and *Brass*. A virtual microstructure composed of fine equiaxed grains is created, and these grains are subsequently assigned rolling textures with banded or random distribution. Grains with orientations of *Copper*, *S*, and *Brass* (with a maximum random deviation of 5° from the standard orientation) account for around 40%, 30%, and 30% of the total respectively, as Fig. 1 shows (all the IPF maps in this paper are with respect to the ND direction). The physical dimensions of computational domains are set as $600 \times 5 \times 600$ μm, with a cell size of 1 μm. The expansion of nuclei in the TD direction is disregarded, much fewer cell layers are thus set in this direction considering computational time cost.

The site saturation nucleation hypothesis is adopted in this paper, and a preset number and orientation of nuclei are artificially implanted at the beginning of annealing. The dislocation density in the deformed matrix is set to 5×10^{15} m^{-2}, whereas that in the nuclei is 1×10^{10} m^{-2}. The focus of this study is on the anisotropy of nuclei growth, regardless of the decrease of driving force caused by static recovery, although it will lead to the reduction of growth rate in the later stage of annealing. In this study, four examples with different annealing condition are calculated individually, as shown in Table 1. The setting for each annealing condition is described in detail in Sect. 3.

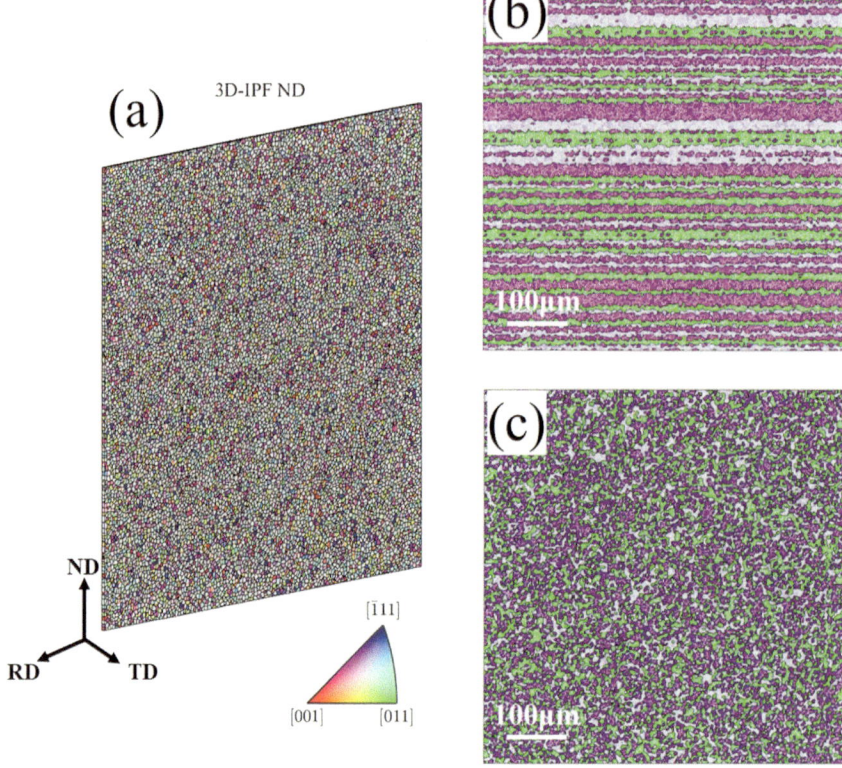

Fig. 1 The initialized cold-rolled microstructure. **a** Random orientations; **b** banded distribution of rolling textures; **c** random distribution of rolling textures

Table 1 Different settings of annealing condition

Case no	Grain shape	Dispersoid	Temp (°C)	Number of nuclei	Nuclei orientation
1	Fibrous	No	400	200	A[a]
2	Equiaxed	No	400 °C	200	A[a]
3	Fibrous	Yes	400 °C	200	A[a]
4	Fibrous	Yes	400 °C	30	B[b]

[a] A-10% cube, 10% Σ7, 10% rolling, 70% random
[b] B-0% cube, 100% Σ7, 0% rolling, 0% random

3 Results and Discussion

In this investigation, the pre-factor M_0 is 7.41×10^{-5} m·s^{-1}·Pa^{-1}, R is 8.314 J·mol^{-1}·K^{-1}, θ_m is 15°, μ is 20.8 GPa in 400 °C, b is 2.86×10^{-10} m [17], γ_m is 0.324 J/m^2 [16], and rolling strain ε is 3 ($\varepsilon_1 = 3$, $\varepsilon_2 = 0$, $\varepsilon_3 = -3$). The volume fraction of dispersoid, if present, is set to 0.7%, and the average radius is 60 nm, with an average axial ratio of 1.2. It is assumed that nuclei occur at the beginning of annealing and the simulation stops as the volume fraction of recrystallization reaches 100%. Figures 2 and 3 present the final recrystallized microstructures of the four cases. It is evident that varying annealing conditions lead to different recrystallized grain sizes, shapes, and textures.

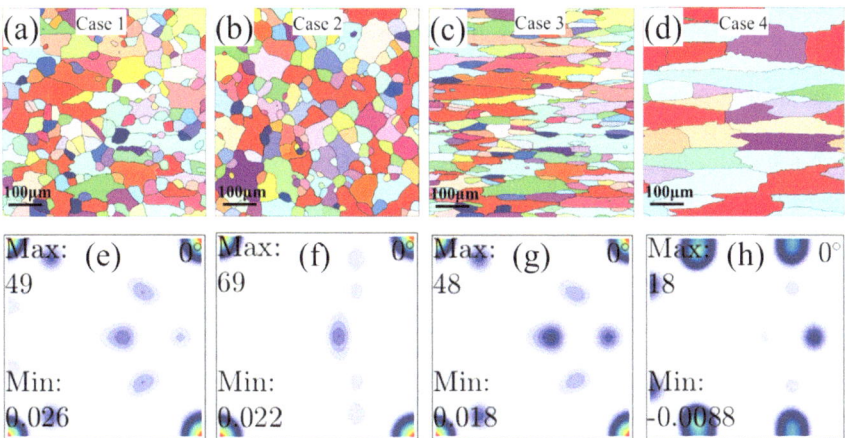

Fig. 2 The simulated recrystallized microstructures in different annealing conditions. **a–d** IPF maps; **e–h** ODF sections ($\varphi_2 = 0°$)

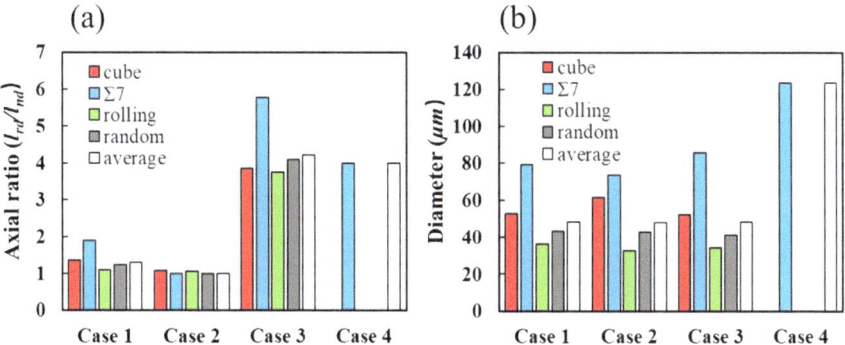

Fig. 3 Recrystallized grain morphology. **a** Axial ratio; **b** diameter

3.1 Case 1 and 2: Oriented Growth Effect

Figure 2a reveals the recrystallized grains in Case 1, in which no dispersoid presents and 200 nuclei are given with orientations of 10% *Cube*, 10% Σ7 (with a misorientation of 40° <111> to the matrix), 10% *rolling* and 70% *random*. It can be seen from Figs. 2a and 3a that four different kinds of nuclei end up with distinct sizes, and the average size of nuclei with *Σ7* orientation is the largest, following by *Cube*, *random* and *rolling*. The microstructures in Case 1 at recrystallization volume fractions of 9%, 25%, 75%, and 100% are depicted in Fig. 4. All types of nuclei can grow isotropically in the early stage of recrystallization when the grain size is small, although the growth rates are quite different. However, as the nuclei continue to grow, those having specific orientation relationship (Σ7) with neighboring matrix contact the original mother grain boundary in the ND direction, and the growth rate gets slower, while along the RD direction they can keep growing rapidly until encountering other nuclei, while there are a large number of smaller grains, still maintaining slow and isotropic growth. ODF sections in Fig. 4e–h show the textures changing from *rolling* to a common recrystallized pattern.

A same annealing condition is set in Case 2, except that a random distribution of rolling textures is applied. Although the grain morphology in the fully recrystallized state (Figs. 2b and 3) is not significantly different from that in Case 1, the recrystallization growth process is quite distinct. The microstructures at recrystallization volume fractions of 25% in Cases 1–3 are shown in Fig. 5. As stated above, under the band-like distributed rolling textures, some nuclei having specific orientation relationship (Σ7) with neighboring matrix can grow quickly along the RD direction, whereas growth along the ND direction is constrained, resulting in recrystallized grain anisotropy. This is not observed in the random distributed rolling textures

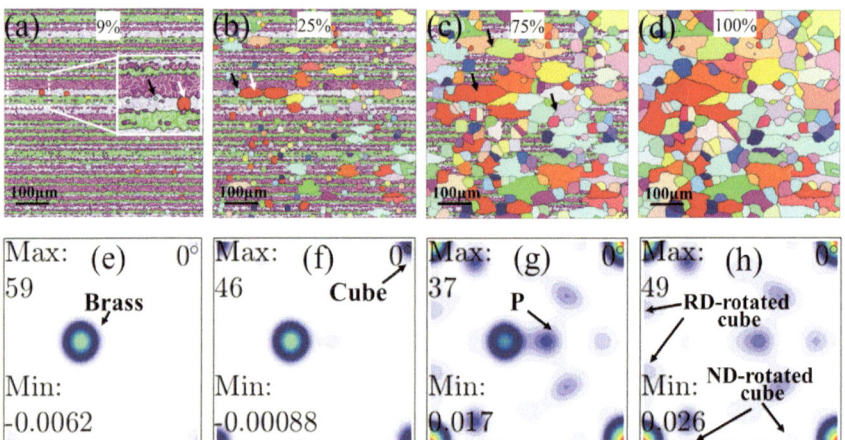

Fig. 4 Simulated evolution of microstructure during annealing in Case 1. **a-d** IPF maps; **e–h** ODF sections ($\varphi_2 = 0°$)

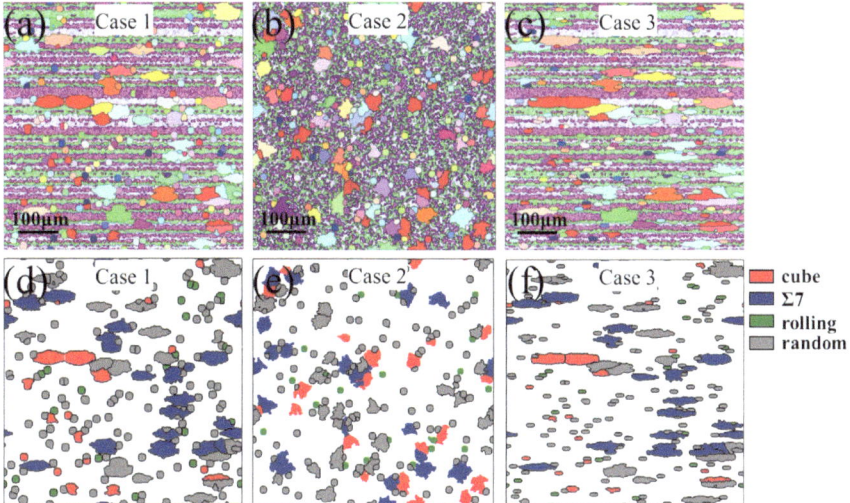

Fig. 5 The growing nuclei when the volume fraction of recrystallization is 25% in Cases 1–3 **a–c** IPF maps; **d–f** nuclei only

because the possibility of growing nuclei encountering specific grain boundaries in all directions is consistent. This further illustrates the effect of oriented growth on the shape and size of recrystallized grains.

3.2 Case 3 and 4: Anisotropic Pinning Effect

In order to study the effect of non-uniformly distributed dispersoid on the growth of recrystallized grains, particles with a constant volume fraction and size are considered in Case 3, on the basis of Case 1. It can be calculated by Eq. (10) that the pinning resistance in RD and ND direction is 0.008 MPa and 0.244 MPa, respectively. In comparison to Case 1, it produces a recrystallized structure compressed in the ND direction and elongated in the RD direction by applying the pinning resistance (Figs. 2c, 5c and f). The axial ratio of recrystallized grains increases (Fig. 3), indicating enhanced growth anisotropy.

In fact, if the homogenization treatment before rolling is insufficient for Al–Mn alloys, evident dynamic precipitation will occur during annealing, considerably inhibiting the quantity of nucleation. Only nuclei with those specific orientations ($\Sigma 7$) can be kept and grow due to their higher mobility and less pinning by particles. Hence, in Case 4, only 30 nuclei are provided, and they are all with $\Sigma 7$ orientations. Surprisingly but reasonably, abnormally coarse and lath-like grain structure is observed (Fig. 2d). Due to the extremely low nucleation rate, competition between nuclei is

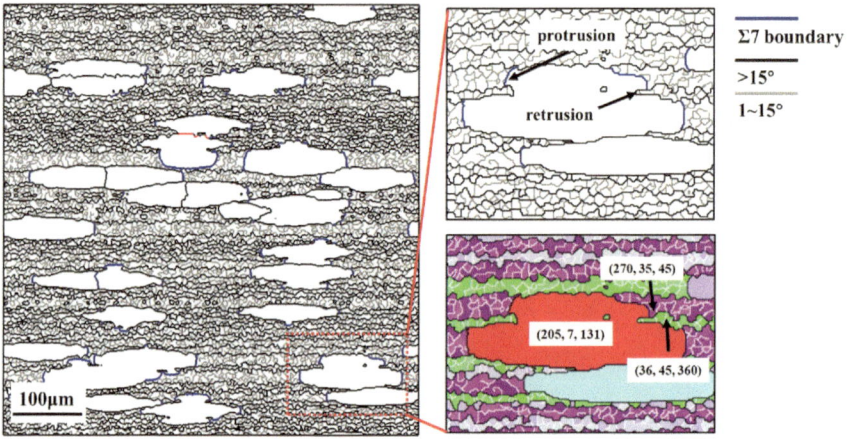

Fig. 6 The boundary map when the volume fraction of recrystallization is 25% in Case 4

very low, and thus there is sufficient space for nuclei to expand. Under the simultaneous effect of oriented growth and anisotropic pinning resistance, abnormally coarse recrystallized grains with high axial ratio can be produced (Fig. 3).

In order to further explain the correlation between oriented growth and deformed structure, Fig. 6 displays the orientation relationship of coarse grains with the surrounding matrix during growth. It can be observed that these coarse grains frequently have a Σ7 relationship with the adjacent parent grains in the RD direction, while they are mostly "pinned" by the adjacent deformed matrix with other orientations in the ND direction. It is also possible to see the protrusions and retrusions of grain boundaries in the RD direction, which is quite compatible with the study of HUANG et al. [1].

3.3 Recrystallized Textures

The most common textures after annealing in cold-rolled aluminum are *Cube, ND-rotated Cube, P, rolling* and *random*. If one rotates the deformed texture by 40° along all eight variants of <111> crystallographic directions, a transformed texture is obtained. Three different transformed textures are generated here by doing this operation for 500 orientations of *Copper, S,* and *Brass*, respectively, as shown in Fig. 7. It is indicated that the *Copper* deformation texture may encourage the development of *P* and *ND-rotated Cube*, while *Cube* and *RD-rotated Cube* after recrystallization can be enhanced by the presence of *S* and *Brass*, respectively. Figure 7d shows the transformed texture obtained by doing the rotation on randomly picked 500 orientations from the initialized rolling texture in Sect. 2.3. A potential recrystallized texture containing *Cube, ND/RD-rotated Cube*, and *P* can be observed. Therefore, it may be

Fig. 7 Textures transformed from: **a** copper; **b** S; **c** brass; **d** randomly-picked rolling orientations

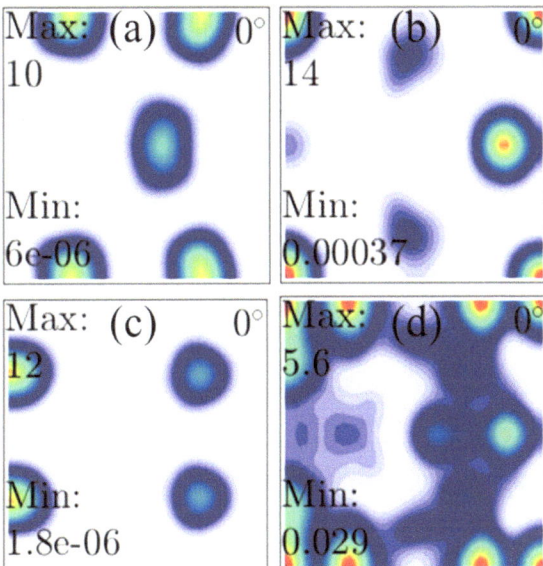

concluded that the recrystallized texture is determined by selective nucleation and oriented growth, given a certain deformed microstructure.

ODF sections at $\varphi_2 = 0$ of all cases are shown in Fig. 2e–h. Strong *Cube, ND-rotated Cube*, and *P* textures are all observed, which are all attributed to the fast growing speed of those nuclei, and certainly, the fact that nuclei of those orientations are preset in this work. A more reasonable nucleation rate model that accounts for the mechanisms of thermal activation and selective nucleation is developed, which will be published in future work.

4 Conclusion

Based on the simulated results, it can be concluded that.

The oriented growth effect, which refers to the high mobility and lower particle pinning for special boundary, has been applied in the present CA model. Nuclei with a misorientation close to 40° <111> with the adjacent matrix can grow faster than those with other orientations. A lath-like grain shape can be obtained at the early stage of recrystallization because of the fibrous rolling microstructure and oriented growth effect, although the axial ratio may decrease afterwards due to the impingement of growing nuclei along RD direction.

The anisotropic pinning effect arises from the anisotropy of the particle distribution and orientation of the ellipsoid axis, resulting in a higher drag force along ND direction and lower along RD direction. A higher axial ratio of recrystallized grains can be obtained when both oriented growth and anisotropic pinning are present, especially for the case that nucleation is hindered.

Acknowledgements The authors acknowledge the financial support from the National Natural Science Foundation of China (No. U2141215, 52105384).

References

1. Huang K, Zhang K, Marthinsen K, Logé RE (2017) Controlling grain structure and texture in Al–Mn from the competition between precipitation and recrystallization. Acta Mater 41:360–373
2. Huang K, Engler O, Li YJ, Marthinsen K (2015) Evolution in microstructure and properties during non-isothermal annealing of a cold-rolled Al–Mn–Fe–Si alloy with different microchemistry states. Mater Sci Eng A 628:216–229
3. Huang K, Li YJ, Marthinsen K (2014) Isothermal annealing of cold-rolled Al–Mn–Fe–Si alloy with different microchemistry states. Trans Nonferrous Met Soc China (English Ed.) 24(8):3840–3847
4. Ånes HW, van Helvoort ATJ, Marthinsen K (2023) Orientation dependent pinning of (sub)grains by dispersoids during recovery and recrystallization in an Al–Mn alloy. Acta Mater 248(1):118761
5. Vertyagina Y, Mahfouf M (2014) A 3D cellular automata model of the abnormal grain growth in austenite. J Mater Sci 50(2):745–754
6. Li Y, Liu C, Chu Z, Li W, Wu Z, Gao S, He W (2020) Grain growth of AZ31 magnesium alloy based on three-dimensional cellular automata. Adv Mater Sci Eng
7. Eivani AR, Zhou J, Duszczyk J (2014) Simulation of transient-state recrystallization of Al-4.5Mg-1Zn alloy after hot deformation. Comput Mater Sci 86:193–199
8. Luan Q, Lee J, Zheng Z, Lin J, Jiang J (2018) Static recrystallization study on pure aluminium using crystal plasticity finite element and phase-field modelling. Procedia Manuf 15:1800–1807
9. Liu R, Li K, Zhou G, Tang W, Shen Y, Tang D, Li D (2022) Simulation of strain induced abnormal grain growth in aluminum alloy by coupling crystal plasticity and phase field methods. Trans Nonferrous Met Soc China 32(12):3873–3886
10. Burke JE, Turnbull D (1952) Recrystallization and grain growth. Prog Met Phys 3
11. Humphreys FJ (1997) A unified theory of recovery, recrystallization and grain growth, based on the stability and growth of cellular microstructures—II. The effect of second-phase particles. Acta Mater 45(12):5031–5039
12. Su F, Liu W, Wen Z (2020) Three-dimensional cellular automaton simulation of austenite grain growth of Fe-1C-1.5Cr alloy steel. J Mater Res Technol 9(1):180–187
13. Kremeyer K (1998) Cellular automata investigations of binary solidification. J Comput Phys 142(1):243–263
14. Read WT, Shockley W (1950) Dislocation models of crystal grainboundaries. Phys Rev 78(3):275–289
15. Humphreys FJ, Ardakani MG (1996) Grain boundary migration and zener pinning in particle-containing copper crystals. Acta Mater 44(7):2717–2727
16. Humphreys FJ, Rohrer GS, Rollett AD (2017) Recrystallization and related annealing phenomena, 3rd edn. Elsevier, Oxford, p 734
17. Chen S, Li D, Zhang S, Han H, Lee H, Lee M (2020) Modelling continuous dynamic recrystallization of aluminum alloys based on the polycrystal plasticity approach. Int J Plast 131

Information Extraction from Time Series in the EDM Drilling Process

Tomasz Jażdżewski⑩, Krzysztof Regulski⑩, Adam Bułka, Pawel Malara, Adrian Czeszkiewicz, and Marcin Trajer

Abstract Electrical discharge machining (EDM) allows to obtain small holes with the high efficiency and high quality. Such features are most common in jet engine turbine airfoils. The main problem of the analysis is detection of a moment when the machine should stop the drilling process—the breakthrough detection. Machine learning applications requires that data and models to be prepared by specialists that can extract the most important information from an input data and choose most suitable Artificial Intelligence (AI) algorithm for particular case. This Article describes an experiment on how to extract valuable information from heterogeneous time series data with various sources (which is popular in an industry 4.0—Internet of Things) from the EDM drilling process. EDM process is conducted by a Computerized Numerical Control (CNC) drilling device that measures the position of the electrode, drilling speed, Additional sensor, monitors impulses current and voltage. The research is focused on classic AI algorithms (decision tree, random forest and eXtreme Gradient Boosting (XGBoost)) which provide fast training and possibility to check more hyperparameters in a time than neural networks algorithms. It is also described how feature extraction can high up AI algorithm predictions.

Keywords Times Series · Feature extraction · EDM drilling · Machine learning

T. Jażdżewski · K. Regulski (✉)
AGH University of Krakow, Kraków, Poland
e-mail: regulski@agh.edu.pl

A. Bułka · P. Malara · A. Czeszkiewicz
Łukasiewicz Research Network – Institute of Aviation, Warsaw, Poland

M. Trajer
Faculty of Material Science and Engineering, Warsaw University of Technology, Warsaw, Poland

J. Kusiak et al. (cds.), *Numerical Methods in Industrial Forming Processes*, Lecture Notes in Mechanical Engineering, https://doi.org/10.1007/978-3-031-58006-2_12

1   Introduction

1.1   Background

Electrical discharge machining is a non-traditional process that is widely used for jet engines turbine airfoils cooling holes. One of the key challenges is to properly detect a moment when the machine should stop the drilling process—when electrode reaches opposite site of the metal wall. This moment is called the breakthrough. Correct detection [1] is critical for parts quality. Literature examples where evaluation of pulse current and voltage was conducted showed that it is possible to identify the fact that a breakthrough took place [2, 3]. Such models make it possible to analytically indicate in the process data record the moment when surface damage occurred [4, 5]. This has allowed the use of machine learning tools in breakthrough detection [6]. However, in each of the mentioned studies, the moment of breakthrough was detected either after or during breakthrough phenomena occurrence. The purpose of this study was to verify whether it is possible to predict a breakthrough moment, and to analyse how far in advance it is possible. Currently evaluated CNC EDM records data and recognizes if a breakthrough has occurred based on analogue algorithms. However, as indicated, higher accuracy alternate solution based on AI models might be considered. It may be possible to develop additional sub-system that allows more accurate evaluation of the drilling process in the future. Development of such AI based system requires learning data properly labelled so based on that data set algorithm can indicate the need to complete drilling before a breakthrough occurs.

1.2   Process Description

EDM drilling process is widely used for small holes manufacturing. Such are made within fuel nozzles in diesel engines, medical equipment and in jet engines high pressure turbine cooled components. Once process is based on set of parameters and numerical control (NC) code parts could be manufactured. There are separate setups for electrical parameters, electrode movement and breakthrough detection. Part is placed in the EDM machine where its position is verified. After that automatic drilling process starts. All elements of the hole drilling process: surface opening, hole drill, breakthrough are NC code controlled. For complex parts such as jet engines elements drilling itself is important but breakthrough is critical. Once part is complete it is withdrawn from the machine for subsequent operations. Test coupon operation is shown in Fig. 1.

Fig. 1 Photo from a process of making a test drills

2 Methods of Feature Extraction from Time Series Data

2.1 Methodology of Analysis

The data acquired in the EDM process have time-series characteristics, which means that they are time-indexed variables. This raises the necessity of using models that can cope with the autocorrelation of parameters, rather than just cross-sectional analysis of the relationships between them [7, 8]. Nevertheless, the methods used in time series modelling are often the same as in cross-sectional analysis. The simpler the methods, the faster the recognition. Hence the conclusion is that it is more reasonable to use the fastest possible algorithms in a situation where fractions of seconds are decisive. Despite the fact that artificial neural networks or support vector machines are a fairly common methodology [6], it was proposed to assess whether traditional methods based on decision trees, fast and easy to interpret the models, could provide satisfactory results. Hence it was decided to focus on three methods based on decision trees: DecisionTree, RandomForest and XGBoost from sklearn and xgboost libraries in Python.

2.2 Decision Trees

In the discussed study, among others, decision trees techniques were used. Using the sklearn library (Scikit Learn [9]), trees with an algorithm based on Classification and Regression Trees (CART) were created. The CART algorithm is one of many decision tree induction algorithms. It is based on hierarchical, binary divisions of a data set for better segregation of cases that are representative of the values of the dependent variable. The algorithm strives for an ideal situation where the division (leaf) created contains cases with the same value of the dependent variable. However, these methods are not as effective in prediction as neural networks or support vector machine. They do not achieve equally precise results, mainly due to the fact of discretization of quantitative variables, and thus forced generalization. Decision trees are graphical representations of rules obtained from the analysis of data structure. The undoubted advantages of tree-based classifiers are their graphical representation, clear and easy to interpret and verify based on domain knowledge, the ability to determine the significance of predictors, insensitivity to noise and outliers, the result in the form of a set of rules that can be used in other applications. A decision tree is a graphical method of decision support. It is a tree structure in which the internal nodes contain tests of attribute values, and the leaves describe decisions on object classification. A decision tree is essentially a structure composed of a series of conditional instructions. Decision trees are an advanced form of knowledge representation that provides a wide range of interpretation possibilities, both at the stage of the knowledge acquisition itself (data mining) and at the stage of its application in the decision-making process. Decision trees are a popular method used in industrial and research applications [10].

2.3 Random Forest

Random Forest (RF, also known as Bagged Decision Trees) is a method based on the principle of decision tree induction [9, 11]. It involves creating complex models consisting of multiple decision trees combined into a single classification model. The trees themselves calculate a value for each successive input, and then the result is averaged. With this approach, the model eventually became independent of outliers, which was previously a major drawback of individual decision trees. A random forest is a set of weak learners that can solve more complex classification problems. The random forest classifier starts by randomizing the training data into several different data sets. Each training data set is then used to create a customized classification tree model. Based on the individual trees, the values of individual attributes are evaluated as predictors by measuring data contamination, entropy. The set of attributes with the highest value will be used to split the data and a random forest will be built combining multiple decision trees.

2.4 XGBoost Classifier

XGBoost model is based on a gradient boosting decision tree (GDDC) algorithm where all separated models are calculated in parallel. It is possible thanks to a novel distributed weighted quantile sketch algorithm which works on a large dataset [12]. That algorithms helps to find proper split point for a given dataset which will be used to create basic decision tree models. Thanks to that it is possible to use the power of simple decision tree algorithms and power of gradient descent learning algorithms in a faster way. Implementation of the error function provides a feature that adds a penalty for giving extra leaves in a tree [13]. Penalty is proportional to the size of the leaf weights. Thanks to that it is less sensitive for overtraining and is more stable with local anomalies triggered by measurement errors. The most important thing is the possibility to calculate the XGBoost algorithm on a graphics processing unit (GPU) processor. That is possible by the process of parallelising the construction of individual trees.

3   Data Description

Data was prepared on a single part of material in which number of holes have been drilled. Pulse parameters were identical for all holes used in this study. Process data has been saved in comma-separated values (CSV) files. Every file represents data from a single hole drill. At first a simple visualization method was used (Figs. 2, 3 and 4). The breakout moment is clearly visible from a CNC machine data. First of all there is a significant change in Z_location data. Change in voltage value is also visible—it occurs in the same place where Z_location disorder has been spotted. Some stationary disorders are at the beginning of a drilling process; however, those disorders are triggered by an initiation process of machine starting the process. That anomaly should not be used in further data evaluation. In some drills the BT_DETECT was set up as true after a local disorder. After consultation with an expert, it was concluded that it must be a delay of a machine in a detecting a breakthrough moment.

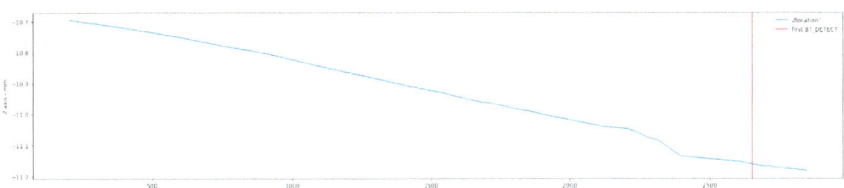

Fig. 2   Plot shows changes in a Z_location in a machine. Z_location represents a head of a machine that is responsible for holding an electrode

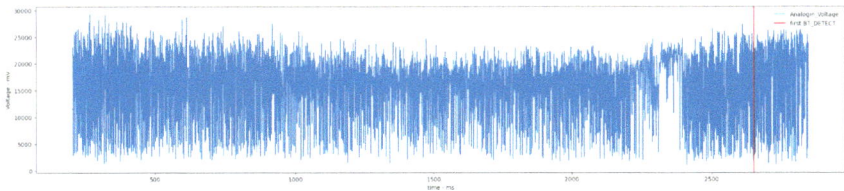

Fig. 3   Plot shows variability of an Analogin_Voltage variable

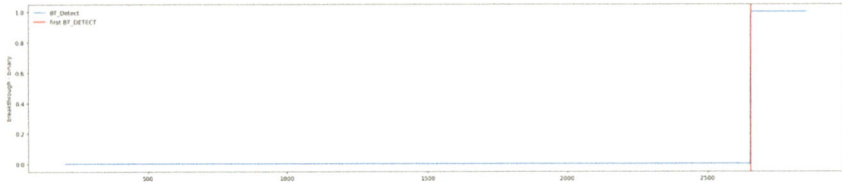

Fig. 4 Plot shows a moment where breakthrough moment is detected (BT_Detetc, value 1 means that machine has detected a breakthrough)

3.1 Feature Extraction

Visualization of data (Figs. 2, 3 and 4) showed that there is a need to extract information regarding stability of a variable in data. Z_location parameter data is similar to a linear function before a breakthrough. It was decided to create a feature which monitors a change of that parameter in a time window. It was achieved by the use of a simple linear function y = x property.

$$X_{n+step} - X_n = step \qquad (1)$$

where step is length between points on X-axis.

For a Z_location step = 300 was used. This value was chosen after initial trial. For selected value output had visible two states: low changes and a high changes. Use of a larger step value makes a feature less sensitive to variation. In contrast a lower value makes a feature too sensitive (Fig. 5).

Extracting information from Analogin_Voltage is quite different because of the fact that the frequency of changing voltage is much higher than a visible process. However it has been decided to select a metric that describes a changing of data in time. That decision was made because it is clearly visible that in the same place like in a Z_location, Analogin_Voltage value is more "stable" (character of changes in data is the same). The first step was to concentrate on a single window and calculate a metric for that window and see how they change in a time. The first basic metric is a mean value in a given window. After calculating a mean for a single window it is visible that in a critical moment there is a place where that value is bigger (Figs. 6 and 7).

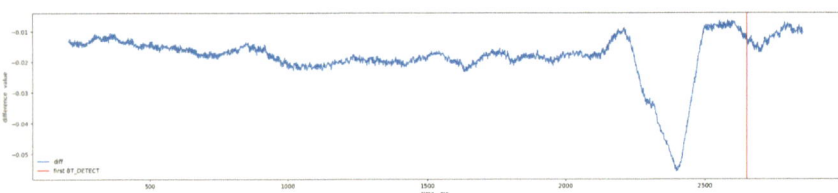

Fig. 5 Plot that describe a difference between every 300 steps

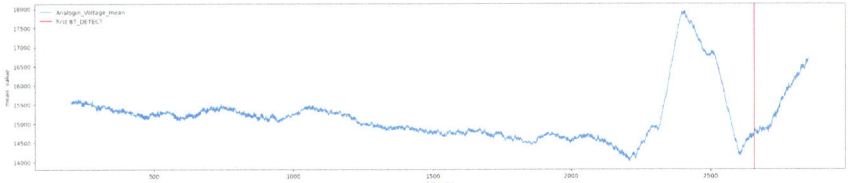

Fig. 6 Plot that describes a mean value for a window equal to 200 steps

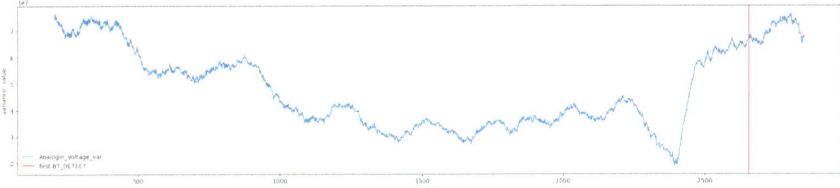

Fig. 7 Plot that describes a various metric for a window equal to 100 steps

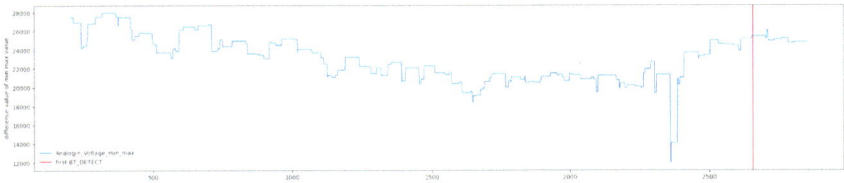

Fig. 8 Plot that describes a subtraction of max value in a window equal to 100 and minimal value for a window equal to 50

Another metric that shows us a change of values is a variance, that is why it has been also used, as another metric, however the change is not significant as in a data based on the mean value.

The last probe was calculating a subtraction between max value in a window and a lowest value in a window (Fig. 8). There is a huge advantage of that method because of the fact that we can control a window from max values and for minimal values.

The method is quite sensitive in this example and the moment of disorder is clearly visible but only for a short time. After the end of this step, our dataset has an extra 3 columns with metrics that describe a change of a value in a time in a current moment.

3.2 Data Preparation

Data preparation was focused on cleaning data and preparing them that can be easily used by machine learning algorithm implementation. The most important thing was to separate data on test, validation and train set. To do that, it has been chosen 3 different groups of drills (one for a training, one for a valid set and one for a test set to calculate final metrics). Data from a machine doesn't have not a number (NaN) values and are pretty well prepared (no visible records missing, etc.). That is why it has been decided to delete only the beginning of series where there is a disorder on a plot triggered by initial strengths. The value of BT_DETECT after detection is always set up as a true value. That is why it has been decided to leave 100 steps after detection and delete the rest of a tail. Thanks to that, machine learning algorithms would not learn wrong patterns (when the anomaly does not exist but the predicted value is still set up as true).

Data from separated drills needed a special preprocessing which allowed them to set up them in one dataframe. Thanks to that it is possible to learn models not only on a single probe. To do it whole preprocessing was generated on a single probe, the main target was to create on a single probe as many vectors that present each state of BT_DETECT in a time as possible. After that all probes were joined, thanks to that only final train set has 26,000 probes.

4 Results

4.1 Hypertuning

To prepare data for machine learning algorithms, it has been decided to use variable values from previous steps to show an algorithm a history of changes. Extra features which are described in Chap. 3 were added in the preprocessing section. It is given in a single vector thanks to the fact that there is no need to use special algorithms like recurrent neural networks.

Set of hyperparameter that has been used—only that parameters were changed in a model:

DecisionTree:

- Max_depth [1;2;3;4;5;6;7;8;9;10]
- Splitter ['best', 'random']
- Class_weight ["balanced"]

RandomForest:

- Max_depth [1;2;3;4;5;6;7]
- N_estimators [10,3-,50,100]
- Max_features ['auto', 'sqrt', 'log2']Class_weight ["balanced"]

XGBoost:

- Max_depth [1;2;3;4]
- Learning_rate[0.01, 0.05, 0.1, 0.15,0 0.2, 0.3, 0.5, 0.75,0.9, 1]

Table 1 Metrics of models accuracy	Model	Test Accuracy	FP rate	FN rate
	Classification tree	0.86	0.019	0.117
	Random forest	0.79	0.000	0.211
	XGBoost	0.96	0.011	0.027

- Colsample_bytree [0.3, 0.5, 0.7, 0.8, 1]
- Reg_lambda [0, 0.1, 0.5, 1, 10, 20, 50, 100]
- Reg_alpha [0, 0.1, 0.5, 1, 2, 5, 10, 20, 50]

To optimize hypertuning, it has been used a random search algorithm. The best models were sought in the parameter space given above. As a metric it has been used as an accuracy score [14]. The train set was used only for trains, and a valid set of data has been used to select the best algorithm. The final result has been calculated on a test data (Table 1).

4.2 Models Comparison

The models found the best accuracy that could be obtained within the assumed range of parameters. The simplest models were created by indirection with the basic CART algorithm, random forests and XGBoost resulted in much more elaborate trees. Examples of graphs can be seen in Fig. 9. A classic problem in publishing large graphs is the difficulty of maintaining readability if we want to transfer the model to a paper, so we present only an illustrative fragment of the XGBoost model. As we can also see, decision trees did not perform very well in mapping complex patterns.

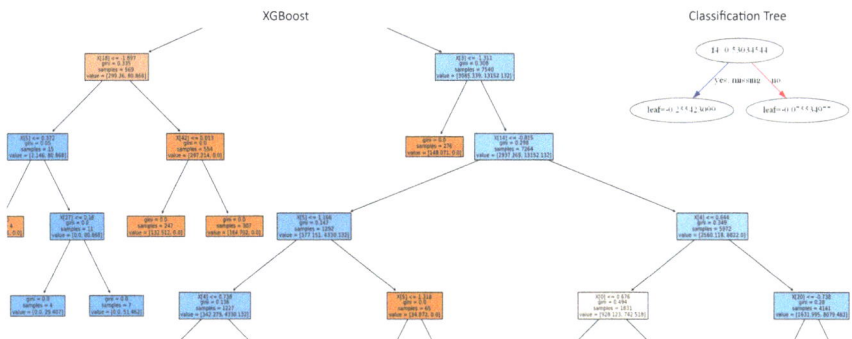

Fig. 9 Sample fragments of models: XGBoost and Classification Tree

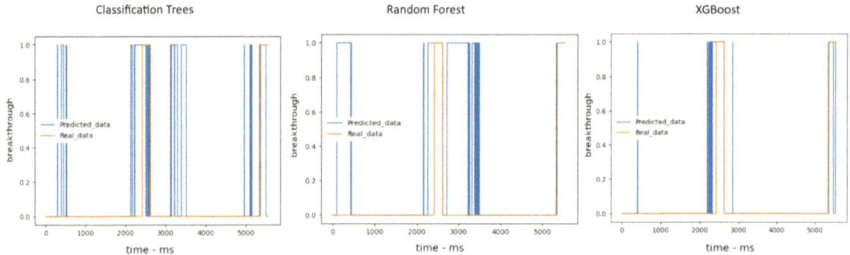

Fig. 10 The results of the BT_Detect prediction

A summary of accuracy metrics and measures of false positives and false negatives are summarized in Table 1. The results of the BT_Detect prediction are shown in Fig. 10.

From the point of view of the usefulness of the models, it is most important that breakthrough detection occur as early as possible before the actual breakthrough, so that the prediction allows the device to respond. As can be seen in Figs. 5, 6 and 7, currently breakthrough detection (by CNC device) occurs with some delay, so the prediction should offset this delay and even get ahead of the event. All models have succeeded in getting ahead of the event, so they can be considered to have made the prediction correctly, but they are not free of flaws. The problem can be a false-positive error, i.e. detecting an event even though it did not happen. The accuracy of the models (Table 1) indicates that this error is negligible, but in further work we will strive to eliminate it completely.

5 Summary

The final result shows that a XGBoost algorithm is the best algorithm to detect anomalies in data which provide information about a breakthrough moment. The main fact is that a decision tree and random forest look like they are too sensitive. That is why there is a lot of FALSE detection at the beginning of a process. However XGBoost algorithm only focused on specific changes that are right before or in a moment of breakout moment.

That information is very useful in in future work and gives a chance for better detection of breakthrough moment in a real time or it can be used to improve actually existing algorithm in a machine which could be very useful because of the fact that better breakthrough detection means better and faster production with a reduction in material (more components will pass a quality test for the first time). For that moment it is important to detect a proper anomaly in a smaller time window which can be modified to an algorithm delay.

Acknowledgements This study was carried out as part of the fundamental research financed by the Ministry of Science and Higher Education, grant no. 16.16.110.663.

References

1. Geng T, Xu Z, Zhang Ch, Ning J (2022) Breakthrough detection in electro-chemical discharge drilling to enhance machining stability. Chin J Aeronaut.https://doi.org/10.1016/j.cja.2022.09.023
2. Bellotti M, Qian J, Reynaerts D (2019) Breakthrough phenomena in drilling micro holes by EDM. Int J Mach Tools Manuf (146):103436, https://doi.org/10.1016/j.ijmachtools.2019.103436
3. Bellotti M, Qian J, Reynaerts D (2020) Self-tuning breakthrough detection for EDM drilling micro holes. J Manuf Process 57:630–640. https://doi.org/10.1016/j.jmapro.2020.07.031
4. Maradia U, Benavoli A, Boccadoro M, Bonesana C, Kliuev M, Zaffalon M, Gambardella L, Wegener K (2018) EDM Drilling optimisation using stochastic techniques. Procedia CIRP (67):350–355. https://doi.org/10.1016/j.procir.2017.12.225
5. Koshy P, Boroumand M, Ziada Y (2010) Breakout detection in fast hole electrical discharge machining. Int J Mach Tool Manuf 50:922–925. https://doi.org/10.1016/j.ijmachtools.2010.05.006.6
6. Weiwen X, Junqi W, Wansheng Z (2018) Break-out detection for high-speed small hole drilling EDM based on machine learning. Procedia CIRP (68):569–574. https://doi.org/10.1016/j.procir.2017.12.115
7. Fu T-c (2011) A review on time series data mining. Eng Appl Artif Intell 24(1):164–181.https://doi.org/10.1016/j.engappai.2010.09.007
8. Olejarczyk-Wożeńska I, Opaliński A, Mrzygłód B, Regulski K, Kurowski W (2022) Bainite transformation time model optimization for Austempered Ductile Iron with the use of heuristic algorithms. Comput Methods Mater Sci 22(3):125–136. https://doi.org/10.7494/cmms.2022.3.786
9. Pedregosa F, Varoquaux G, Gramfort A, Michel V, Thirion B, Grisel O, Blondel M, Prettenhofer P, Weiss R, Dubourg V, Vanderplas J, Passos A, Cournapeau D, Brucher M, Perrot M, Duchesnay E (2011) Scikit-learn: machine learning in Python. J Mach Learn Res (JMLR) 12:2825–2830
10. Baran W, Regulski K, Milenin A (2022) Influence of materials parameters of the coil sheet on the formation of defects during the manufacture of deep-drawn cups. Processes 10:578. https://doi.org/10.3390/pr10030578
11. Gumienny G, Kacprzyk B, Mrzygłód B, Regulski K (2022) Data-driven model selection for compacted graphite iron microstructure prediction. Coatings 12(11):1676. https://doi.org/10.3390/coatings12111676
12. Chen T, Guestrin C (2016) XGBoost: a scalable tree boosting system, arXiv:1603.02754. https://doi.org/10.1145/2939672.2939785
13. Shi L, Qian C, Guo F (2022) Real-time driving risk assessment using deep learning with XGBoost. Accid Anal Preven (178). https://doi.org/10.1016/j.aap.2022.106836
14. Zhang Y, Tang Z, Yang R (2022) Data anomaly detection for structural health monitoring by multi-view representation based on local binary patterns. Measurement (202):111804. https://doi.org/10.1016/j.measurement.2022.111804

Sensitivity Analysis and Formulation of the Inverse Problem in the Stochastic Approach to Modelling of Phase Transformations in Steels

Danuta Szeliga⬮, Natalia Jażdżewska⬮, Jakub Foryś, Jan Kusiak⬮, Rafał Nadolski⬮, Piotr Oprocha⬮, Maciej Pietrzyk⬮, Paweł Potorski⬮, and Paweł Przybyłowicz⬮

Abstract The need for a reliable prediction of the distribution of microstructural parameters in metallic materials after processing was the motivation for this work. The model describing phase transformations, which considers the stochastic character of the nucleation of the new phase, was formulated. Numerical tests of the model, including sensitivity analysis, were performed and the optimal parameters such as time step, kind of the random numbers generator (RNG) and the number of the Monte Carlo points were determined. The validation of the model requires an application of proper coefficients corresponding to the considered materials. These coefficients have to be identified through the inverse analysis, which, on the other hand, uses optimization methods and requires the formulation of the appropriate objective function. Since the model involves stochastic parameters, it is a crucial task. Therefore, in the second part of the paper, a specific form of the objective function for the inverse analysis was developed. In the first approach, an objective function based on measurements of the average parameters was used and primary optimization was performed. Various optimization methods were tested. In the second approach, the hybrid objective function, which combined measured average transformation temperatures with a measure based on histograms, was used. Since, at this stage, we do not have measurements of the distribution of microstructural features, the basic histograms were generated by the model with the coefficients obtained in the first step of the optimization. The capability of finding the optimal solution for different starting points was evaluated and various approaches were compared. The elaborated original stochastic approach to modelling the phase transformations occurring during cooling after hot forming was validated on selected carbon steel.

D. Szeliga (✉) · J. Foryś · J. Kusiak · R. Nadolski · M. Pietrzyk
Faculty of Metals Engineering and Industrial Computer Science, AGH University of Krakow, Al. Mickiewicza 30, 30-059 Kraków, Poland
e-mail: szeliga@agh.edu.pl

N. Jażdżewska · P. Oprocha · P. Potorski · P. Przybyłowicz
Faculty of Applied Mathematics, AGH University of Krakow, Al. Mickiewicza 30, 30-059 Kraków, Poland

161

J. Kusiak et al. (eds.), *Numerical Methods in Industrial Forming Processes*, Lecture Notes in Mechanical Engineering, https://doi.org/10.1007/978-3-031-58006-2_13

Keywords Multiphase steels · Heterogeneous microstructures · Phase transformations · Nucleation · Stochastic model · Identification

1 Introduction

Diverse materials, like metallic alloys, notably multiphase advanced high-strength steels (AHSS), are extensively utilized in modern-day applications [1, 2]. The microstructure of AHSSs typically consists of multiple phases with distinct properties in order to achieve a favourable combination of strength and ductility [3]. Among multiphase steels, dual-phase (DP) and complex-phase (CP) steels are the two widely used in car body applications [4]. Both these steels have good strength and global formability represented by elongation in the tensile tests, but local formability is much better for the CP steels. It is due to the complexity of the CP microstructure, which is characterized by smoother gradients of properties compared to the DP steels. Despite the growing application of CP steels, the correlation between the CP microstructure and its unique mechanical properties has not been fully understood so far. It is expected that advanced numerical models can support the investigation of heterogeneous microstructure in the CP steels. Thus, numerical tools, which can predict distributions of various parameters in heterogeneous materials, are intensively searched for. Mean-field and full-field material models have been distinguished in the literature during the last few decades. The latter have much wider predictive capabilities which, however, go side by side with high computing costs. The need for a reliable and fast prediction of the distribution of microstructural parameters in metallic materials after processing was the motivation for our work. The objective was to develop mean-field (fast) model capable of predicting distributions (histograms) of microstructural features instead of their average values. The model describing the evolution of dislocation populations and grain size, which considers the stochastic aspects of phenomena occurring during hot forming, was formulated in [5]. The coefficients in this model were identified on the basis of experimental data in [6, 7] and the validated version of the model with examples of applications is presented in [8]. Because product properties develop during the cooling process following hot forming, we opted to expand the model's scope by incorporating phase transformations. Utilizing the dislocation density and grain size histograms derived from the stochastic hot deformation model, we employed them as inputs for simulating phase transformations in the initial method outlined in the publication [9], the deterministic model incorporated this data to calculate phase composition histograms. The model's evolution involves considering the stochastic nature of new phase nucleation as the subsequent stage. In the present paper, the stochastic phase transformation model is proposed. The model calculates distributions (histograms) of microstructural features after cooling of products after hot forming. In the present paper, we focused on the numerical tests of the model and the sensitivity analysis. Formulation of the inverse problem dedicated to the identification of the coefficients in the model was the next objective. A specific form of the objective function for the inverse analysis was proposed using a measure of the distance between two histograms, measured and calculated ones.

2 Stochastic Model of Phase Transformations

The developed model simulates phase transformations occurring in steels during controlled cooling after hot deformation processes. The main equations of the model are described below.

2.1 State of the Equilibrium

The equilibrium state of the metallurgical system is described by the thermodynamics. The phase transformation model describes the kinetics of the transient state between the two equilibrium states. The equilibrium state for steels is characterized by the phase equilibrium diagram Fe–Fe$_3$C. Approximation of the two important lines (GS, ES) in this diagram gives the following equations:

Equilibrium carbon concentration in the austenite (at the γ/α interface)

$$c_{\gamma\alpha} = c_{\gamma\alpha0} + c_{\gamma\alpha1} T(t) \tag{1}$$

Maximum carbon concentration in the austenite (at the $\gamma/cementite$ interface)

$$c_{\gamma\beta} = c_{\gamma\beta0} + c_{\gamma\beta1} T(t) \tag{2}$$

In Eqs. (1) and (2), temperature T is in °C and $c_{\gamma\alpha0}$, $c_{\gamma\alpha1}$, $c_{\gamma\beta0}$ and $c_{\gamma\beta1}$ are coefficients, which were determined using ThermoCalc software by researchers from the Silesian Institute of Technology, see joint publication [6]. The current average carbon content in the austenite c_γ is another crucial parameter, which determines whether a phase transformation may occur

$$c_\gamma = \frac{c_0 - F_f c_\alpha}{1 - F_f} \tag{3}$$

where c_0—carbon content in steel, F_f—ferrite volume fraction with respect to the whole volume and c_α—carbon content in ferrite.

Other parameters in the model are A_{e3}—theoretical temperature of the beginning of phase transformations, A_{e1}—theoretical temperature of the end of ferritic transformation and c_{eut}—equilibrium carbon concentration in the austenite at the eutectic temperature, which is defined by a cross point of lines (1) and (2):

$$c_{eut} = c_{\gamma\alpha0} + c_{\gamma\alpha1}\frac{c_{\gamma\alpha0} - c_{\gamma\beta0}}{c_{\gamma\beta1} - c_{\gamma\alpha1}} \tag{4}$$

The equilibrium volume fraction of ferrite at the temperature A_{e1} is calculated from the following equation:

$$F_{eut} = \frac{c_{eut} - c_0}{c_{eut} - c_\alpha} \tag{5}$$

Equilibrium parameters determined by Eqs. (1)–(5) were used as boundary conditions for the model, which describes the kinetics of phase transformations.

2.2 Nucleation of a New Phase

The nucleus of a new phase may appear only if the condition for starting a given transformation is met. These conditions for ferrite, pearlite, bainite and martensite are listed in Table 1, where B_s, M_s—bainite and martensite start temperatures, a_{25}, a_{31}, a_{32}—coefficients. The ferritic transformation cannot start if the pearlitic, bainitic or martensitic transformation has already started, the pearlitic transformation cannot start if the bainitic or martensitic transformation has already started and the bainitic transformation cannot start if the martensitic transformation has already started.

As it has been mentioned above, the phase transformations introduce stochastic elements in the model connected with the random character of the nucleation of the new phase. In the first approach, we will perform a solution assuming Poisson homogenous nucleation. The statistical approach is based on the fundamental knowledge regarding nucleation [10]. The deterministic nucleation rate equation is replaced by an equation with a stochastic variable, which accounts for the stochastic character of the nucleation. The parameter $\xi(t_i)$ representing this stochastic variable, satisfies

$$\mathbf{P}[\xi(t_i) = 0] = \begin{cases} p(t_i) & \text{if } p(t_i) < 1 \\ 1 & otherwise \end{cases} \tag{6}$$

$$\mathbf{P}[\xi(t_i) = 1] = 1 - \mathbf{P}[\xi(t_i) = 0]$$

In Eq. (6), $p(t_i)$ is a function, which bounds together the probability that the material point becomes a critical nucleus in a current time step and present state of

Table 1 Conditions determining the beginning of subsequent transformations

Transformation	Condition to start
Ferrite	$T < A_{e3}$
Pearlite	$c_\gamma > c_{\gamma\beta}(T)$
Bainite	$T < B_s = a_{25}$
Martensite	$T < M_s = a_{31} - a_{32}c_\gamma$

the material. This probability is based on the following knowledge about nucleation sites:

- Nucleation rate increases with an increase of the undercooling below temperature specific for given transformation, A_{e3}, A_{e1} and $B_s = a_{25}$ for ferrite, pearlite and bainite transformations, respectively.
- Grain boundaries and shear bands in the deformed microstructure are the privileged locations of the nuclei. Therefore, the probability of nucleation should increase with an increase of dislocation density and a decrease of the austenite grain size.

Based on this knowledge and assuming Poisson homogenous nucleation, the following equations are used:

- Ferrite

$$p(t_i) = a_1 D^{-a_2} \rho^{a_3} [A_{e3} - T(t)]^{a_4} \Delta t \tag{7}$$

- Pearlite

$$p(t_i) = a_{11} D^{-a_{12}} \rho^{a_{13}} [A_{e1} - T(t)]^{a_{14}} \Delta t \tag{8}$$

- Bainite

$$p(t_i) = a_{21} D^{-a_{22}} \rho^{a_{23}} [a_{25} - T(t)]^{a_{24}} \Delta t \tag{9}$$

where D—austenite grain size, ρ—dislocation density, a_1, a_2, a_3, a_4, a_{11}, a_{12}, a_{13}, a_{14}, a_{21}, a_{22}, a_{23}, a_{24}, a_{25}—coefficients.

In each time step of calculations, a random number within the range [0,1] is generated and compared with the probability $p(t)$. If the value of the function $p(t)$ in a given time step is greater than the generated random number, $\xi(t_i) = 0$ and the nucleus appears. In the case of martensite, this random number is not generated and martensitic transformation occurs always when temperature drops below the M_s temperature. However, according to Table 1 M_s depends on the current carbon concentration in the austenite, which, in turn, depends on the progress of earlier transformations, see Eq. (3). In consequence, the stochastic component is introduced into the martensitic transformation as well. In Eqs. (7), (8) and (9), grain size and dislocation density are stochastic variables (in the form of histograms), which are calculated by the hot forming model described in [8], see selected example of such results in Fig. 1.

In the present solution, hot forming model and phase transformation model are connected. Calculations for each MC point begin after heating before hot deformation and go through the whole process of hot deformation and cooling of samples.

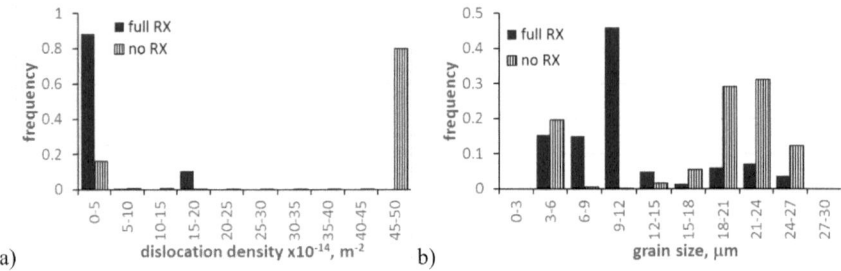

Fig. 1 Selected examples of the calculated histograms of the dislocation density **a** and the grain size and **b** at the beginning of phase transformations for the fully recrystallized and not recrystallized material

2.3 Kinetics of Transformations

After the appearance of a nucleus of a new phase, it starts to grow until another transformation begins or until all phases excluding austenite occupy the whole volume, which marks the end of the simulation. We assumed that phase growth in the model is deterministic in nature, which corresponds to the real phase growth process. To avoid problems which occur when the temperature varies during the process, we selected the upgrade of the Leblond model [11], which describes the kinetics of the growth of the new phase using the differential equation with respect to time. In consequence, it does not need application of the additivity rule when the temperature varies during the process. It is the main advantage of this approach. The original Leblond model [11] assumes that the rate of the transformation is proportional to the distance from the thermodynamic equilibrium in a given temperature:

$$\frac{dX}{dt} = k\left(X_{eq} - X\right) \qquad (10)$$

where t—time, X—volume fraction of a new phase, X_{eq}—equilibrium volume fraction of the new phase at the temperature T, k—coefficient

Coefficient k for each phase depends on the temperature and it is defined by a modified Gauss function, see Table 2, where $a_6, a_7, a_8, a_9, a_{10}, a_{30}, a_{16}, a_{17}, a_{18}, a_{19}, a_{20}, a_{26}, a_{27}, a_{28}$ and a_{29} are coefficients.

The Eq. (10) is solved using explicit Euler method:

Table 2 Formulae describing coefficient k in Eq. (10) for ferrite, pearlite and bainite

Ferrite	$k_f(T(t_i)) = \frac{a_6}{D^{a_{10}}} \exp\left(-\left(\frac{|T(t_i)-a_7|}{a_8}\right)^{a_9}\right)$	(11)
Pearlite	$k_p(T(t_i)) = \frac{a_{16}}{D^{a_{20}}} \exp\left(-\left(\frac{|T(t_i)-a_{17}|}{a_{18}}\right)^{a_{19}}\right)$	(12)
Bainite	$k_b(T(t_i)) = \frac{a_{26}}{D^{a_{30}}} \exp\left(-\left(\frac{|T(t_i)-a_{27}|}{a_{28}}\right)^{a_{29}}\right)$	(13)

Table 3 Formulae describing equilibrium volume fractions X_{eq} in Eq. (10) and volume fractions F for each phase

Transformation	Equilibrium volume fraction	Volume fraction with respect to the whole volume
Ferrite	$$\left.\begin{array}{l} X_{eq}(T) = \frac{F_{f\max}(T)}{F_{eut}} \\ F_{f\max}(T) = \frac{c_{\gamma\alpha}(T) - c_0}{c_{\gamma\alpha}(T) - c_\alpha} \end{array}\right\} \text{ for } T > A_{e1}$$ $X_{eq}(T) = X_{eut} = 1$ for $T \le A_{e1}$	$F_f = X_f F_{eut}$
Pearlite	$X_{eq}(T) = 1$	$F_p = X_p(1 - F_f)$
Bainite	$X_{eq}(T) = 1$	$F_b = X_b(1 - F_f - F_p)$
Martensite		$F_m = (1 - F_f - F_p - F_b)$

$$X(t_i) = X(t_{i-1}) + k(T(t_i))\big(X_{eq}(T(t_{i-1})) - X(t_{i-1})\big)\Delta t \tag{14}$$

where Δt—time increment.

Volume fraction X in Eq. (14) is calculated with respect to the maximum volume fraction of the considered phase at the temperature $T(t)$, whereas the final result of the model calculations is F, which is the volume fraction of the considered phase with respect to the whole volume. Thus, X for each phase is within a range [0,1] and the sum of F of all phases equals 1 because it corresponds to occupancy of the whole volume of the sample. As it has been mentioned, grain grows until another transformation starts or until all phases, excluding austenite, occupy the whole volume. In the case of martensite, it is assumed that this phase occupies the whole volume of the austenite, which remained after ferrite, pearlite and bainite transformations at the temperature M_s. Formulae describing equilibrium volume fractions X_{eq} in Eq. (10) and volume fractions F for each phase are listed in Table 3. In this table, $F_{f\max}(T)$ is the equilibrium volume fraction of the ferrite in steel at the current temperature T. Maximum ferrite volume fraction in steel is equal to the equilibrium ferrite volume fraction at the eutectic point F_{eut}, which is defined by Eq. (5).

Coefficients in the developed model are grouped in the vector $\mathbf{a} = \{a_1, ..., a_{32}\}^T$ and they are determined by the inverse analysis for the experimental data. The model with optimized coefficients predicts distributions of such parameters as volume fractions of phases. The parameters will be used to predict local gradients of properties, which influence formability [12].

3 Numerical Tests and Validation of the Stochastic Model

The objectives of the numerical tests were twofold. The first was to evaluate the influence of selected numerical parameters in the model on the accuracy of simulations and on the computing times. The following parameters were investigated: (i) type of the random number generator (RNG), (ii) maximum time step and temperature change during the time step and (iii) a number of the Monte Carlo points. The optimal values of these parameters were proposed having in mind **a** balance between accuracy and computing times. Sensitivity analysis of the model's output with respect to the coefficients was the second objective of the numerical tests.

All the numerical tests were carried out for the model coefficients obtained in the primary identification, which was performed using inverse analysis based on the comparison of the measured and calculated average output parameters. In all the tests the output histograms were divided into 10 bins.

3.1 Selection of the Optimal Numerical Parameters

The tests that were designed to check the influence of the random number generator (RNG) on the model output were performed first. As a basis, we considered the RNGs from the C++ Standard Library. In the tests, we performed computations with 20,000 Monte Carlo points with the default RNG (default_random_engine from <random> C++ header) and with few other RNGs available in the library: mt19937, mt19937_64, minstd_rand0, minstd_rand, ranlux24_base, ranlux24, ranlux48_base and ranlux48, knuth_b. The results obtained for various RNGs were compared with the ones obtained for default RNG. The comparison consisted of computing Earth Mover's Distance (EMD) between the output histograms. According to this metric, the distance between histograms H_1 and H_2 is calculated as follows:

$$d(H_1, H_2) = EMD = \sum_{i=1}^{n} |EMD_i|$$

$$EMD_i = \sum_{j=1}^{i} \left[H_1(j) - H_2(j) \right] \tag{15}$$

The EMD, which is explained in detail in [13], was used in our earlier publication [10] and it gave good results in the comparison of histograms arising in applications considered there. The analysis of the results of the test allowed to conclude that the influence of the used RNG on the model output is negligible. Additionally, due to the stochastic nature of the model, the model output varies slightly in each simulation run, even when the same RNG is used. So, the received positive distances may have occurred only because of generated random numbers and not because of the RNG itself. It is also worth noting that the model works as intended because in each simulation run it gives a similar output (small values of EMD), but slightly different due to its stochastic nature.

The maximum temperature change per time step and the maximum time step are other model parameters, which were considered. They should be constant for all the simulations for all the materials. When both are calibrated, the first one affects mainly simulations with high cooling rates and the second one affects mainly simulations with low cooling rates. The influence of these parameters on the model output was investigated. Obviously, the shorter the time step and the smaller the temperature change per time step, the better the accuracy of simulations, but it is paid with an increase in the computational cost. Thus, our objective was to find a balance between accuracy and computing costs. We took into consideration many values of these parameters and compared the average output parameters with the ones obtained in the experiments. We took into consideration every cooling speed that was included in the experimental data, which included both high and low cooling speeds. After these comparisons, we decided to set the maximum temperature change per step equal to 0.1 °C and the maximum time change per step equal to 0.5 s.

Searching for the optimal number of the Monte Carlo points was the next objective of the numerical tests. This number is a crucial model parameter. Since the model is stochastic in nature, the output parameters for individual points may be different. The number of the Monte Carlo points determines the result stability, i.e. how much the results differ between simulation runs. If the number of the Monte Carlo points is too small, the model output is too random. The stability of the results is paid with computational cost, because, for each Monte Carlo point, the computations have to be performed separately. The tests were designed to find the least number of the Monte Carlo points for which the results are stable. The simulations were performed with exactly the same parameters, excluding the number of Monte Carlo points, which took all values from the set {100, 200, ..., 19,900, 20,000}. For every number of Monte Carlo points, the simulation was repeated five times. Then all the results were compared with the basic result, which was obtained for the 20,000 Monte Carlo points. In comparison, we took into consideration the difference between the average output parameters and the Earth Mover's distance between the output histograms. Then the largest value (in each compared item) for each number of Monte Carlo points was selected and these values are shown in Fig. 2. In the case of the average output parameters, there are no values for the pearlite because, for the investigated range of the parameters, the pearlitic transformation has not occurred for a majority of the Monte Carlo points in all simulations. We assumed that the solution for the average output parameter is stable when it differs from the basic result by less than 1oC for temperatures and by less than 0.003 for volume fractions. It was fulfilled for at least 7500 Monte Carlo points. However, it can be seen in Fig. 2c that the curve for the histogram distances flattens out above 10,000 Monte Carlo points. Thus, we decided that the least number for which the results are stable is 10000. Performing a simulation for more points is futile because this increases the computational cost and does not improve the result. Examples of a comparison of the output histograms for 100, 1000 and 10,000 points are shown in Fig. 3. It is seen that indeed the output histograms differ noticeably for a small number of the Monte Carlo points, and that for 10,000 Monte Carlo points or more, the difference stabilizes.

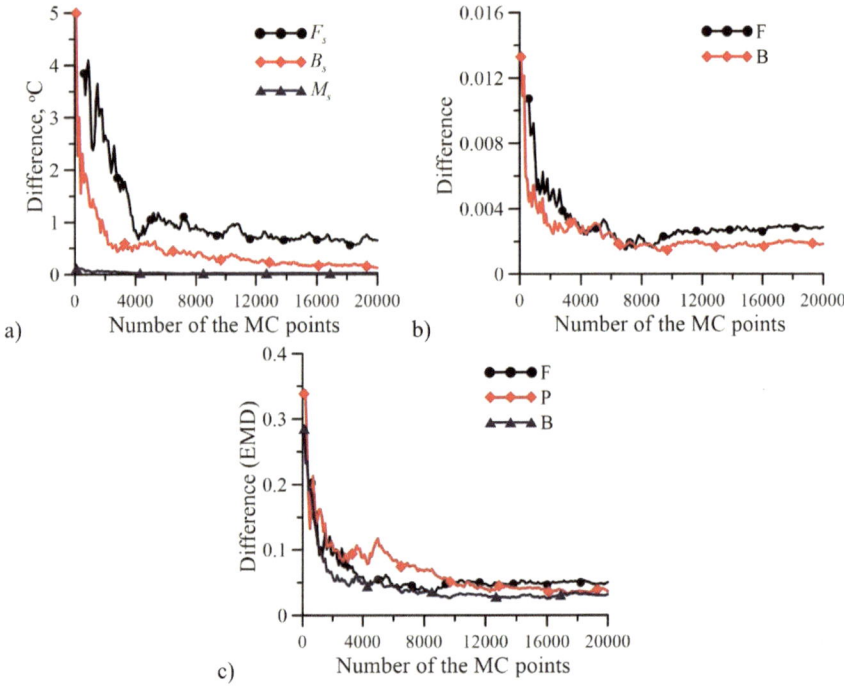

Fig. 2 Maximum difference from 5 simulation runs between the result with a given number of Monte Carlo points and the basic result with 20,000 Monte Carlo points for the average output temperatures (**a**), the average output volume fractions (**b**) and the histograms of the volume fractions (EMD between histograms) (**c**). Notation: F_s, B_s, M_s—start temperatures for ferrite, bainite and martensite, respectively, F, B, M—volume fractions of ferrite, bainite and martensite, respectively

3.2 Sensitivity Analysis

Identification of the model with 32 coefficients is time-consuming and problems with the uniqueness of the solution can be encountered. To avoid these problems, sensitivity analysis (SA) [14] was applied prior to the inverse analysis. The goal of the SA was to find the coefficients which influence the output most and to identify these coefficients in the first step of the IA. The effect of the change of the ith coefficient (Δa_i) on the solution (χ_i) at the time t is given below:

For the average output parameters:

$$\chi_i(\mathbf{a}) = \frac{a_i}{\Delta a_i} \frac{y_c(a_1, \ldots, a_{i-1}, a_i + \Delta a_i, a_{i+1}, \ldots, a_k) - y_c(\mathbf{a})}{y_c(\mathbf{a})} \tag{16}$$

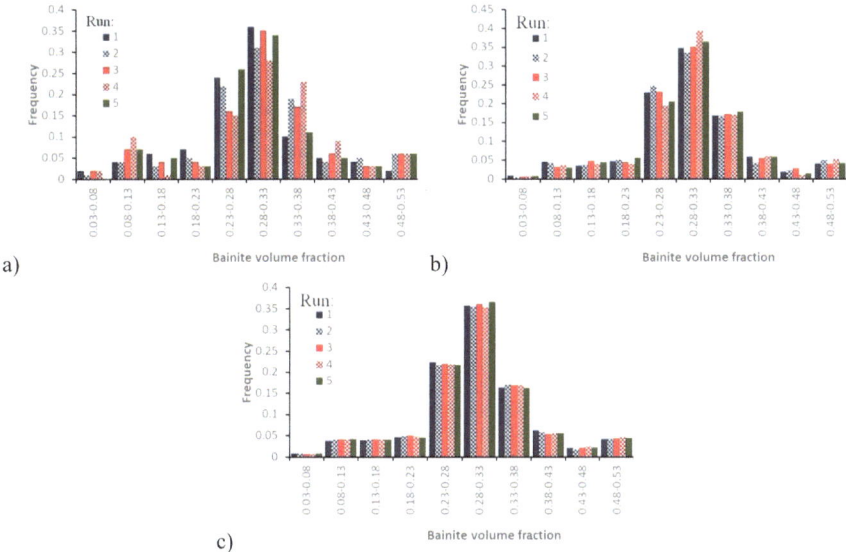

a)

b)

c)

Fig. 3 A comparison of the output histograms for 100 (**a**), 1000 (**b**) and 10,000 (**c**) MC points—an example for bainite volume fraction

For the output histograms:

$$\chi_i(\mathbf{a}) = a_i \frac{d(H_1, H_2)}{\Delta a_i} \tag{17}$$

where Δa_i—small increment of the ith parameter, $y_c(\mathbf{a})$—model output, H_1—the basic histogram calculated for the coefficients $\mathbf{a}$, H_2—histogram obtained after small disturbance of a_i.

The Earth Movers Distance, see (15), was used to calculate the distance between histograms H_1 and H_2 in the Eq. (17).

During the SA the simulation was performed for 10,000 Monte Carlo points, and the Δa_i was set to 10% of a_i. The SA determined the model parameters, which contribute the most to the model output and those which are not significant [15]. The SA preceded identification of the model using inverse analysis and the SA results were used to design the best optimization strategy. The coefficients a_3, a_{13} and a_{23} were not taken into consideration in the SA because they are responsible for the influence of a dislocation density which currently is not taken into account in the model. The coefficients a_5 and a_{15} also were not taken into consideration in the SA because, currently, they do not appear in the model. In the case of the average output parameters, there are no values for the pearlite because, for the investigated range of the parameters, this transformation has not occurred for a majority of the

Monte Carlo points in all simulations. In the whole SA, the pearlite temperature end
and the bainite temperature end were not taken into consideration because, for the
investigated range of the parameters, these output values have not appeared in any
Monte Carlo point. The results of the sensitivity analysis are shown in Fig. 4. The
analysis of the results of the SA brought us to the conclusion that for the simulation
result, the most crucial coefficients are a_2, a_4, a_{10}, a_{25}, a_{30} and a_{31}. For the individual
output parameters, the following coefficients have the biggest influence:

- ferrite start temperature: a_2, a_4
- pearlite start temperature: a_2, a_{10}, a_{12}, a_{25}
- bainite start temperature: a_{22}, a_{25}
- martensite start temperature: a_{31}
- ferrite volume fraction: a_2, a_4, a_{10}, a_{25}
- pearlite volume fraction: a_2, a_{10}, a_{12}, a_{25}
- bainite volume fraction: a_2, a_{10}, a_{25}, a_{30}, a_{31}
- martensite volume fraction: a_2, a_{10}, a_{30}, a_{31}

Coefficients a_1, a_2, ..., a_9, a_{10} appear in the ferrite transformation in the model,
coefficients a_{11}, a_{12}, ..., a_{19}, a_{20} in the pearlite transformation, coefficients a_{21},
a_{22}, ..., a_{29}, a_{30} in the bainite transformation, and coefficients a_{31} and a_{32} in the
martensite transformation. A coefficient may significantly influence the respective
temperature or volume fraction in the model output and it is quite obvious. However,
the coefficients that appear in one transformation can influence the volume fraction of
other as well. Since the volume fractions of all phases sum up to 1, if one phase takes
significantly more/less volume (because of an earlier/later start of the transformation
or a faster/slower growth of the phase), it will also affect other phases. That change
in the volume does not affect the temperatures at the start of the transformations,
though—in Fig. 4a, it is clearly seen that the main influence on the temperatures
comes from the coefficients that appear in the respective transformation.

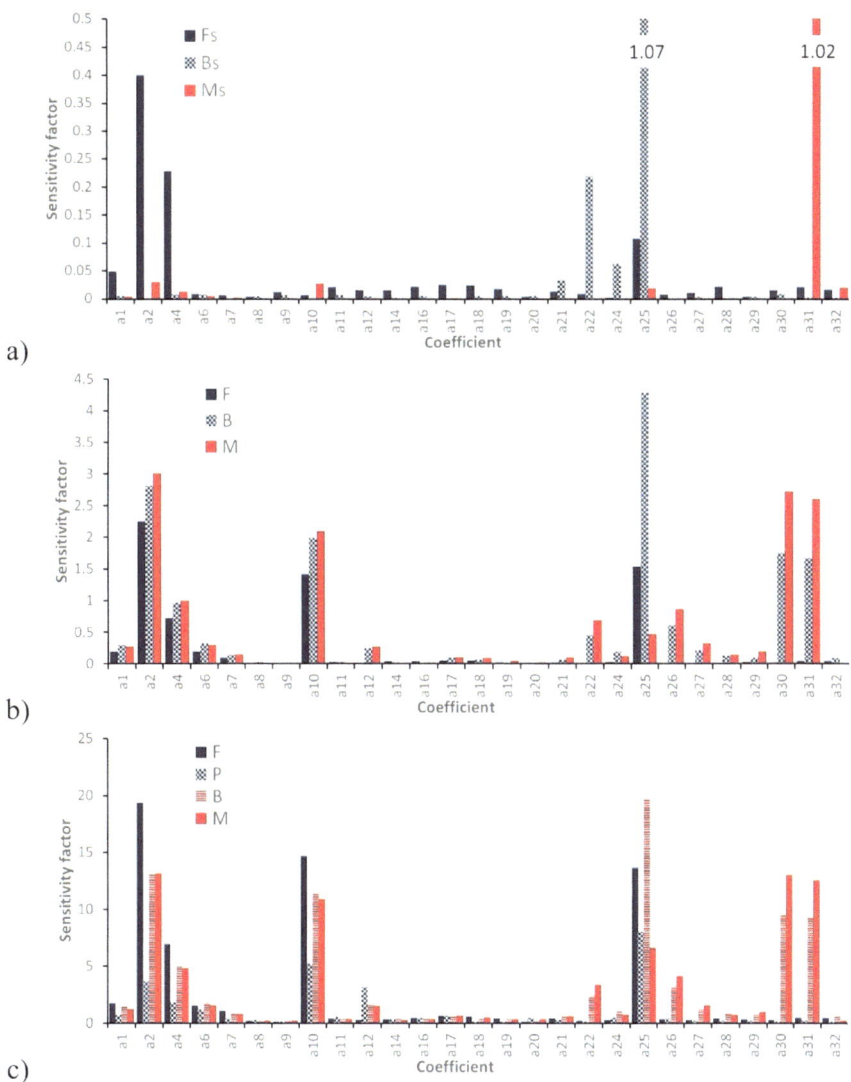

Fig. 4 Results of the sensitivity analysis: influence of the model coefficients on the output parameters: temperatures of the beginning of transformations (**a**), average volume fractions of phases (**b**) and histograms of the volume fractions of phases (**c**). Symbols in the legend are explained in the Fig. 2 caption

4   Identification of the Coefficients in the Stochastic Model

As it has been mentioned, the stochastic model of phase transformations contains several coefficients, which must be determined for each specific material. Identification of the coefficients is performed using inverse analysis for the experimental data. These data contain both measurements of the average values of some microstructural parameters and measurements of histograms of other parameters.

4.1   Formulation of the Inverse Problem for the Stochastic Experimental Data

The problem of the identification of the coefficients in material models is well-known and widely discussed in the scientific literature as an inverse problem [16, 17]. The algorithm for the stochastic inverse problem is described in detail in [18]; we repeat it briefly below for the completeness of the paper. Since our model is non-linear, there is no analytic solution for the inverse problem. Therefore, we reformulated the inverse problem as an optimization task with the coefficients in the model **a** becoming the state variables. The aim of the inverse analysis was finding the optimal values of coefficients **a**, which are determined by searching for a minimum of the following objective function:

$$\Phi(\mathbf{a}) = d(y_c(\mathbf{a}), y_m) \tag{18}$$

where $y_c(\mathbf{a})$—outputs calculated for the model coefficients **a**, y_m—measurements in the experimental tests, d—metric in the output space Y.

The optimization task defined for the deterministic model by Eq. (18) was redefined in [5] for the stochastic variable model and the following objective function was proposed:

$$\Phi(\mathbf{a}) = d(H_c(\mathbf{a}), H_m) \tag{19}$$

where $H_c(\mathbf{a})$—histogram obtained by several calculations of $y(\mathbf{a})$, H_m—measured histogram (from the experiment), d—a ranking function comparing two histograms.

The stochastic model solution is in the form of a histogram, approximating the real distribution of phase fractions. Therefore, it was necessary to compare the model outputs for particular sets of coefficients, taking into account that the random variable $\xi(t_i)$ in the Eq. (6) and stochastic nature of D and ρ (input parameters for the model) can lead to completely different single solutions for the same starting values. Similarly, as it was done in the sensitivity analysis (Sect. 3.2), EMD was used as the metrics $d(H_c(\mathbf{a}), H_m)$. It ensured good convergence in the optimization.

As we have said before, for the identification task, the objective function (19) should be minimized with respect to the model coefficients **a**. To be able to apply metrics (15), the experimental data should include information on distributions of the temperatures of phase transformations and phase fractions. Since the measurement of histograms of temperatures is not physical, we decided to compare only histograms of phase fractions and average start and end temperatures of transformations. Thus, the objective function (19) was reformulated to a hybrid form as follows:

$$\Phi(\mathbf{a}) = \Phi_T(\mathbf{a}) + \Phi_F(\mathbf{a}) \tag{20}$$

The components of the objective function are calculated as follows:

$$\Phi_T(\mathbf{a}) = d\left(T^c(\mathbf{a}), T^m\right) \tag{21}$$

$$\Phi_F(\mathbf{a}) = d\left(H^c(\mathbf{a}), H^m\right) \tag{22}$$

where $T^c(\mathbf{a})$—expected average value of the start/end temperature of transformation calculated for model coefficients $\mathbf{a}$, T^m—the average start/end temperature of transformation determined from the dilatometric tests [6], $H^c(\mathbf{a})$—distribution (histogram) of the phase fraction after cooling calculated for the model using coefficients $\mathbf{a}$. Superscripts m and c refer to measurement and calculations, respectively.

The distance $d(T^c(\mathbf{a}),T^m)$ in Eq. (21) is defined as the sum of the mean square root errors (MSRE) between measured and calculated average start/end temperatures of transformations, that is

$$d\left(T^c(\mathbf{a}), T^m\right) = \sqrt{\frac{1}{Ne}\sum_{i=1}^{Ne}\frac{1}{Nt_i}\sum_{j=1}^{Nt_i}\left(\frac{T_{ij}^c(\mathbf{a}) - T_{ij}^m}{T_{ij}^m}\right)^2} \tag{23}$$

where Ne—number of the tests, Nt_i—number of the temperatures measured in the ith test.

Recall that as before, the distance between histograms in the Eq. (22) is calculated as the sum of the EMDs, which are defined in the Eq. (15).

4.2 Optimization with the Objective Function Based on the Measured and Calculated Average Values of Microstructural Parameters

The correct definition of the objective function is a crucial factor from the point of view of the quality of the solution obtained from the optimization. In the case of the reverse analysis carried out in this work, the coefficients $\mathbf{a} = \{a_1,..., a_{32}\}^T$ for the analysed stochastic model of phase transformations have to be determined. As it is shown in Chap. 2, the coefficients **a** directly influence modelling of the investigated process, which is the cooling of the steel components after hot forming. In this process, the model predicts the start and end temperatures of phase transformations (T) and volume fractions of structural components (F) after cooling. The inverse analysis described in Sect. 4.1 is used to determine the coefficients of the model based on the experimental data. The experiments were composed of dilatometric tests performed with a cooling rate in the range of 0.1 °C/s–100 °C/s. Two austenitization temperatures were used, in consequence, two different austenite grain sizes prior to transformations (17 μm and 24 μm) were obtained. The material used in the experiments was steel containing 0.12%C and 1.3%Mn [6, 19]. In the first approach, the experimental data provided information on the transformation temperatures, as well as average values of volume fractions for ferrite, pearlite, bainite and martensite. Thus, the primary objective of our work was the identification of the model coefficients for the average values of the output parameters. For this purpose, the objective function (22) was omitted and the function (23) was extended by including average volume fractions of phases, as follows:

$$\Phi = \sqrt{\frac{1}{Ne} \sum_{i=1}^{Ne} \left[\frac{w_T}{Nt_i} \sum_{j=1}^{Nt_i} \left(\frac{T_{ij}^m - T_{ij}^c}{T_{ij}^m} \right)^2 + \frac{w_F}{Nf_i} \sum_{k=1}^{Nf_i} \left(\frac{F_{ik}^m - F_{ik}^c}{F_{ik}^m} \right)^2 \right]} \quad (24)$$

where Ne—number of the tests, Nt_i—number of the temperatures measured in the ith test, Nf_i—number of the phase fractions measured in the ith test, T—start or end temperature of the phase transformation, F—phase fraction after cooling, w_T, w_F—weights for temperatures and phase fractions, respectively. Superscripts m and c refer to measurement and calculations, respectively.

Since the influence of temperatures and phase fractions on the objective is different, a selection of weights is extremely important. In the first approach, the following weights were used: $w_T = 0.5$ and $w_F = 0.5$. The results, however, definitely differed from the experimental values that were sought, and the objective function oscillated around 0.0424 (Fig. 5). It was then decided to increase the weights for volume fractions and decrease the weights for temperatures, as it was expected to improve the fit. This resulted in a significant decrease in the value of the objective function to values around 0.00552. However, in some of the tests, the values from the model still differed from the measurements.

We observed that the use of constant weights for volume fractions of all phases has the disadvantage that very small volume fractions generate large errors, even when the absolute difference between measurements and calculations is small. To eliminate this effect, we decided to test a solution with varying weights, which depended on the value of the parameter the weight is dedicated to. In consequence, a similar level of the error is generated regardless of the value of the volume fraction. In the case of temperatures, it was not so important and constant weights were used. This approach made it possible to find results that closely match the experimental data.

By using the objective function (24) in the inverse analysis, it was possible to determine the model coefficients for the analysed steel. The Self-adaptive Differential Evolution method [20] was used in the optimization. The target coefficients were divided into several categories: coefficients responsible for the ferrite start temperature (a_1, a_2, a_3, a_4), pearlite start temperature ($a_{11}, a_{12}, a_{13}, a_{14}$), bainite start temperature ($a_{21}, a_{22}, a_{23}, a_{24}, a_{25}$), martensitic transformation start temperature (a_{31}, a_{32}), ferrite volume fraction ($a_6, a_7, a_8, a_9, a_{10}$), pearlite volume fraction ($a_{16}, a_{17}, a_{18}, a_{19}, a_{20}$) and bainite volume fraction ($a_{26}, a_{27}, a_{28}, a_{29}, a_{30}$). It was decided to divide the optimization based on the measurements of average parameter values into several stages. In the first stage, the temperatures for each of the phase transitions were optimized in order to find appropriate ranges for the coefficients responsible for the temperatures. The next step was to perform similar optimizations in terms of finding the appropriate ranges for the coefficients responsible for the size of the phase fractions. These initial optimizations were performed multiple times to determine the possible ranges for the coefficients. The reason for such initial optimizations was the exclusion from the search area of the ranges for coefficients in which there were no good solutions.

Changes in the objective function in subsequent iterations of the optimization for various weights in the Eq. (24) are shown in Fig. 5. The justification for the selection of different weights has been described earlier. As expected, better results were obtained with the increase in the number of iterations. However, as it is seen in Fig. 5, the plot of the objective function flattens out significantly after around 100–300 iterations, depending on weights. The existence of a significant number of local minima allows for a slight improvement of the solution from time to time, but in the long run, it is difficult to have a significant decrease in the objective function. The quality of the experimental data is also of great importance. The measurement data could also contain deviated data resulting from incorrect sensor measurements, which translated into the quality of optimization.

Fig. 5 Plot of the objective function versus the number of iterations during differential evolution optimization with the objective function (24) for various weights

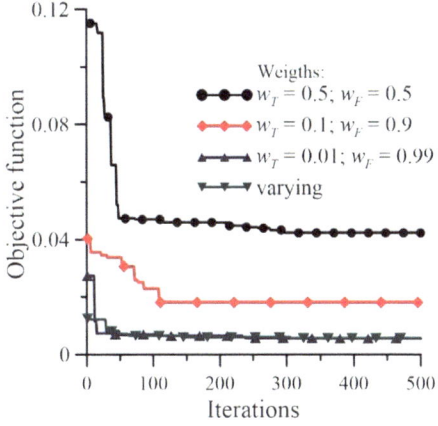

Table 4 Optimal values of the coefficients in the model

a1	a2	a3	a4	a5	a6	a7	a8
8.64×10^{-11}	1.06	0	5.454	0	3.777	539.7	534.37
a9	**a10**	**a11**	**a12**	**a13**	**a14**	**a15**	**a16**
1.2065	1.0272	1.49×10^{-24}	2.15	0	11.92	0	0.295
a17	**a18**	**a19**	**a20**	**a21**	**a22**	**a23**	**a24**
662.88	541.27	15.699	0.1586	3.9785	2.5181	0	2.0849
a25	**a26**	**a27**	**a28**	**a29**	**a30**	**a31**	**a32**
556.35	5.2359	496.2	608.19	5.8933	2.1735	388.7	20.357

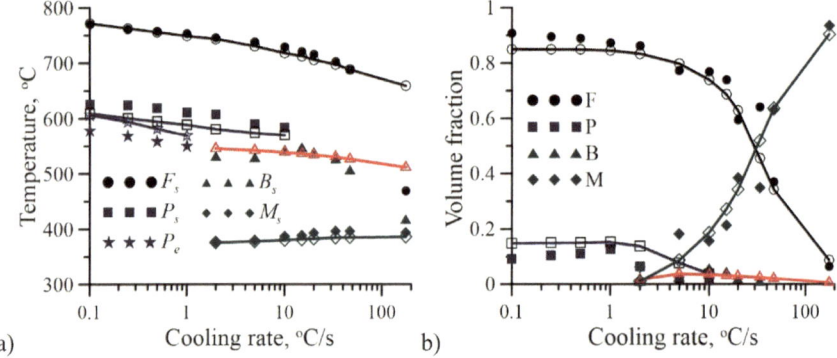

Fig. 6 Comparison of the measured (full symbols) and calculated for the optimal coefficients (open symbols with lines) temperatures of phase transformations (**a**) and volume fractions of structural components (**b**)

The optimal coefficients obtained during optimization for the experimental data are presented in Table 4. The graphs showing a comparison of measured and calculated parameters are shown in Fig. 6.

4.3 Numerical Tests of the Inverse Analysis for the Stochastic Experimental Data

Optimization of the microstructure parameters on the histograms allows for more accurate identification of the model coefficients by using the entire available frequency distribution. However, this also translates into a much higher computational effort, and thus a much longer optimization.

At this stage of the project, we do not have histogram measurements of the output parameters. The objective of the numerical tests described below was to evaluate the capability of the inverse analysis to determine coefficients in the model when such measurements of the histograms are available. Thus, we have calculated the histograms using the model with the coefficients in Table 4 and we considered these histograms as experimental data. Following this, the values of the coefficients were disturbed and optimization was performed. The hybrid objective function (20) was used, but it was reformulated having in mind the specifics of the dilatometric tests:

$$\Phi = \sqrt{\frac{1}{Ne} \sum_{i=1}^{Ne} \left[\frac{w_T}{Nt_i} \sum_{j=1}^{Nt_i} \left(\frac{T_{ij}^m - T_{ij}^c}{T_{ij}^m} \right)^2 + \frac{w_F}{Nf_i} \sum_{k=1}^{Nf_i} \text{EMD}_{ik} \right]} \qquad (25)$$

Symbols in the Eq. (25) are explained below the Eq. (24). The objective function (25) combines measurements of the average temperatures of transformations with measurements of histograms of phase fractions. The earth mover's distance is a metric of the distance between histograms, and it is defined in the Eq. (15) with $H_1 = H_m$ and $H_2 = H_c$—measured and calculated histograms of phase fractions for the phase k in test i.

5 Results

The calculations in this section were carried out using the data obtained from optimization on average values of parameters for the investigated steel (Table 5). The number of 200 Monte Carlo points was used for the initial optimizations, which were performed using the differential evolution method. This method proved to be the most efficient in the optimization of the average values of the measured parameters, see Sect. 4.2. The generated solutions for 3 process parameters were placed in histograms containing 10 bins. In order to reduce the required number of calculations, it was decided to limit the number of the used cooling rates to 7, selecting those that are the most important, as they are located on the edges of phase transformations.

The change in the value of the objective function in subsequent iterations is shown in Fig. 7. As expected, the objective function decreases with successive optimization iterations. The computational costs of optimization are negligible, but generating subsequent results from the model with a high number of Monte Carlo points was cumbersome. It was decided to use a small population (10 individuals) and a number of iterations of the optimization method limited to at most 300 because increasing neither the population nor iterations brought significant benefits, which was confirmed also by other publications [5].

We used the same approach as in the case of the optimization with average values. As the first step, the ranges for the coefficients have been narrowed, and then, acceptable values for the coefficients of + -10% relative to the best point from the narrowing were adopted. The implemented optimizations allowed to achieve the value of the objective function at the level of 0.47. This value of the function enabled a good mapping between the measured data and the response from the model. There were

Fig. 7 Changes in the value
of the objective function in
subsequent iterations during
optimization based on
histograms with the objective
function (25)

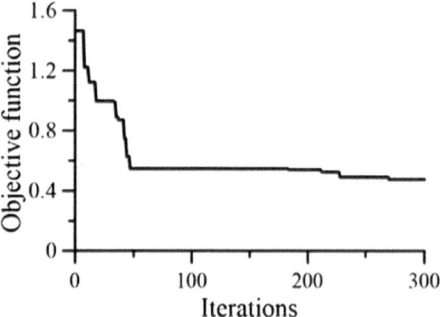

no visual differences compared to the plots in Fig. 6. Therefore, these data are not presented.

The conclusions drawn from the performed optimizations are that by performing subsequent optimizations, it is possible to find coefficients that match the experiment well. Differences in volume fractions are not significant for the newly found coefficients, even though they were different from the original ones. This fact raises questions about the uniqueness of the solution, which will be explored in our future works.

The key aspect in the optimization of histograms, as well as in the optimization of average values, was to narrow the ranges for the coefficients to the ranges in which acceptable results occurred. Subsequent optimizations for specific temperatures and fractions of volume allowed to find the best ranges for the coefficients. Such earlier examination of the ranges allowed perform a more focused search in the optimization leading to a significant reduction of the computational cost during the optimization itself.

Figure 8 shows an example of a comparison between the measured and calculated histograms for the volume fractions of ferrite, pearlite and bainite. The calculated histograms were obtained from the model with coefficients, which gave the lowest value of the objective function (25). Histograms were generated for 1000 Monte Carlo points.

The histograms obtained from the model coincide with those of the experiments. The differences in the histograms are negligible, which also translates into a lower value of the objective function. Through the applied approach, it was possible to find coefficients close to the experimental ones.

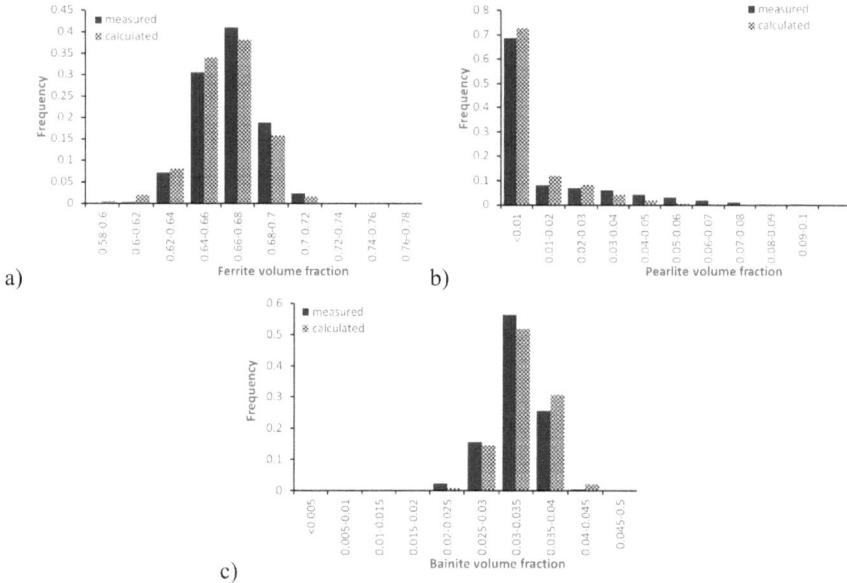

a)

b)

c)

Fig. 8 Selected examples of comparison of measured and calculated histograms of the volume fractions of ferrite (**a**), pearlite (**b**) and bainite (**c**) for the cooling rate of 15 °C/s

6 Conclusions

The stochastic phase transformations model, which accounts for a random character of phase transformations during the cooling of steels and calculates distributions (histograms) of microstructural features, was proposed. The tests of the model, with the objective of selecting the best numerical parameters, were performed and the following conclusions were drawn:

1. As expected, the use of a random factor results in generating diverse output histograms, which allows to characterize the heterogeneous microstructures. Also, due to an introduction of a random factor, the model reflects a stochastic nature of nucleation of a new phase in steels.
2. Differences between the results from the stochastic model obtained for various random number generators are negligible.
3. Setting the maximum temperature change per step to 0.1 °C and the maximum time change per step to 0.5 s results in a good balance between accuracy and computing costs.
4. Numerical tests have shown that simulations are stable above 10,000 Monte Carlo simulations of the individual trajectories. The application of more Monte Carlo simulations does not lead to much better results, but at the same time highly increases the computational cost.

5. Performed sensitivity analysis allowed to identify which coefficients are the most crucial for the model output as a whole and for the individual output parameters.
6. A number of optimization methods were analysed, of which genetic algorithms turned out to be the best in solving the investigated problem. The differential evolution method and its upgrades performed particularly well, decreasing the value of the objective function fastest and usually reaching the lowest value.
7. When defining the objective function, it was observed that proper selection of weights leads to better performance in optimizing the average values. Focusing on matching the weight values for volume fractions, they should have a correspondingly higher value in relation to temperatures. In the case of volume fractions, it was a good idea to use variable weights. This approach allowed a better fit to the experimental data because small fractions do not generate high values of the objective function. During the optimization of the histograms, the values of the weights were not so important anymore.
8. The optimization itself is not computationally expensive, but the numerical model of phase transformations requires considerable computational effort.
9. When optimizing histograms, a valuable approach turned out to be the use of a hybrid objective function, where differences between volume fractions were calculated using the Earth Movers Distance (EMD). This approach allowed to find coefficients that coincide with the experiments reasonably well. The calculated histograms sufficiently reflected the experimental data for volume fractions of all phases, which allowed to conclude that the adopted methodology is correct.
10. It turned out to be a good idea to initially narrow the ranges during global optimization for each temperature and volume fraction separately. This allowed for the initial determination of potential ranges for the coefficients, in which there were good solutions. This multi-stage approach to optimization proved to be the most valuable in terms of obtained results.

Acknowledgements Financial support of the National Science Foundation in Poland (NCN), project no. 2021/43/B/ST8/01710, is acknowledged.

References

1. Kok Y, Tan XP, Wang P, Nai MLS, Loh NH, Liu E, Tor SB (2018) Anisotropy and heterogeneity of microstructure and mechanical properties in metal additive manufacturing: a critical review. Mater Design 139:565–586
2. Chang Y, Lin M, Hangen U, Richter S, Haase C, Bleck W (2021) Revealing the relation between microstructural heterogeneities and local mechanical properties of complex-phase steel by correlative electron microscopy and nanoindentation characterization. Mater Design 203:109620
3. Fonstein N (2015) Advanced high strength sheet steels. Springer International Publishing, Cham

4. Singh MK (2016) Application of steel in the automotive industry. Int J Emerg Technol Adv Eng 6:246–253
5. Klimczak K, Oprocha P, Kusiak J, Szeliga D, Morkisz P, Przybyłowicz P (2022) Inverse problem in stochastic approach to modeling of microstructural parameters in metallic materials during processing. Math Problems Eng, Article ID 9690742
6. Poloczek Ł, Kuziak R, Pidvysotsk'yy V, Szeliga D, Kusiak J, Pietrzyk M (2022) Physical and numerical simulations to predict distribution of microstructural features during thermomechanical processing of steels. Materials 15:1660
7. Szeliga D, Czyżewska N, Klimczak K, Kusiak J, Kuziak R, Morkisz P, Oprocha P, Pidvysotsk'yy V, Pietrzyk M, Przybyłowicz P (2022) Formulation, identification and validation of a stochastic internal variables model describing the evolution of metallic materials microstructure during hot forming. Int J Mater Forming 15:53
8. Szeliga D, Czyżewska N, Klimczak K, Kusiak J, Kuziak R, Morkisz P, Oprocha P, Pietrzyk M, Poloczek Ł, Przybyłowicz P (2022) Stochastic model describing the evolution of microstructural parameters during hot rolling of steel plates and strips. Arch Mech Civil Eng 22:239
9. Szeliga D, Czyżewska N, Klimczak K, Kusiak J, Morkisz P, Oprocha P, Pietrzyk M, Przybyłowicz P (2023) Accounting for the random character of some metallurgical phenomena and uncertainty of process parameters in modeling phase transformations in steels. Canad Metall Quart. (accepted for publication)
10. Clouet E (2009) Modeling of nucleation processes. In: Furrer DU, Semiatin SL (eds) ASM Handbook Vol. 22A, Fundamentals of modeling for metals processing, pp. 203–219. ASM International
11. Leblond JB, Devaux J (1984) A new kinetic model for anisothermal metallurgical transformations in steel including the effect of austenite grain size. Acta Metall 32:137–146
12. Madej Ł, Chang Y, Szeliga D, Bleck W, Pietrzyk M (2021) Criterion for microcrack resistance of multi-phase steels based on property gradient maps. CIRP Ann Manuf Technol 70(1):243–246
13. Rubner Y, Tomasi C, Guibas LJ (1998) A metric for distributions with applications to image databases. In: IEEE International conference on computer vision, pp 59–66. IEEE, Bombay
14. Saltelli A, Chan K, Scot EM (2000) Sensitivity analysis. Wiley, New York
15. Pietrzyk M, Madej Ł, Rauch Ł, Szeliga D (2015) Computational materials engineering: achieving high accuracy and efficiency in metals processing simulations. Butterworth-Heinemann, Elsevier, Amsterdam
16. Gavrus A, Massoni E, Chenot JL (1996) An inverse analysis using a finite element model for identification of rheological parameters. J Mater Process Technol 60:447–454
17. Szeliga D, Gawąd J, Pietrzyk M (2006) Inverse analysis for the identification of rheological and friction models in metal forming. Comput Methods Appl Mech Eng 195:6778–6798
18. Szeliga D, Czyżewska N, Klimczak K, Kusiak J, Morkisz P, Oprocha P, Pietrzyk M, Przybyłowicz P (2021) Sensitivity analysis, identification, and validation of the dislocation density-based model for metallic materials. Metall Res Technol 118:317
19. Poloczek Ł, Kuziak R, Foryś J, Szeliga D, Pietrzyk M (2023) Accounting for the random character of nucleation in modeling phase transformations in steels. Comput Methods Mater Sci 23(2):17–28
20. Brest J, Greiner S, Bošković B, Mernik M, Zumer V (2006) Self-adapting control parameters in differential evolution: a comparative study on numerical benchmark problems. Evolut Comput IEEE Trans 10(6):646–657

Cold Forging of Hollow Shafts with Variable Wall Thickness by Means of an Adjustable Forming Zone

Alexander Weiß and Mathias Liewald

Abstract Hollow shafts are widely used due to their high weight saving potential and only slight reduction of their bending and torsional stiffness compared to solid shafts. Additional weight savings can be achieved by a specific material allocation within the hollow shaft, which can be achieved by varying their wall thickness. The production of such hollow shafts by conventional forming processes and multiple forming stages appears quite costly today. Alternatively, incremental forming processes could be used, but those are very time-consuming. In order to overcome these disadvantages, this paper proposes a special cold forging process characterized by a locally adjustable forming zone, which allows the production of hollow shafts with variable wall thickness within one single press stroke. In previous numerical investigations, the potential of this cold forging process for the production of customized hollow shafts with predefined wall thickness locations or internal splines has already been demonstrated. However, experimental validation has not yet been carried out. In this paper, the first experimental results on hollow shafts produced with wall thickness changes along their longitudinal direction are presented. The experimental setup and the results of the tests carried out so far are described. Furthermore, the experimental results are compared with the numerical calculations of this new forming process.

Keywords Cold forging · Hollow shafts · Adjustable forming zone

1 Introduction

Hollow shafts are widely used in many technical fields due to lightweight design, resource efficiency and the necessity of the integration of functionalities. Their weight saving potential can be explained by comparing the section modulus of a solid and a hollow shaft. With an increasing internal diameter of the hollow shaft, the section modulus decreases only slightly compared to a solid shaft with an identical outer

A. Weiß (✉) · M. Liewald
Institute for Metal Forming Technology, University of Stuttgart, 70174 Stuttgart, Germany
e-mail: alexander.weiss@ifu.uni-stuttgart.de

© The Rightsholder, under exclusive licence to [Springer Nature Switzerland AG], part of Springer Nature 2024
J. Kusiak et al. (eds.), *Numerical Methods in Industrial Forming Processes*, Lecture Notes in Mechanical Engineering, https://doi.org/10.1007/978-3-031-58006-2_14

diameter. This means that the spared volume has only a minor impact on the resistance to bending and torsion of the shaft [1]. Therefore, hollow shafts are often used, e.g., for torque transmission in gearboxes, while the saved volume also offers additional assembly space [2]. Regarding the high torsional load in such cases, cold forging processes appear appropriate for manufacturing of such shafts due to the strain hardening of the material [3].

Rotary swaging and axial forming (for splined components) are often used cold forming processes for manufacturing tailored hollow shafts for lightweight applications [4]. Those processes offer a certain flexibility and therefore are recognized for manufacturing such tailored components. However, the flexibility of rotary swaging requires higher cycle times compared to non-incremental processes. There are also some investigations regarding drawing and axial forming processes using a movable mandrel for manufacturing hollow components with different cross sections along their length [5, 6]. Further approaches for manufacturing hollow profiles with varying wall thickness have been carried out in hot extrusion of round [7, 8] and rectangular [9] aluminum tubes.

At the Institute for Metal Forming Technology at the University of Stuttgart, a novel cold forging process has been developed which allows for manufacturing hollow shafts with varying cross sections in one single press stroke by use of an adjustable forming zone [10]. The process is based on a conventional hollow cold forging process while the mandrel comprises at least two different cross sections which can be moved relatively to the die. By means of this relative position adjustment, it is possible to adjust the forming zone during the process. Until now, numerical investigations have been carried out regarding this process [11, 12] and also a tool concept for the required tool kinematics has been developed [13]. The main objective of this present study is the experimental proof of the feasibility of the process in general. Furthermore, the numerically calculated and the experimentally determined part geometries were compared regarding the radial underfilling at the outer surface of the workpieces.

2   Material and Methods

At first, a material characterization and further numerical investigations for the material AA6082 were carried out. Afterward, experimental tests were conducted and the results were compared to the numerical results by means of optical measurement of the part geometry. The material characterization, developed numerical model and experimental setup are described below.

2.1 Flow Curves and Numerical Model

The flow curves of the test material AA6082 were determined by compression tests using a Gleeble 3800c thermomechanical simulator and are depicted in Fig. 1. Cylindrical specimens having a height of 15 mm and a diameter of 10 mm were used for the compression tests according to [14]. To consider the rate and temperature dependence of the deformation behavior of the material flow curves for three different strain rates (0.1 s^{-1}, 1 s^{-1}, 10 s^{-1}) and four different temperatures (20 °C, 100 °C, 200 °C, 300 °C) were obtained. These temperatures were chosen since the material heats up during deformation which leads to a softening of the material. During the compression tests, copper paste and graphite foil were used as lubricants to keep friction low and reduce barreling of the specimens to a minimum. Nevertheless, very slight barreling of the specimens was observed.

For the numerical investigations, the FEM software DEFORM was used. Figure 2 shows the numerical model for the 2D and 3D simulations. Since the investigation of radial underfilling at the outer surface of the workpieces was one of the objectives of this study, the workpiece was modeled as an elastic–plastic object (flow curves depicted in Fig. 1, constant Young's modulus of 70 GPa and constant Poisson's ratio of 0.33).

Regarding the tool components, only the die was modeled as an elastic object (constant Young's modulus of 207 GPa and constant Poisson's ratio of 0.3). Mandrel, hollow punch and ram-side ejector were modeled as rigid objects. The hollow punch is considered as stationary and the die is moved in the -z direction at a constant velocity of 30 mm/s. The ram-side ejector is pushed in the upward direction during the forming process. At the same time, a very low counterforce is applied on the workpiece. The respective load–displacement curve of the ram-side ejector was determined experimentally for this numerical investigation.

The heat energy due to deformation was calculated as a product of the mechanical energy and a constant value of 0.9 according to the standard setting in DEFORM.

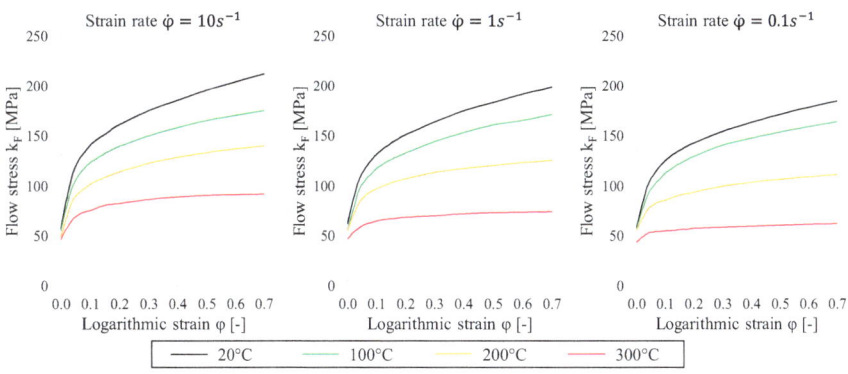

Fig. 1 Experimental determined flow curves of the test material AA6082

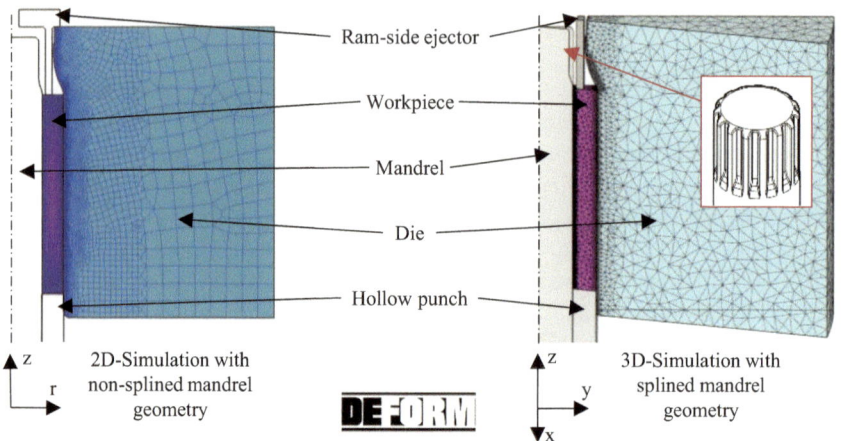

Fig. 2 Numerical model for the 2D and 3D simulations

Regarding friction, the shear friction model with a constant factor of m = 0.2 was used based on experience gained in various research projects at the institute. Mesh windows were used to refine the mesh in relevant sections, e.g., for the internal splined sections of the workpiece. In the 2D simulation, 4,000 quadrilateral elements were used for the mesh of the workpiece comprising a minimum edge length of 0.14 mm. In the 3D simulation, 300,000 tetrahedral elements were used for the mesh of the workpiece comprising a minimum edge length of 0.05 mm.

Besides the tip of the mandrel, the geometry of all other components was the same for the 2D and 3D simulations. The length of the workpiece was chosen to be 58 mm, its outer diameter was 31.9 mm and its internal diameter 19.05 mm. The internal diameter of the die was initially 32 mm and was reduced to 26 mm in the forming zone using a radius of 15 mm at the end of the forming zone. The outer diameter of the die was 160 mm with a height of 85 mm. In the 2D simulation, the diameter of the mandrel was chosen to be 19 mm in its lower section and 17 mm at the tip of the mandrel. In the 3D simulation, a splined mandrel with 15 teeth, a teeth height of 1 mm, a tooth tip diameter of 19 mm and a flank angle of 30° was used. Only half of a tooth was calculated in this case, so a 12° segment of the forming process was investigated. Due to the boundary conditions of machining process of the splined mandrel, all radii at the teeth were chosen to be 0.32 mm, while the tooth thickness corresponds to the tooth gap at medium tooth height.

The movement of the die was directly coupled with the movement of the ram. The mandrel kinematic was implemented as a position control in relation to the ram position. Figure 3 shows the position of the mandrel and die in relation to the absolute value of the ram position. At the beginning, the mandrel and die move synchronously with the same velocity (1). Then the mandrel stops, causing a reduction of the cross section in the forming zone (2). For a certain amount of time, the mandrel remains in this position. In order to return to the initial relative position with respect to the

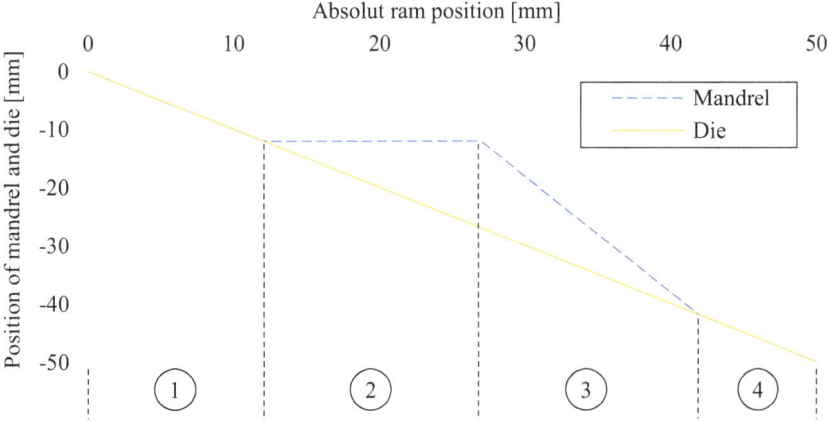

Fig. 3 Position of the mandrel and die in relation to the absolute value of the ram position

die and thus to increase the cross section in the forming zone again, the mandrel is moved at a higher velocity compared to the die (3). Finally, the movement of the mandrel and the die is synchronized when they reach the same position (4).

2.2 Experimental Setup

For the experimental investigations, a hydraulic press SMG HPZUI/300/300–1300/1000 with a maximum capacity of 6,000 kN was used. The used tool consists of an upper, a lower and a base frame (Fig. 4). The base frame includes a hydraulic cylinder with a maximum force of 563 kN during extension and a maximum force of 343 kN during retraction at 280 bar. The hydraulic cylinder is driven by a separate hydraulic unit. The mandrel is connected to the rod of the hydraulic cylinder. The lower frame holds the hollow punch and is connected to the press table via the base frame. Thus, the hollow punch does not move during the process. The upper frame contains the die and the ram-side ejector. Zinc stearate was used as a lubricant and was applied homogeneously on the surface of the billets.

Load cells based on strain gauges were used to measure the tool load at mandrel, hollow punch and die. Furthermore, the displacements of ram (die) and hydraulic cylinder (mandrel) were measured. After the process, the outer surface of the pressed parts was measured by using the optical measurement system GOM ATOS Compact Scan 5 M and the radial deviation compared to a cylinder was determined.

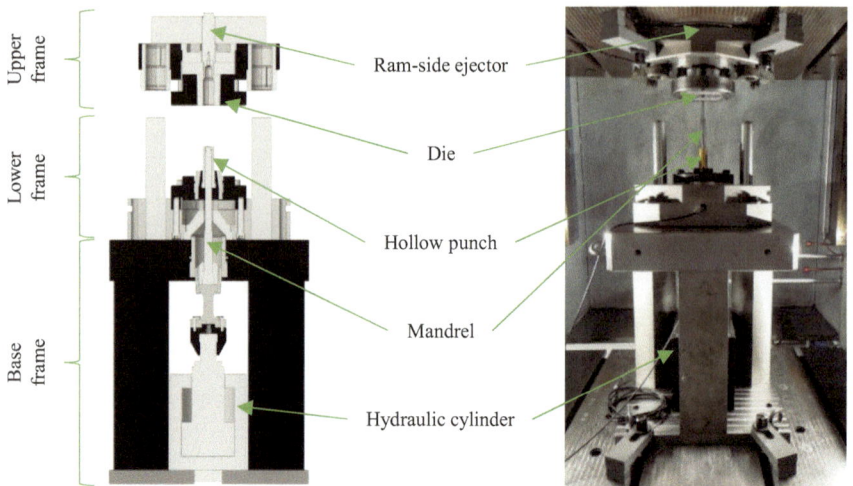

Fig. 4 Sliced CAD model and real tool with labeling of main tool components

3 Results and Discussion

First of all, it can be stated that the feasibility of the process could be successfully demonstrated with the developed tool. At first, the experiments with an increased wall thickness toward the ends of the hollow shaft were carried out. The underfilling at the outer surface of the pressed parts, which was already observed in the numerical investigations, also occurs on the real parts. Figure 5 depicts sliced pressed parts made of AA6082 and their main dimensions.

Figure 5a shows a sliced hollow shaft with an increased wall thickness toward the ends of the shaft by using the mandrel kinematic shown in Fig. 3. In a further experiment, the mandrel kinematic was varied in such a way that a hollow shaft with an increased wall thickness in the center of the shaft could be pressed (Fig. 5b). These results demonstrate the high flexibility of the presented process with regard to the variation of radial dimensions within the axial proportions of the pressed parts. Besides the experiments with the non-splined mandrel, tests with a splined mandrel were also carried out. By using the same mandrel kinematic as shown in Fig. 3, it was possible to manufacture a hollow shaft with local internal splined sections within one single press stroke. A respective pressed part with such a geometry is depicted in Fig. 5c.

Figure 6 shows the numerical and experimental determined tool load onto the hollow punch and the mandrel versus the absolute value of the ram displacement for the splined part geometry.

The maximum tool load onto the hollow punch was measured to 174 kN in the experiment and calculated to 152 kN in the simulation. Regarding the mandrel, a minimum load of -42 kN was measured in the experiment and a minimum load of -22 kN was calculated in the simulation. These deviations can be explained

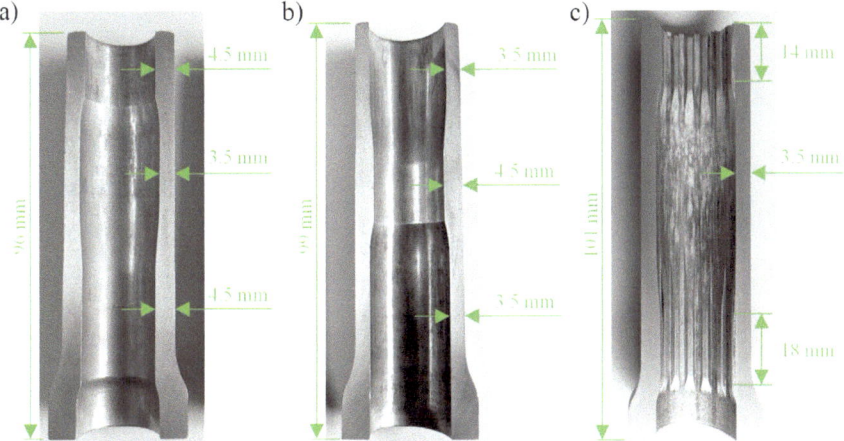

Fig. 5 Sliced pressed parts made of AA6082 and main dimensions: **a** increased wall thickness toward the ends of the shaft, **b** increased wall thickness in the center of the shaft due to a varied mandrel kinematic and **c** local internal splined sections toward the ends of the shaft

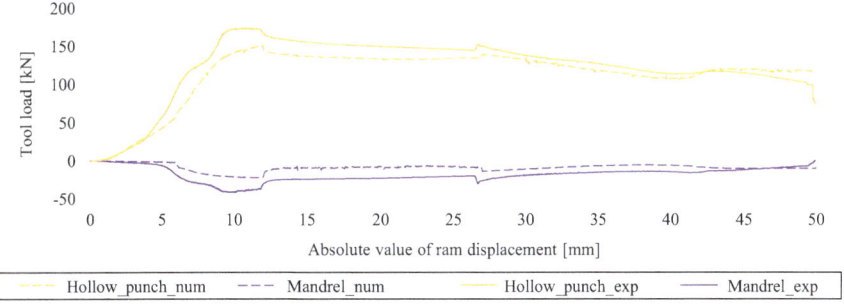

Fig. 6 Numerical and experimental determined tool load onto the hollow punch and the mandrel versus the absolute value of the ram displacement for the splined part geometry

by the friction modeling in the simulation on the one hand and the rigid modeling of the mandrel and the hollow punch on the other hand. However, the qualitative characteristic of the experimental and numerical load–displacement curves appear to be very similar.

Figure 7 shows a comparison of the optical measurements of the pressed part and the numerical results in terms of the radial deviation of the part geometry compared to an ideal cylindrical shape. Figure 7a shows the comparison for the non-splined part and Fig. 7b for the internally splined part. It can be obtained that the experimentally determined radial deviation matches the numerically determined radial deviation quite well with respect to the quantitative values. The underfilled section has a minimum radial deviation of −0.13 mm in the optical measurement and − 0.12 mm in the simulation.

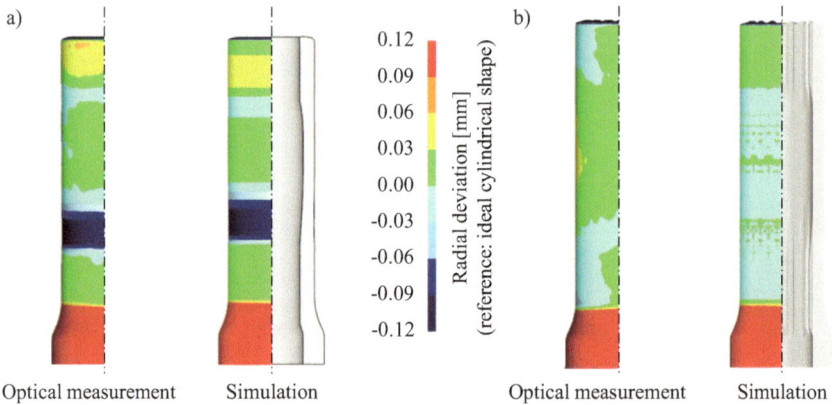

Fig. 7 Comparison of optical measurements and numerical results in terms of radial deviation of the part geometry compared to an ideal cylindrical shape: **a** non-splined part and **b** splined part

However, a slight axial misalignment regarding the axial position of the radial deviations was found which can be seen in Fig. 7a. This can be explained by three aspects, namely the modeling of the mandrel in the simulation as a rigid body, the characteristics of the displacement sensor of the hydraulic cylinder and the delay of the hydraulic control unit. An elastic modeling of the mandrel will have an impact on the ram-position-related change of the forming zone due to the elastic elongation of the mandrel. When comparing the sensor signal with gauge blocks and a dial indicator, a slight nonlinear sensor behavior of the sensor for measuring the axial position of the mandrel was detected. Since this sensor signal is the reference signal for the control unit of the mandrel, the kinematics require a correction before being transmitted to the control unit. For further experiments, this correction curve for compensation of the nonlinear sensor behavior has already been determined based on the current results. In addition, a delay in the hydraulic control unit has been identified. This results in a small deviation of the defined and the actual position of the mandrel during the process. It has been found that the delay depends on the load at the mandrel.

Figure 7b in particular proves that a continuous back pressure between the workpiece and the mandrel induced by the tooth tips of a splined mandrel leads to a significant reduction of the underfilling. This hypothesis has already been stated in the numerical investigation in [10].

In addition to the experimental investigations conducted with AA6082, also a few tests were carried out with the low-carbon steel DIN/EN-1.0303. The gained results showed that the developed tool can also be used for cold forging of steel parts with a variable wall thickness.

4 Conclusion and Outlook

In this contribution, the first experimental results regarding a special cold forging process characterized by a locally adjustable forming zone are presented. The locally adjustable forming zone of this process allows for the production of hollow shafts having a variable wall thickness within one single press stroke. At first, the material data and the numerical model are presented. A splined and a non-splined workpiece geometry were investigated and the respective used numerical models (3D and 2D) were described. Additionally, relevant geometric details and the developed tool and used hydraulic press are presented.

The experimental tests were carried out successfully with the developed tool presented in this paper. Thus, the feasibility of the proposed process for cold forging of hollow shafts having a variable wall thickness or local internal splined sections within one single press stroke has been proven. Although the main focus was on the material AA6082, the tool was also capable of forming low-carbon steel DIN/EN-1.0303. Regarding the radial underfilling at the outer surface of the workpieces, it was found that the deviation between the numerical and experimental results is quite low with a maximum radial deviation of −0.13 mm in the experiment and −0.12 mm in the numerical investigation. Furthermore, it could be proven that a continuous back pressure between the workpiece and the mandrel induced by the tooth tips of a splined mandrel leads to a significant reduction of the underfilling.

In order to further improve the numerical model and to reduce the kinematic and geometrical deviations between the experiment and the simulation, further adjustments to the numerical model will be carried out, such as the use of the actual mandrel kinematics from the experiment and the elastic modeling of the mandrel and the hollow punch. In the next step, an experimental investigation with the case-hardening steel AISI 5115 will be carried out.

Acknowledgements The research project "Development of a cold forging process for manufacturing hollow parts with variable wall thicknesses" (fund number ZF4012808LP9) in cooperation with LS-Mechanik GmbH is funded by the Federal Ministry of Economic Affairs and Energy through the German Federation of Industrial Research Associations (AiF) as part of the Central Innovation Programme for SMEs (ZIM) based on a decision of the German Bundestag. The authors would like to thank the German Federation of Industrial Research Associations for the financial support and the project partner LS-Mechanik GmbH for the collaboration.

References

1. Liewald M, Felde A, Weiss A et al (2019) Hollow shafts in lightweight design—state of the art and perspectives, 34. Jahrestreffen der Kaltmassivumformer, VDI
2. Kiel E (2008) Drive solutions—mechatronics for production and logistics. Springer, Berlin, Heidelberg, ISBN: 9783540767053
3. Lange K, Kammerer M, Pöhlandt K et al (2008) Fließpressen—Wirtschaftliche Fertigung metallischer Präzisionswerkstücke, Springer, Berlin, Heidelberg, ISBN: 9783540309093

4. Lüftl S, Degischer HP (2012) Leichtbau—Prinzipien, Werkstoffauswahl und Fertigungsvarianten, Wiley, ISBN: 9783527659869
5. Boutenel F, Delhomme M, Velay V et al (2018) Finite element modelling of cold drawing for high-precision tubes. Comptes Rendus Mécanique 346(8):665–677. https://doi.org/10.1016/j.crme.2018.06.005
6. Walter Henrich GmbH (2019) Verfahren und Vorrichtung zum axialen Umformen eines Rohres, patent number DE102019103926A1
7. Makiyama T, Murata M (2002) Controlling inside diameter of circular tube by extrusion. Mater Sci Forum 396–402(1):513–518. https://doi.org/10.4028/www.scientific.net/msf.396-402.513
8. Negendank M, Müller S (2018) Strangpressen von Aluminiumhohlprofilen mit axial variabler Wandstärke, 10. Ranshofener Leichtmetalltage—Hochleistungsmetalle und Prozesse für den Leichtbau der Zukunft, ISBN: 9783902092106.
9. Selvaggio A, Haase M, Khalifa NB et al (2014) Extrusion of profiles with variable wall thickness. Procedia CIRP 18:15–20. https://doi.org/10.1016/j.procir.2014.06.100
10. Weiß A, Liewald M (2019) Numerical investigation of a cold forging process for manufacturing hollow shafts with variable wall thickness. NUMIFORM 2019: the 13th international conference on numerical methods in industrial forming processes
11. Weiß A, Liewald M (2020) Fließpressen hohler Wellen mit Wanddickenvariation—Numerische Untersuchung zum Hohl-Vorwärts-Fließpressen mit einstellbarer Umformzone, wt Werkstattstechnik online 110(10):684–688
12. Weiß A, Arny M, Liewald M (2022) Impact of a splined mandrel geometry on die filling in a cold forging process with adjustable deformation zone. Key Eng Mater 926:612–620. https://doi.org/10.4028/p-aynwq8,ISBN:9783035717594
13. Weiß A, Liewald M (2021) Flexible Fertigung maßgeschneiderter Hohlwellen—Werkzeug und simulative Untersuchung zur Fertigung von Hohlwellen mit Wanddickenvariation, wt Werkstattstechnik online 111(10)
14. NN (2014) Material testing and modelling. ICFG Document No. 24/14, Meisenbach Verlag Bamberg, ISBN: 978-3-87525-366-5

Numerical and Experimental Analysis of the Influence of Manufacturing Parameters in Additive Manufacturing SLM-PBF on Residual Stress and Thermal Distortion in Parts of Titanium Alloy Ti6Al4V

M. O. Santos, A. S. F. R. Maiolini, F. Miranda, A. Farias, V. Seriacopi, E. C. Bordinassi, and G. F. Batalha

Abstract Additive manufacturing (AM) has become popular in recent years due to its integration with Industry 4.0 and for enabling the production of complex and optimized geometries. However, the process leads to disadvantages, such as thermal distortions and the generation of residual stresses due to thermal gradients during the parts production. Based on these undesirable characteristics, this work aims to model AM numerically through the Finite Element Method (FEM), besides obtaining experimental results that relate the impact of different process parameters, such as laser power, speed, and hatch distance on the residual stress and distortions distribution, to parts generated by Selective Laser Melting (SLM)—Powder Bed Fusion (PBF)—of titanium alloy Ti6Al4V. The experimental tests, as well as the numerical analyses, followed a factorial design. The results for thermal distortions are excellent with differences below 4.5% between the manufactured and simulated parts. As for the results of residual stresses measured at a single point on the lateral face, for most of the results, the difference was less than 4%; however, there was a large dispersion of results. When the values of average residual stresses were considered in the simulation, the results had statistical relevance and showed greater robustness. It was found that the average diameter of Ti6Al4V titanium parts was greatly influenced by the laser power followed by the hatch distance. The increase in power increased the average diameter of the samples, while the increase in speed and hatch decreased their diameter. In turn, increasing laser power decreased residual stress, while increasing speed and hatch tended to increase the mean residual stress.

M. O. Santos (✉) · A. S. F. R. Maiolini · A. Farias · V. Seriacopi · E. C. Bordinassi
Department of Mechanical Engineering, University Center of Maua Institute of Technology, São Caetano do Sul, SP, Brazil
e-mail: marcelo.santos@maua.br

M. O. Santos · F. Miranda · G. F. Batalha
Department of Engineering Mechatronics and System Mechanics, University of São Paulo, Polytechnic School of USP, São Paulo, SP, Brazil

195
J. Kusiak et al. (eds.), *Numerical Methods in Industrial Forming Processes*, Lecture Notes in Mechanical Engineering, https://doi.org/10.1007/978-3-031-58006-2_15

Keywords Additive manufacturing SLM · Titanium alloy Ti6Al4V · Numerical simulation · Thermal distortions · Residual stresses

1 Introduction

Industry 4.0 is the result of the twenty-first-century demand for innovation toward a revolution in products and production chains. This concept has become widely spread throughout society and is evidenced by the transformations that manufacturing processes have undergone. According to Dilberoglu [3], this new industry format integrates physical and digital systems, giving rise to smart factories. Among the concepts regarding the physical component, manufacturing is a fundamental part; conversely, it also acts as a limiting agent to the capacity of these factories. Therefore, this scenario makes the development of new non-traditional manufacturing methods a vital factor, among which additive manufacturing (AM), widely known as "3D printing", can be highlighted.

According to Chryssolouris [2], this manufacturing technology has the potential to meet the current demand for flexibility in industries—being able to print customized products with complex geometries that are difficult to manufacture using conventional techniques. It still allows integration with digital systems—using CAD (Computer-Aided Design), CAE (Computer-Aided Engineering), and CAM (Computer-Aided-Manufacturing) software.

Metallic materials are widely used in various fields of engineering. There is also the emergence of AM technology by the fusion of metal powder bed FLPM (MPBF—Metal Powder Bed Fusion), [7]. Among these metallic materials, titanium alloy Ti6Al4V is the material focused on by this research.

Although previous works indicate the importance of understanding the generation of residual stress and distortions in Selective Laser Melting (SLM) processes, the underlying mechanisms for the generation of residual stress remain poorly understood. In order to better determine the factors that influence the accumulation of residual stress as well as the prevention of distortions, delamination, and fractures [5], a combination of parameters is sought for an ideal process window (one that meets the requirements product performance). Therefore, thermomechanical models for finite element simulation (FEM) of SLM are potentially valuable, although they are a challenge due to the complexity of the physics involved in the process [6].

A significant problem associated with SLM components is the development of high internal residual stress [1]. The repeated cycles of heating and cooling successive layers of powdered raw material during the SLM construction process produce high cooling rates and high-temperature gradients associated with the process, resulting in residual stress buildup in SLM components. Parts may fail during SLM construction or later in service due to these high internal residual stresses [4].

Therefore, the present work aims to contribute to understanding how some process parameters such as laser power, scanning speed, and hatch affect the generation of residual stresses and distortions in the printed part—both numerically (using the

Fig. 1 OmniSint-160
additive manufacturing
machine

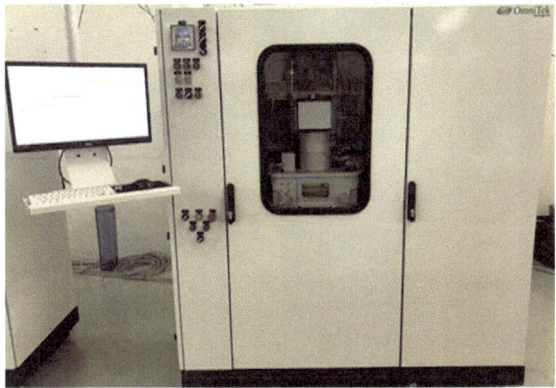

commercial software Simufact Additive 2020 FP1 from MSC Softwares) and experimentally. A factorial design is used to select the variation of parameters, and the analysis of variance relates the results (residual stress and distortion) with the selected variables.

2 Materials and Methods

2.1 Equipment

The manufacture of Ti6Al4V titanium alloy samples via additive manufacturing technique by selective laser melting was carried out in an OmniSint-160 SLM equipment, ytterbium fiber laser module, Rycus source 500 W of nominal power, from OMNITEK (see Fig. 1).

Measurements of residual stresses of the titanium parts were carried out using the RIGAKU brand X-ray diffraction equipment, model Ultima IV (Fig. 2a) located at the Nuclear Research Institute (IPEN). To measure the thermal distortions, the optical measuring equipment ATOS Core 80—CP40/MV100 was used, with a resolution of 5 megapixels. Data was processed by the GOM Inspect 2021 software—in a partnership with the company Vtech (Fig. 2b).

2.2 Experimental Methodology

For evaluating the impact of the variation of the SLM manufacturing process parameters, such as the distance between beams (hatch), scanning speed, and laser power,

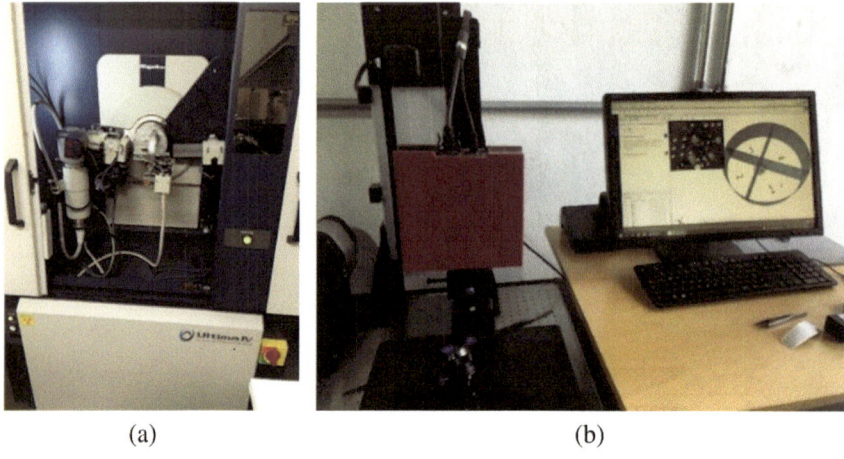

(a)   (b)

Fig. 2 **a** RIGAKU diffractometer, Ultima IV model, **b** ATOS Core 80 optical meter

on the appearance of residual stresses and thermal distortions, 16 cylindrical specimens with 11.3 mm diameter and 10 mm height were used, according to a factorial
design with the following variable factors:

- Power (100 W–200 W)
- Speed (500 mm/s–1500 mm/s)
- Hatch distance (50 μm–90 μm)

    Table 1 presents the order of the tests and the parameters considered for each part
manufactured by the SLM manufacturer.

    Figure 3 shows the 16 samples of Ti6Al4V titanium printed on the printing mat.

    The parts were cut from the printing base using wire EDM, which greatly
minimizes the possibility of generating residual stresses during cutting.

2.3 Computational Methodology

Numerical analysis was performed using the Simufact Additive 2020 FP1 software
from Hexagon MSC Software, which uses CAD models to apply the same manufacturing conditions, defining a thermomechanical simulation. Such settings are established directly by the graphical user interface and the software calibration process.
To minimize the complexity—caused by the wide range of data that influence the
process—it's necessary a pre-calibration step that uses the inherent strain method,
marked by the definition of the volume expansion factor (VEF). The material's properties can be selected in Simufact Material 2020 FP1 that will supply our model
with standard data for that material or can be defined manually using the datasheet
provided by the supplier or data collected experimentally.

Table 1 Manufacturing parameters during the experimental stage

Test order	Power	Speed	Hatch
	W	mm/s	mm
1	150	1000	50
2	150	1000	90
3	100	1000	70
4	150	500	70
5	200	1000	70
6	150	1500	70
7	200	1500	90
8	150	1000	70
9	200	500	50
10	100	1500	50
11	150	1000	70
12	200	1500	50
13	100	1500	90
14	100	500	90
15	200	500	90
16	100	500	50

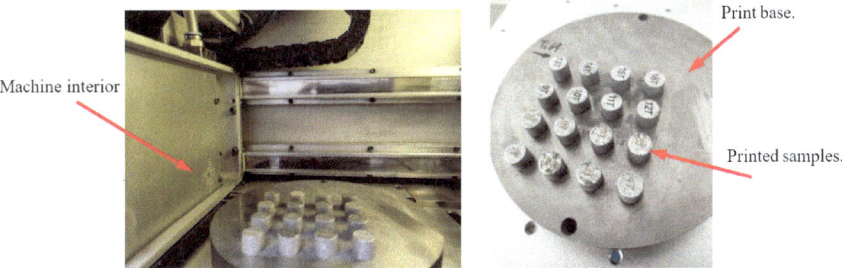

Fig. 3 Printed Ti6Al4V samples

VEF is responsible for correcting the effects of thermal expansion and contraction, since the element generated by the discretization of the domain in voxels (hexahedral finite elements—Fig. 4) consists of more than one layer of material. For selecting this factor, a geometry pre-selected by the application is generally used, the cantilever (which facilitates the characterization of the state of deformations). However, as such specifications would already be provided by the experimental study, a convergence analysis of the factor was carried out for an arbitrary sample, through the relationship with the data obtained numerically and experimentally—aiming at the approximation of the thermal distortions, so that, if there is an equivalence in the displacements, it

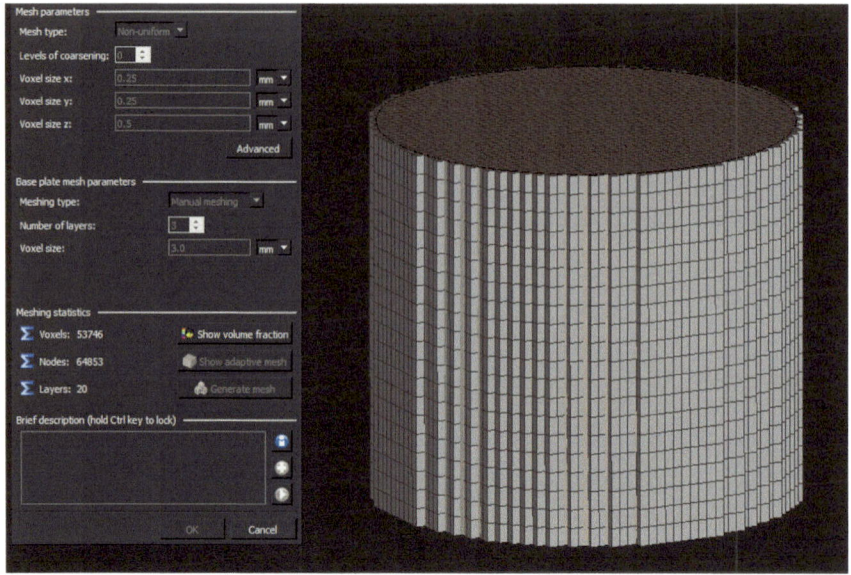

Fig. 4 Discretization of the continuous domain of geometry

can be stated that the thermal deformation was correctly captured by the software and the residual stresses can thus be calculated, since they derive from deformations caused by the temperature gradient.

When defining the metal powder bed melting process, the stages of the process were also defined, namely: construction of the part and separation of the base. The base separation was set up straight away. For the construction of the part, it is necessary to insert the constant parameters (the width of the laser beam and the scanning strategy) and variables for the study, namely: laser power, scanning speed, and distance between scanning vectors (hatch)—being the same parameters varied for the experimental part.

3 Results and Discussion

3.1 Experimental Thermal Distortions

Initially, the sample (Fig. 5a) is subjected to scanning in the ATOS Core 80 equipment of its entire external surface (Fig. 5b) resulting in the geometry of Fig. 5c, which is treated using the GOM Inspect software.

After scanning each printed piece, the GOM Inspect software is used to perform the analysis of metrological parameters. Initially, the geometry scanned by a cylinder is superimposed with the nominal dimensions of the part, which allows evaluating

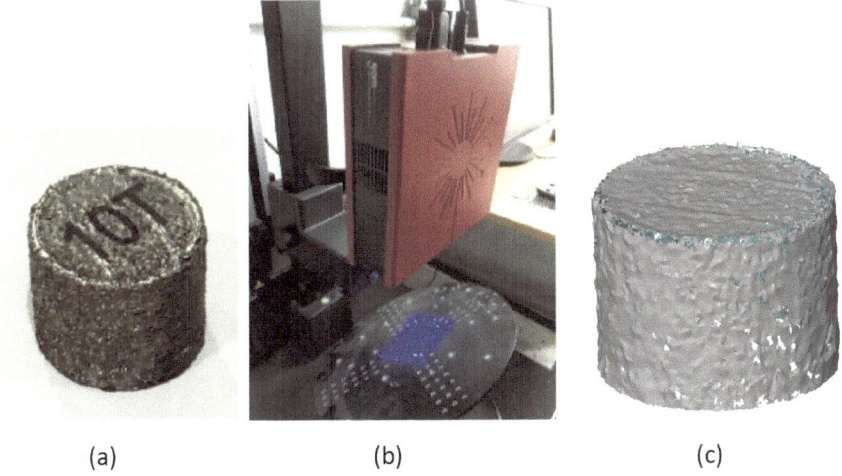

(a)   (b)   (c)

Fig. 5 **a** 10 T printed sample, **b** optical measurement process, **c** scanned geometry of the sample

the distortions resulting from the manufacturing process. For this research, only the mean diameter was defined as the distortion response parameter.

Figure 6 shows a dimensional comparison of the entire scanned surface of sample 10T in relation to the nominal cylindrical body, with nominal dimensions of 10 mm height by 11.3 mm diameter. Warm colors indicate larger dimensions and cool colors indicate smaller dimensions than the nominal value.

3.2 Experimental Residual Stresses

In this step, the samples had their residual stresses measured. Measurements were taken on the side of each piece. In Fig. 7a, b, it is possible to visualize the part being measured by the RIGAKU diffractometer, model Ultima IV, in the lateral position.

3.3 Numerical Simulations

After the printing parameters setup steps according to factorial planning for each sample, it was possible to measure the thermal distortions and calculate the residual stresses. Distortions were measured at a total of 15 points distributed on the cylindrical surface, 5 points at a height of 2.5 mm from the base, 5 points at a height of 5.0 mm, and finally, 5 points at a height of 7.5 mm Sample. Figure 8 shows some of the points measured in the simulation stage. The measurement of the residual stress was made by a single point of the same coordinate that was considered in the

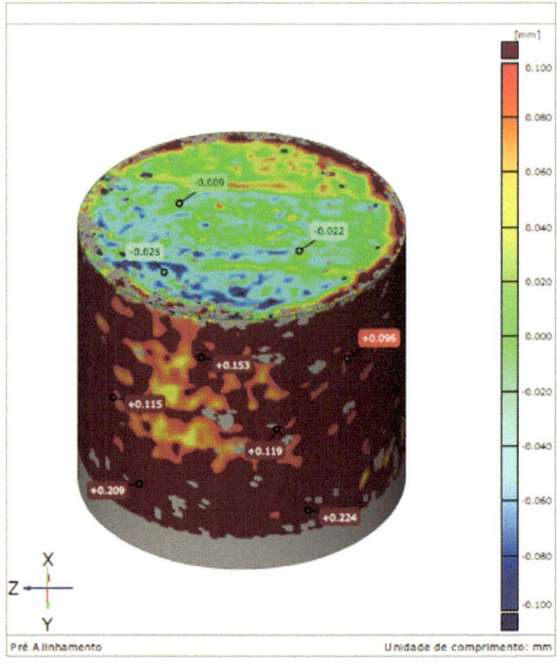

Fig. 6 Representation of distortions of the 10 T printed sample

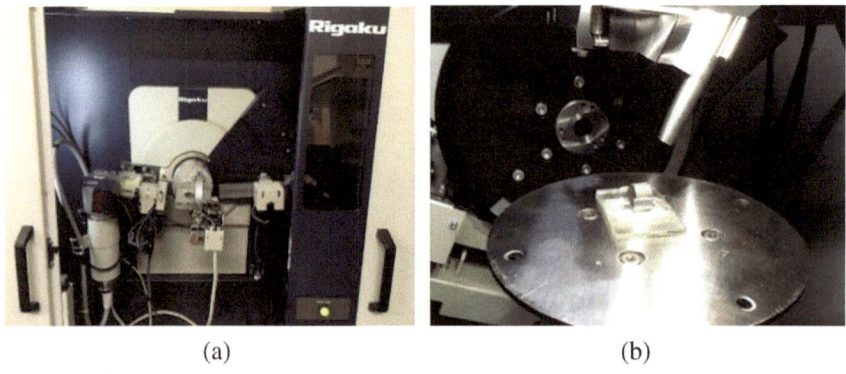

| (a) | (b) |

Fig. 7 Measurement of the 10 T sample by diffraction **a** external view of the machine, **b** measurement position

experimental stage in the first phase, and later the other points were considered as explained later.

Table 2 presents the distortion and residual stress results from the experimental stages and simulations for each of the 16 specimens with the respective relative percentage deviations between the values obtained.

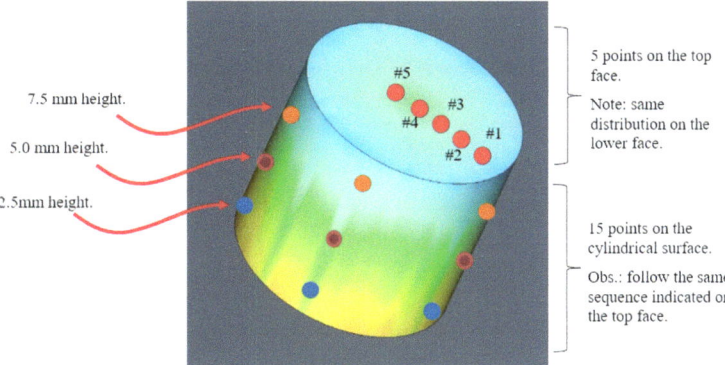

Fig. 8 Distortion measurement points in the simulation stage

Table 2 Results of experimental and simulated distortions and residual stresses

Test	Simulated Diameter [mm]	Experimental Diameter [mm]	Deviation %	Residual Stress (MPa) Simulated	Residual Stress (MPa) Experimental	Deviation %
1	11.185	11.561	3.3	536.8	538.0	0.2
2	11.203	11.383	1.6	650.2	629.2	−3.3
3	11.200	11.346	1.3	553.8	528.9	−4.7
4	11.183	11.471	2.5	588.0	586.3	−0.3
5	11.197	11.542	3.0	345.3	345.1	−0.1
6	11.205	11.463	2.3	281.8	268.2	−5.1
7	11.203	11.430	2.0	493.0	495.5	0.5
8	11.199	11.479	2.4	502.3	502.8	0.1
9	11.178	11.670	4.2	471.5	463.8	−1.7
10	11.203	11.415	1.9	360.8	353.2	−2.1
11	11.199	11.459	2.3	498.7	518.9	3.9
12	11.221	11.623	3.5	273.0	280.5	2.7
13	11.209	11.281	0.6	514.3	513.1	−0.2
14	11.188	11.298	1.0	435.9	442.1	1.4
15	11.180	11.495	2.7	449.5	453.8	1.0
16	11.188	11.409	1.9	447.1	436.2	−2.5

3.4 Analysis of Variance

Figure 9a, b show the result of the analysis of variance for the experimental stage. Figure 9a shows the Pareto chart of the mean diameter with the reference line for statistical significance ($\alpha = 0.05$). The factors that present values lower than the reference line (to the left of the line) do not present statistical significance in the

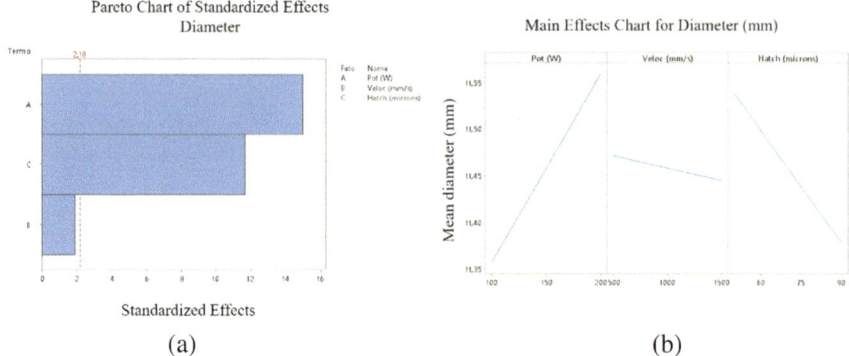

Fig. 9 **a** Pareto chart for mean diameter, **b** Main effects chart for mean diameter—experimental stage

result of the response studied. Figure 9b presents the main effects of the factors on the mean diameter.

The result of the analysis indicated that the power and hatch distance factors, respectively, influence the variation in the average diameter of the printed parts at the 95% confidence level. The velocity factor was not statistically significant, at an adequate confidence level (95%), for any discussion or correlation with the result of the mean diameter variation.

Analyzing Fig. 9b, a trend toward an increase in diameter is observed with increasing power, which may be related to an increase in local temperature at the time of fusion of the powder in the layer. However, with increasing scanning speed and hatching (hatching at greater intensity), the mean diameter decreased. The variation of these two factors changes the cooling rate of the layer, which may have caused a reduction effect on the final diameter dimension.

Figure 10 shows the Pareto chart of the residual stress measured at a single point on the side of the printed parts in the experimental stage with the reference line for statistical significance ($\alpha = 0.05$).

The result of the analysis indicated that none of the power, velocity, and hatch factors statistically influence the residual stress on the side of the part at the 95% confidence level. This is most likely due to the difficulty in obtaining a homogeneous stress distribution condition in such a small geometry piece.

To allow a more reliable verification, once the relationship between the printed part and the simulated part was proven for the measured point shown in Table 2, we decided to measure more points along the radius and also at the height of the part, which allowed working with an average value of residual stress (Fig. 8).

The average residual stress values can be seen in Table 3 and compared with the stress values measured at a single point. A great difference is observed, which makes the computational simulation step even more important in this sample characterization process. The residual stresses on the side presented their lowest values, when

Fig. 10 Pareto chart for the
lateral residual
stress—experimental stage

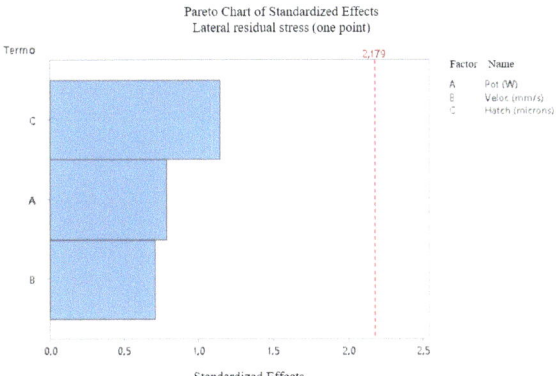

sufficiently close to the bottom and top. These results made it possible to perform
the analysis of variance, observing this time significance in the results.

Figure 11a, b show the Pareto chart of the residual stress obtained on the side face
of the parts simulated in the numerical step considering the average value measured.
Figure 11b shows the graphs of the main effects of the parameters on the residual
stress result.

Table 3 Comparative results between average and "one point" residual stress

Test	Residual Stress (MPa) Simulated (Average)	Residual Stress (MPa) Simulated (1 point)	Deviation %
1	631.7	536.8	−17.7
2	691.3	650.2	−6.3
3	721.3	553.8	−30.2
4	625.1	588.0	−6.3
5	639.3	345.3	−85.2
6	691.7	281.8	−145.5
7	671.9	493.0	−36.3
8	679.8	502.3	−35.3
9	569.6	471.5	−20.8
10	724.1	360.8	−100.7
11	679.4	498.7	−36.2
12	645.7	273.0	−136.5
13	716.6	514.3	−39.3
14	689.3	435.9	−58.1
15	603.7	449.5	−34.3
16	667.4	447.1	−49.3

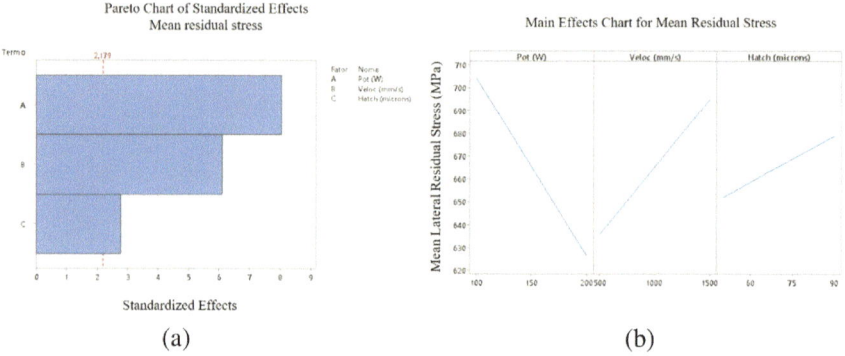

Fig. 11 **a** Pareto plot for mean lateral residual stress, **b** Main effects plot for mean lateral residual stress—numerical step

The result of the analysis indicated that all three factors, laser power, speed, and hatch distance, statistically influence the residual stress of the part at the 95% confidence level.

Analyzing Fig. 11b, it is possible to verify a decreasing tendency of the residual voltage on the lateral face with the increase of the power. However, by increasing sweep speed and hatch distance, the residual stress of the side face increased.

4   Conclusions

After carrying out the studies, the modeling of the distortions showed to present good repeatability while the residual stresses and their measurements in a single point present significant uncertainty. The software can thus be concluded to be able to accurately predict the distortion and, qualitatively, the residual stresses measured on the surfaces studied.

It was also found that the average diameter of Ti6Al4V titanium parts was greatly influenced by the laser power followed by the hatch distance. The increase in power increased the average diameter of the samples, while the increase in speed and hatch decreased their diameter.

The residual stress measured on the lateral face through a single point showed great variation when comparing the experimental and numerical results. This is due to imprecision in the measured coordinate and also to difficulties in stabilizing the X-ray focus during the acquisition of measurements in the experimental stages. The reduced dimensions of the printed pieces and high porosity are believed to have generated this difficulty in measurement.

In turn, the simulated mean residual stress measured on the lateral face of the Ti6Al4V titanium parts showed a statistical behavior, being greatly influenced by the laser power, followed by the sweep speed and the hatch distance. Increasing laser power decreased residual stress, while increasing speed and hatch tended to increase the mean lateral residual stress.

Acknowledgements The authors would like to thank Omnitek and Doctor Eng. Marcello Mergulhão for making the samples, MSC Software for providing software, IPEN and Vtech for making resources available and carrying out measurements, and Maua Institute of Technology for providing infrastructure and laboratory resources.

Responsibility for Information The authors are solely responsible for the information included in this work.

References

1. Casavola C, Campanelli SL, Pappalettere C (2008) Experimental analysis of residual stresses in the Selective Laser Melting process. In: Proceedings of the XIth international congress and exposition, Orlando, Florida, USA
2. Chryssolouris G et al (2009) Digital manufacturing: history, perspectives, and outlook. Proc Inst Mech Eng Part B J Eng Manuf 223(5):451–462
3. Dilberoglu UM et al (2017) The role of additive manufacturing in the era of industry 4.0. Procedia Manuf 11:545–554
4. Elambasseril JSF, Bringezu M, Brandt M (2012) Influence of process parameters on selective laser melting of Ti-6Al-4V components, RMIT University School of Aerospace, Mechanical and Manufacturing Engineering (SAMME)
5. Kuşhan C, Poyraz M, Uzunonat Ö, Orak YS (2018) Systematical review on the numerical simulations of laser powered bed additive manufacturing. 2018. Sigma J Eng Nat Sci 36 (4):1197–1214
6. Parry L, Ashcroft IA, Wildman RD (2016) Understanding the effect of laser scan strategy on residual stress in selective laser melting through thermo-mechanical simulation. Addit Manuf 12:1–15. ISSN 2214 8604
7. Volpato N et al (2017) Manufatura aditiva: Tecnologias e aplicações da impressão 3D. São Paulo: Blücher, 400 p. ISBN 978-85-212-1150-1

Material Parameters Identification of Thin Fiber-Based Materials Using the Method of Machine Learning Exploiting Numerically Generated Simulation Data

Cedric Wilfried Sanjon⬛, Yuchen Leng⬛, Marek Hauptmann, Jens-Peter Majschak, and Peter Groche

Abstract The determination and validation of material parameters required for finite element simulation of the forming processes of fiber-based materials such as paperboard can be accomplished by strain-based loading of a specimen in combination with a simulation-based reverse engineering approach. Due to the complexity of the material itself, such as anisotropy, the development of such approaches can be very time-consuming and requires programming skills as well as expertise in FEM analysis and optimization. Machine learning methods offer a practical alternative to optimization, parameterization, and reverse engineering approaches, assuming that the data is fully known, generalized, and learned by the machine learning model. More specifically, a machine learning model can compute the material parameters required for a finite element simulation directly from the experimental measurements, if the hypothetical mapping function in this case is learned from the numerical study between material parameters and deformation behavior. In this paper, such data generated by numerical studies are used to train the machine learning model and, based on this, to determine elastic (e.g., Young's modulus), plastic, and Hill's parameters.

Keywords Finite element analysis · Materials analysis · Material parameters · Material identification · Machine learning

C. W. Sanjon (✉) · M. Hauptmann · J.-P. Majschak
Fraunhofer Institute for Process Engineering and Packaging IVV, Heidelberger Str. 20, 01189 Dresden, Germany
e-mail: cedric.sanjon@ivv-dd.fraunhofer.de

Y. Leng · P. Groche
TU Darmstadt Institute for Production Engineering and Forming Machines PtU, Otto-Berndt-Str. 2, 64287 Darmstadt, Germany

209
J. Kusiak et al. (eds.), *Numerical Methods in Industrial Forming Processes*, Lecture Notes in Mechanical Engineering, https://doi.org/10.1007/978-3-031-58006-2_16

1   Introduction

Nowadays, industrial products have to meet both technical and environmental requirements, making sustainability a key factor in product design and material selection. Sustainability in the form of recycling, degradability, and circularity has also become very important in the packaging industry. Plastic packaging, which has been used extensively in the past, clearly does not meet this requirement, so paper and paperboard are gaining ground as an alternative that can meet sustainability requirements at low cost. However, since paper is made from natural fibers, the material itself is more complex in nature and much less ductile than plastic products, so there is still much to be investigated about the forming process of paper and paperboard. In particular, an accurate material model is essential for the numerical approach, which offers a great opportunity to improve both the quality and efficiency of production.

Paper consists mainly of natural fibers plus fillers and additives, and the natural fibers form the basic structure of the material. The small fibers, bonded together by hydrogen bonds, form the fiber network, which exhibits heterogeneity, structural disorder, and anisotropy. For this reason, it is difficult to make accurate predictions about the mechanical behavior of paper and paperboard as a fiber-based material [1]. Obviously, the first consideration is modeling at the microscopic level, such as fibers or fiber networks. Despite the gradual maturation with the advancement of optical measurement techniques, especially Computed Tomography (CT), the microstructure measurement of paperboard remains labor-intensive and costly due to the random and dense arrangement of the fibers and the paperboard itself consisting of multiple compressed layers. In addition, the computation of microscopic models is also very time-consuming and has limited use in microstructural studies. Multiscale mechanical modeling of paperboard has also been proposed in recent years, but still relies heavily on X-ray CT measurements and is not yet widely used for the simulation of forming processes. [2]. It can be seen that sheet scale modeling using continuum models is still dominant due to the convenience of modeling and the practicality of computation. In this approach, paperboard can be considered as a homogeneous anisotropic material. To account for the anisotropic nature, three main directions are usually defined as machine direction (MD), cross direction (CD), and thickness direction (ZD).

Various continuum models have been proposed to describe the anisotropic properties of fiber-based materials using an orthotropic yield function that accounts for different yield stress values in different preferential directions [3]. The most commonly used formulation for modeling paperboard is that of Hill [4], which includes six yield stress values. It has been successfully used to study the forming process and was extended by Hoffman to include tension–compression properties [5]. Another well-known method was later proposed by Xia et al. [6], where the yield surface can be defined as the sum of six sub-surfaces, four for tension and compression in MD and CD, while two correspond to shear behavior. Li et al. [7] proposed a modification of Xia's model by reducing the material parameters from 23 to 10, which includes both isotropic and kinematic hardening behavior. Xia's model was also extended by Borgqvist et al. [8] to account for the plastic response of the ZD.

As can be seen from these typical continuum models, there are always many parameters involved in modeling paper materials, which means that many experiments are required. The resulting problem is that specialized equipment for testing material properties and tools developed specifically for paper materials are always in demand. In addition, due to the inhomogeneity of paper, it is not easy to obtain reliable experimental data, and more replicates are needed than for homogeneous materials. Thanks to advances in full-field optical measurements, more and more inverse methods are being used for material characterization [9]. There are currently two main approaches. One is Finite Element Model Updating (FEMU), which determines the exact values of material parameters by building a finite element model of the characterization experiment and using simulation iterations. One of the best-known FEMU methods is the inverse parameter identification method, where the corresponding FE simulations are based on the results of material property tests. The calculated strain distribution is compared with the measured strain distribution until the deviation reaches a convergence criterion so that the simulated strain distribution is as close as possible to the measured distribution (see Fig. 1).

The second method is the fitting procedure or direct evaluation of the measured strain field using the equilibrium equation, and the Virtual Field Method (VFM) is currently the most widely used method based on this approach. However, these methods are difficult, costly, and time-consuming for users.

Machine learning (ML) approaches, as part of artificial intelligence techniques, which build a model based on sample data, known as training data, in order to make predictions or decisions without being explicitly programmed to do so, can simplify this process [10]. This approach provides a new and efficient way for the measurement, calibration, and validation processes to find the material parameters.

In this work, ML models are proposed to determine the material parameters obtained directly from experiments and required for FE simulation. The data generated by purely numerical studies used to establish the hypothetical function between material parameters and deformation behavior is used to train the ML model. From this, the material parameters (e.g., Young's modulus, yield stress, Hill's parameters, etc.) of the fiber-based material required for the FEA can be determined.

2 Materials and Methods

2.1 Material

Paperboard used in this study (thickness: 0.3 mm; grammage: 250 g/m^2) is an uncoated recycled paper made from 100% recycled fibers, which is on the borderline of the definition of paper and paperboard and is commonly used for printing or forming.

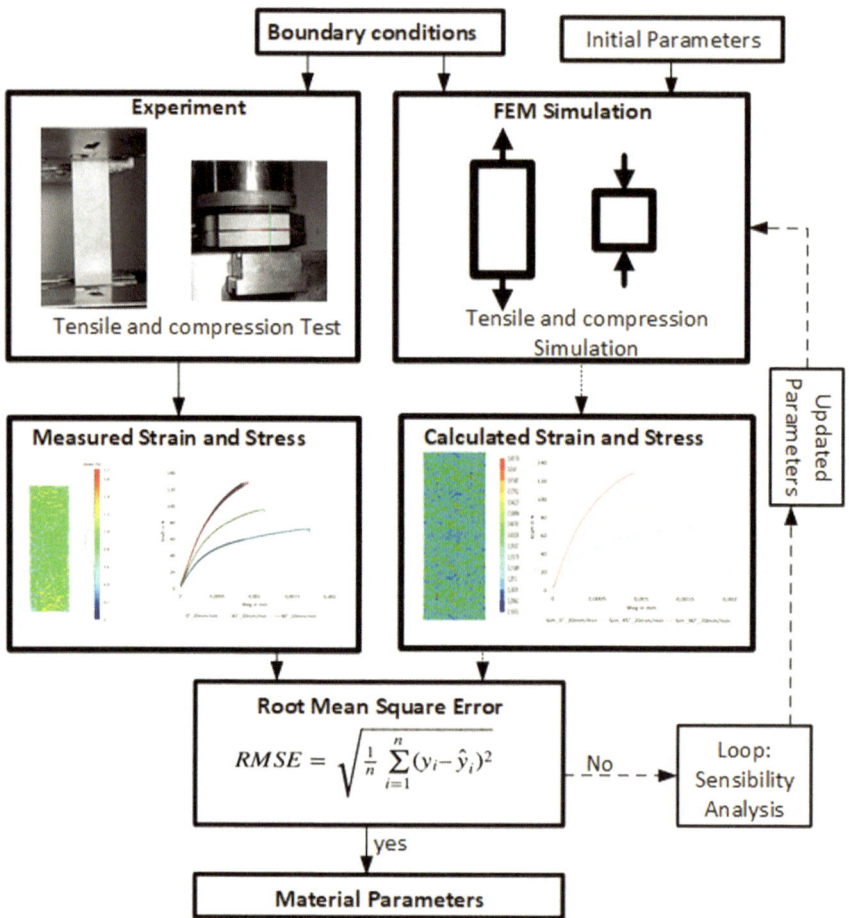

Fig. 1 Principle procedure of the inverse parameter identification method

2.2 Experimental Material Characterization

The tensile test is conducted in five directions of the paper samples, namely MD or x-direction (0°), 22.5°, 45°, 67.5°, and CD (90° or y-direction), as shown in Fig. 2. The test is strain-controlled and the test speed is 20 mm/min. A Zwick/Roell Z100 materials testing machine equipped with a high-resolution video extensometer was used to perform the experiments. GOM Aramis 5M, an optical system for 3D deformation analysis, provided strain distribution information during the uniaxial tensile test, and Zeiss Quality Suite software was used to analyze the captured images. All tests are performed in a controlled laboratory conditioned at 23 °C and 50% relative humidity (RH).

Fig. 2 Tensile samples in five directions

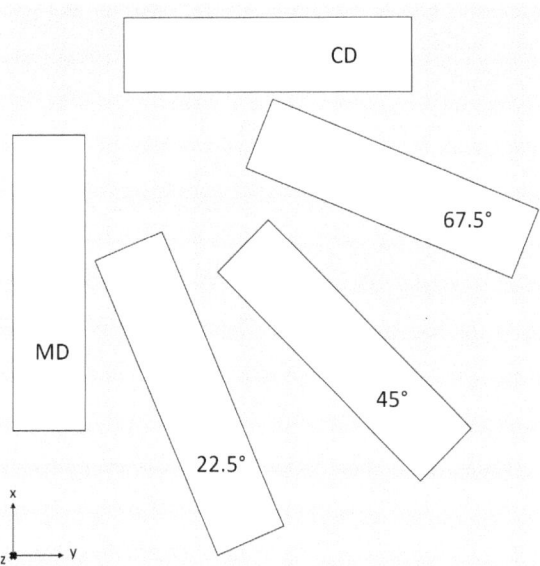

Since the material behavior in the z-direction (ZD) affects the shear behavior in the planar direction (MD-CD), it is necessary to determine the material behavior in the ZD. Due to the delamination in the tensile direction, the compression test was performed to determine the material behavior in the ZD [11]. The compression test is performed under the same conditions as the tensile test, and the machines used are the same. The simple square specimen with a side length of 30 mm is compressed to 20 kN by the Zwick machine and the displacement is recorded by both video extensometer and GOM Aramis. The test speed is 0.36 mm/min.

2.3 Paperboard Material Model and Parameters

Paperboard is an anisotropic material that can be approximated as an orthotropic material. The material exhibits different behavior in response to tension and compression. For this reason, a 2D plane stress state with Hill's yield criterion [4] and multilinear isotropic hardening law is used.

2.3.1 Elastic Part

The elastic part of the paperboard is described using Hooke's law $\epsilon = C\sigma$. The elastic model is defined by three elastic moduli E_x, E_y and E_z, three Poisson's ratios ν_{xy}, ν_{xz}, and ν_{yz}, and three shear moduli G_{xy}, G_{xz}, and G_{yz}, where x represents MD, y represents CD, and z represents ZD (also see in Fig. 2). It is assumed that Poisson's ratios in thickness direction are zero, i.e.,

$$v_{xz} = v_{yz} = 0, \tag{1}$$

and

$$G_{xz} = G_{yz} = G_{xy}, \tag{2}$$

The orthotropic elastic material properties are reduced to:

$$\begin{bmatrix} \epsilon_{xx} \\ \epsilon_{yy} \\ \epsilon_{xy} \end{bmatrix} = \begin{bmatrix} \frac{1}{E_x} & \frac{-v_{yx}}{E_y} & 0 \\ -\frac{v_{xy}}{E_x} & \frac{1}{E_y} & 0 \\ 0 & 0 & \frac{1}{2G_{xy}} \end{bmatrix} \begin{bmatrix} \sigma_{xx} \\ \sigma_{yy} \\ \sigma_{xy} \end{bmatrix} \tag{3}$$

The in-plane shear modulus G_{xy} is given by the empirical expression [12]:

$$G_{xy} = \frac{1}{\frac{4}{E_{45}} - \frac{1}{E_x} + 2\frac{v_{xy}}{E_x} - \frac{1}{E_y}} \tag{4}$$

and Poisson's ratio v_{xy} is then calculated following equation [13]:

$$v_{xy} = 0.293 \sqrt{\frac{E_x}{E_y}} \tag{5}$$

To determine the E_x, E_y, E_{45} and E_z for Young's modulus and the initial yield stress σ_{xx}^0, σ_{yy}^0, σ_{45}^0, and σ_{zz}^0 parameters, a linear regression is used that allows to determine the linearity at the beginning of the stress–strain curve. Another possibility would be to use the method of reverse engineering for the determination of the parameters.

2.3.2 Plastic Part

The constitutive models for elastic–plastic behavior can be written with a decomposition of the total strain into elastic and plastic parts.

$$\epsilon = \epsilon^{el} + \epsilon^{pl} \tag{6}$$

The stress is proportional to the elastic strain ϵ^{el}

$$\sigma = D\epsilon^{el} \tag{7}$$

Table 1 Material constants and evaluated values

E_x [MPa]	E_y [MPa]	E_z [MPa]	E_{45} [MPa]	σ_0 [MPa]	a	b	c	d	R_{xx}	R_{xy}
5349	2350	65	3160	7.64	−80.52	18.33	464.28	1.564	2.208	1.781

and the evolution of plastic strain ϵ^{pl} is a result of the plasticity model. The plastic behavior of the material is assumed to follow Hill's yield criterion [4], which is expressed on the basis of the ratio R_{ij} of the yield stress in direction ij with respect to a reference direction. If the y-axis is chosen as the reference direction, i.e., $R_{yy} = 1$, the 2D-Hill's criterion is given by

$$f(\sigma, \sigma_y) = \sqrt{\sigma^T M \sigma} - \sigma_y = 0 \tag{8}$$

where **M** is Hill's orthotropic coefficient matrix given by

$$M = \begin{bmatrix} \frac{1}{R_{xx}^2} & -\frac{1}{2R_{xx}^2} & 0 \\ -\frac{1}{2R_{xx}^2} & 1 & 0 \\ 0 & 0 & \frac{3}{R_{xy}^2} \end{bmatrix} \tag{9}$$

and σ_y is the yield stress that can, in general, evolve as a function of some material internal variables.

$$\sigma_y = a(1 - e^{-b\hat{\epsilon}_{pl}}) + c\hat{\epsilon}_{pl}^{1/d} + \sigma_0 \tag{10}$$

where a, b, c, and d are plastic parameter constants, σ_0 is the initial yield stress, and $\hat{\epsilon}_{pl}$ is the equivalent plastic strain. It should be noted that Eq. (10) is fitted to the response in the y-direction since $R_{yy} = 1$, the shear parameter in planer direction is determined with the properties behavior in 45° tensile direction and the plastic behavior in the ZD was assumed to have the same value as the y-direction $R_{zz} = 1$.

The following material parameters were obtained for the material from the reverse engineering method, as shown in Table 1.

The properties of this paperboard in different directions, MD, 22.5°, 45°, 67.5°, and CD, are studied and compared numerical experimental data with numerical experimental data. The results can be seen from Fig. 3, which shows a good agreement.

2.4 Machine Learning (ML) for Parameters Identification

This work demonstrates that machine learning can be used to predict the parameters and thus the material properties. In the material model presented in the previous section, the values of the following material parameters are variables (as in Table 2),

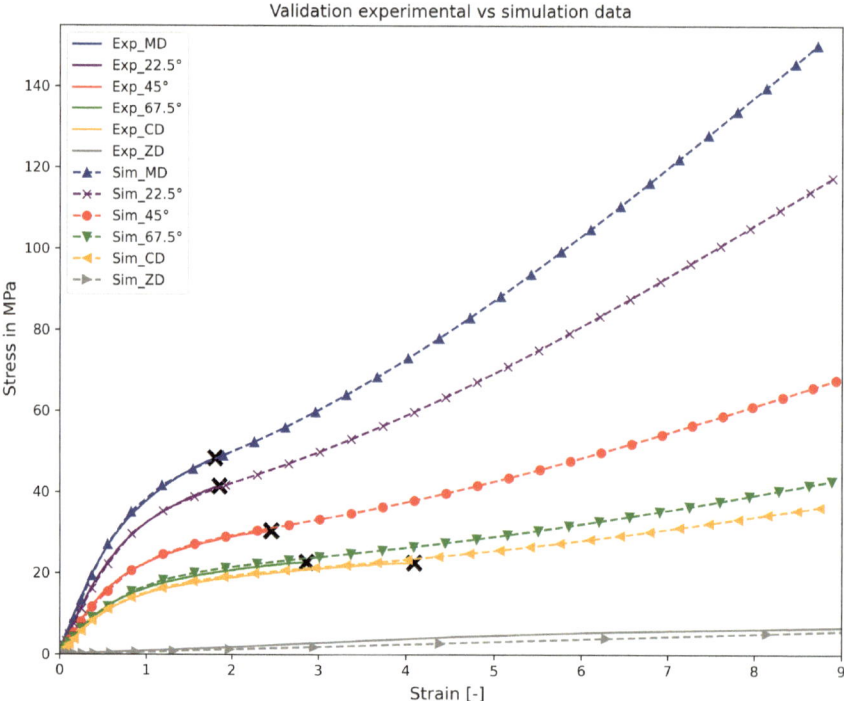

Fig. 3 Comparison of the numerical material model (Sim) and the experimental data (Exp) for the paperboard

while their ranges are defined in such a way that the computation load is controllable and the generated datasets contain enough information to perform a feasibility study of the method. In order to determine the material parameters for the forming process of thin-walled fiber-based materials, a total of 11 parameters are required. Using the parameter ranges from Table 2, 10000 data were generated with the advanced Latin hypercube sampling. The strength of many fiber-based materials is highest in the MD direction, followed by 45° and then CD direction. We have applied the following condition $E_x \geq E_{45}$, $E_{45} \geq E_y$, $\hat{\epsilon}_{pl}$ (The minimal maximum plastic strain has been set to 1.5) and Eq. 4 should always be positive. In total, 2526 generated stress–strain sets were obtained. Both the input (material parameters) and the output (strain values) of the FEM simulation were stored in a database, which is necessary for the subsequent training of the ML model. For the prediction of parameters, the machine model of the SKlearn module in Python was focused due to the fixed number of generated data and the determination of important features for the prediction of material parameters. If the determination of important features was not necessary in the workflow, the Artificial Neural Network (ANN) model would be used for the prediction of material parameters.

Table 2 Range of material parameters for numerical simulations

Material parameter	Unit	Symbol with direction	Ranges of value
Elastic modulus	MPa	E_x, E_y, E_{45}	1–10
	MPa	E_z	0.001–10
Hardening function	MPa	a, b	−1–1
	MPa	c	0–1
	–	d	1–10
Yield Stress	MPa	$\sigma_{yy}^0 = \sigma_0$	0.001–0.8
Hill's parameters	–	R_{xx}, R_{xy}	0.75–4

To perform the FEM study, the Ansys 2021R2 solver was used in a quasi-static mode under implicit analysis. The use of Optislang software was also used to perform the DOE analysis.

The process models of tensile test and compression test, including boundary conditions, loads, and mesh, are shown in Fig. 4. The tensile test uses a shell model, and due to orthogonal symmetry, only a quarter of the samples are simulated to save computation time. The size of the simulated sample is $45 \times 15 \text{ mm}^2$, and the mesh size is 1 mm. The entire sample is evaluated as in the experiments. In the compression test, a solid model is used to investigate the behavior of the ZD, here only one eighth of the specimens are simulated for the same reason as before. The size of the simulated sample is $15 \times 15 \times 0.3 \text{ mm}^3$. The mesh size in the planar direction is also 1 mm,

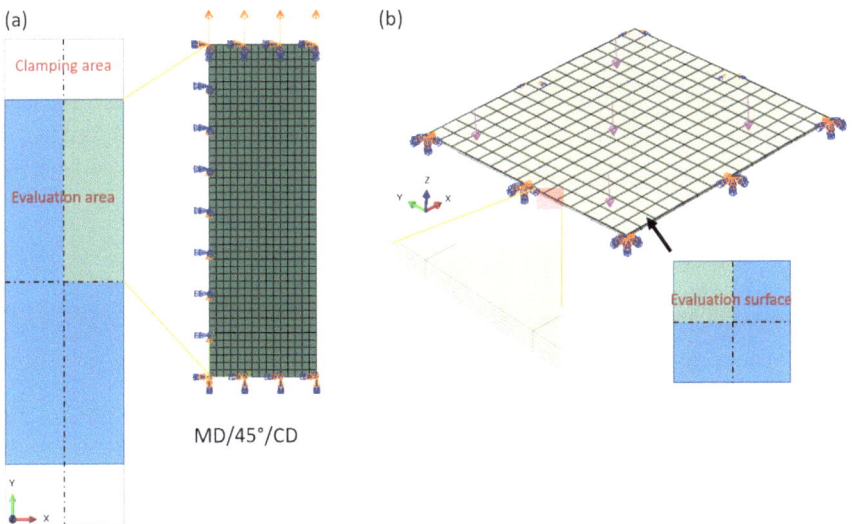

Fig. 4 Process models with boundary conditions, loads and mesh of **a** tensile test and **b** compression test

however, there are five elements in the thickness direction. The material response in each direction (average stress and strain) of the sample surface is evaluated.

3   Results and Discussion

3.1   *ML Modeling*

In order to predict the material parameters, a large study was conducted in which the material parameters were varied and a total of 2526 different stress–strain curves were generated (see Fig. 5). The generated stress–strain curve (ML-Input) was used as input to identify the material parameters (ML-Output), thus reducing the effort to determine the material parameters. Three different ML models were developed to determine the material parameters. One model to determine the elastic material parameters and the yield point for the reference direction y, one model to determine the plastic material properties for the reference direction y, and one model to determine the Hill's parameters. To develop the three ML models, a dataset with each having a data length of 50 was used as input (the data length for the ML input is $8 \times 50 = 400$): $\sigma_{yy}, \epsilon_{yy}, \sigma_{xx}, \epsilon_{xx}, \sigma_{zz}, \epsilon_{zz}, \sigma_{45}, \epsilon_{45}$ (see Fig. 6). It should be mentioned that 0 was always inserted as the first data for the stress and strain in each direction.

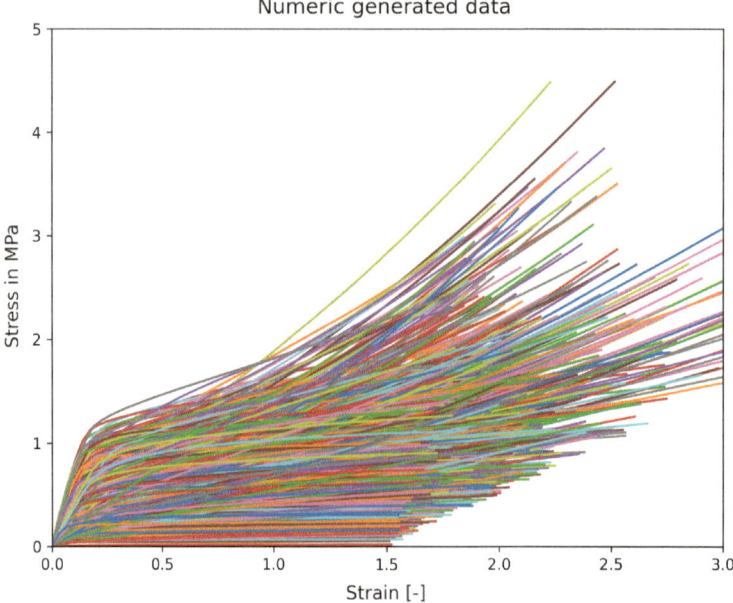

Fig. 5   Numerically generated stress–strain curves that can be used for characterization

Table 3 Scaling of the material parameters for the ML model

$Real_{parameter}$	E_x	E_y	E_z	E_{45}	σ_0	a	b	c	d	R_{xx}	R_{xy}
$Scaled_{parameter}$	E_x/S	E_y/S	E_z/S	E_{45}/S	σ_0/S	a/S	b/E	$(c/S)(1/E)^{1/d}$	d	R_{xx}	R_{xy}

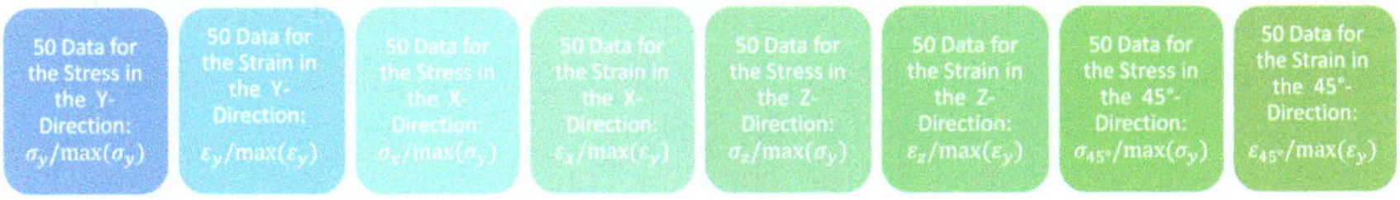

Fig. 6 Sequence of values for the ML-Input

To ensure that the method for predicting material parameters is general and works for all curves, the stresses are divided by the maximum stress and all strains are divided by the maximum strain in the main direction (here is y-axis).

$$Scaled_{\sigma_{ii}} = \sigma_{ii}/max(\sigma_y) \tag{11}$$

$$Scaled_{\epsilon_{ii}} = \epsilon_{ii}/max(\epsilon_y) \tag{12}$$

The following scaling was performed for the material parameters (see Table 3) where S = max(σ_y) and E = max(ϵ_y).

For the prediction of the material parameters, different ML models from the Sklearn library of Python 3.9 were investigated.

Depending on whether the parameters are elastic, plastic, or Hill's parameters, three different ML models have been developed according to the high accuracy.

Elastic Parameters: For predicting the elastic parameters, the ML model gradient boosting regressor with multi-output regressor was found to give the best results. The accuracy of the ML model is 97.88%. The structure of this model is as follows: MultiOutputRegressor (GradientBoostingRegressor (learningrate = 0.1, nestimators = 500, subsample = 1, criterion = friedmanmse, maxdepth = 3))

Plastic Parameters: The ML model multi-output regressor with ExtraTreesRegressor was used to determine the plastic parameters. The model has an accuracy of 92.38%.

To build the model, the following parameters are needed:
MultiOutputRegressor (ExtraTreesRegressor (nestimators = 200, criterion = squarederror, minsamplessplit = 2, minsamplesleaf = 1, minweightfractionleaf = 0.0))

Table 4 Material constants and evaluated values

Parameters	E_x [MPa]	E_y [MPa]	E_z [MPa]	E_{45} [MPa]	σ_0 [MPa]	a	b	c	d	R_{xx}	R_{xy}
Real	5349	2350	65	3160	7.64	−80.52	18.33	464.28	1.564	2.208	1.781
Predicted	5630.985	2279.129	65.021	3696.203	7.51	−80.190	18.382	464.29	1.564	2.208	1.798

Hill's Parameters: The linear regressor from Sklearn is the best model for predicting Hill's parameters.

The parameters for the model are shown below:

LinearRegression(fitintercept = True, copyX = True, positive = False) was used. The model has an accuracy of 95.87%.

3.2 Scaling of Parameters and Verification with Experimental Results

In order to predict the material parameters, four tests should be performed from the developed ML method: tensile test in MD, 45°, CD, and compression test in ZD. Afterward, the experimental data are interpolated with a data length of 49 and null was inserted as the first data point since the generated data or experimental data frequently do not start with null.

The material parameters predicted from the ML models are shown in Table 4 and compared with the real experimental values.

The predicted stress–strain curves in different directions are also shown in Fig. 7, which shows a good agreement with the experimental parameters as well.

3.3 Discussion

The materials used in the forming process are not all homogeneous and isotropic, nor do they possess an ideal stress–strain curve and a simple yield equation. For these materials, there is a similar need for a material model that can accurately describe the material properties for further studies such as forming limits or process optimization. To this end, in this work, the ML method has been developed to determine the parameters using only numerically generated data, which makes this process more available and efficient. Taking the paperboard material as an example, different ML models were tested and compared in terms of accuracy for the elastic part, plastic part, and Hill's parameters, and it was shown that they all found suitable models with satisfying accuracy.

From this workflow, it can be seen that it is absolutely applicable to other fiber-based materials, composite materials, and orthotropic materials as well. This also greatly

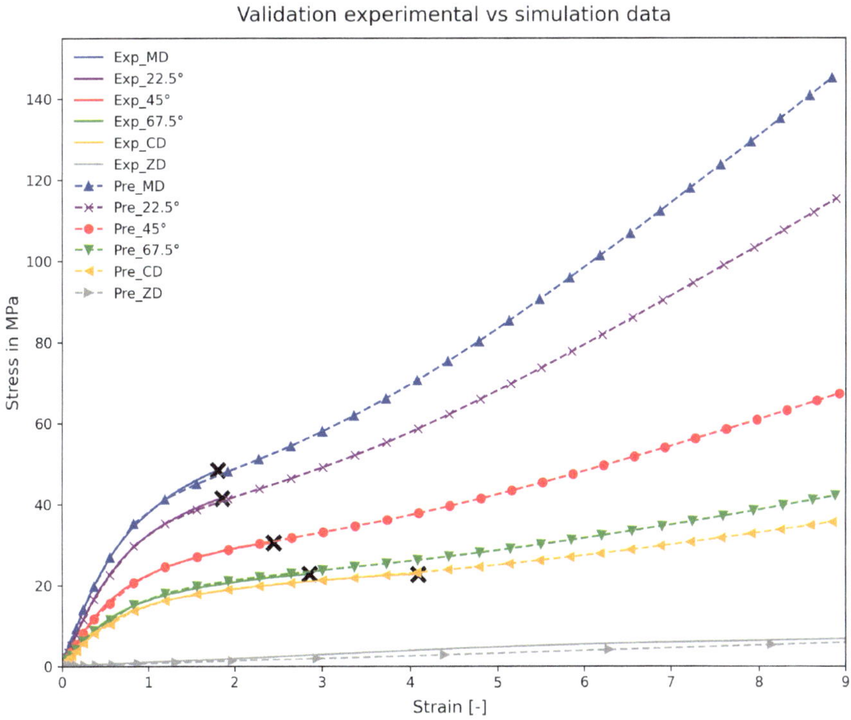

Fig. 7 Comparison of the predicted material parameters (Pre) and the experimental data (Exp) for the paperboard

simplifies the calculation process from experimental data to the parameters needed for simulation and provides the possibility to reduce the number of experiments. Not only that but also other materials used in the forming process, such as aluminum alloys which are currently being widely used due to their lightweighting advantages, an accurate material model including properties such as anisotropy, temperature dependence, etc., is also a demand. Even some more complex intrinsic structure relationships, or other yield functions, can take this approach to make the building of material models more accurate and efficient.

4   Conclusion

For materials with anisotropy, such as fiber-based materials, building an accurate material model in numerical simulations is not an easy task, partly because of the large number of experiments and partly because of the calculation from experimental data to material parameters. The ML method simplifies this process using numerically generated data and offers the possibility to determine material parameters efficiently and accurately, and can also be applied to modeling other relatively complex materials. In the future, ML will be used to reduce the number of experiments needed to determine material parameters, and the developed method will be validated for other types of materials. In addition, increasing the amount of simulation data used to train the ML model will improve the accuracy of the model in predicting plastic parameters.

References

1. Alzweighi M, Mansour R, Tryding J et al (2022) Evaluation of Hoffman and Xia plasticity models against bi-axial tension experiments of planar fiber network materials. Int J Solids Struct 238:111358. https://doi.org/10.1016/j.ijsolstr.2021.111358
2. Boman G (2022) Multiscale mechanical modeling of paperboard
3. Simon JW (2021) A review of recent trends and challenges in computational modeling of paper and paperboard at different scales. Arch Comput Methods Eng 28(4):2409–2428. https://doi.org/10.1007/s11831-020-09460-y
4. Hill R (1948) A theory of the yielding and plastic flow of anisotropic metals. Proc Royal Soc Lond Ser A Math Phys Sci 193(1033):281–297. https://doi.org/10.1098/rspa.1948.0045
5. Hoffman O (1967) The brittle strength of orthotropic materials. J Compos Mater 1(2):200–206. https://doi.org/10.1177/002199836700100210
6. Xia QS, Boyce MC, Parks DM (2002) A constitutive model for the anisotropic elastic-plastic deformation of paper and paperboard. Int J Solids Struct 39(15):4053–4071. https://doi.org/10.1016/S0020-7683(02)00238-X
7. Li Y, Stapleton SE, Reese S, Simon J-W (2016) Anisotropic elastic-plastic deformation of paper: in-plane model. Int J Solids Struct 100:286–296. https://doi.org/10.1016/j.ijsolstr.2016.08.024
8. Borgqvist E, Wallin M, Ristinmaa M, Tryding J (2015) An anisotropic in-plane and out-of-plane elasto-plastic continuum model for paperboard. Compos Struct 126:184–195. https://doi.org/10.1016/j.compstruct.2015.02.067
9. Karadogan C, Cyron P, Liewald M (2021) Potential use of machine learning to determine yield locus parameters. IOP Conf Ser Mater Sci Eng 1157:012064. 10.1088/1757-899X/1157/1/012064
10. Koza JR, Bennett FH, Andre D, Keane MA (1996) Automated design of both the topology and sizing of analog electrical circuits using genetic programming. In: Artificial intelligence in design. Springer, Dordrecht. https://doi.org/10.1007/978-94-009-0279-4_9
11. Lindberg G, Kulachenko A (2021) Tray forming operation of paperboard: a case study using implicit finite element analysis. Packaging Technol Sci 35. https://doi.org/10.1002/pts.2619
12. Nygårds M (2008) Experimental techniques for characterization of elasticplastic material properties in paperboard. Nordic Pulp Paper Res J 23(4):432–437. https://doi.org/10.3183/npprj-2008-23-04-p432-437
13. Baum GA, Habeger CC, Fleischman EH (1981) Measurement of the orthotropic elastic constants of paper. The institute of paper chemistry, Appleton, Wisconsin 26

Numerical Analysis of Inhomogeneous Parameters of Paperboard Using Tensile Tests

Yuchen Leng[ID]**, Cedric Wilfried Sanjon**[ID]**, Peter Groche, Marek Hauptmann, and Jens-Peter Majschak**

Abstract For pure metals, typically a homogeneous distribution of material properties is assumed. This assumption reduces the complexity of the models significantly. For inhomogeneous materials like paperboard, however, this assumption is questionable. Experimental findings indicate that the structural inhomogeneity can lead to variations in mechanical properties, which in turn reduce the robustness of processes and require actions to control the product quality. In this work, we introduce an approach to modeling the local material structure in numerical simulations and investigate the material response to an uniaxial tensile test. The effect of various inhomogeneities, e.g., distribution of mass, density, and fiber orientation, on material properties was investigated, and it was found that fiber orientation has the greatest effect in most cases, while the effect of density is usually the least.

Keywords Numerical simulation · Inhomogeneous material · Material behavior · Distribution

1 Introduction

As a material that can be used for forming processes, paperboard has very special properties. It is the material with the highest added value within the paper and paperboard system and is most frequently used for packaging and high-quality printed products.

Paperboard is a fibrous material that can be designed by a targeted fiber orientation during production, resulting in varying degrees of anisotropic behavior. Fibers are

Y. Leng (✉) · P. Groche
TU Darmstadt Institute for Production Engineering and Forming Machines PtU, Otto-Berndt-Str, 2, 64287 Darmstadt, Germany
e-mail: yuchen.leng@ptu.tu-darmstadt.de

C. W. Sanjon · M. Hauptmann · J.-P. Majschak
Fraunhofer Institute for Process Engineering and Packaging IVV, Heidelberger Str. 20, 01189 Dresden, Germany

J. Kusiak et al. (eds.), *Numerical Methods in Industrial Forming Processes*, Lecture Notes in Mechanical Engineering, https://doi.org/10.1007/978-3-031-58006-2_17

225

mainly dispersed in the in-plane direction, and the degree of anisotropy between the machine direction (MD) and the cross-machine direction (CD) is usually from 1 to 5, while it may differ by 50–100 between the through-thickness direction (ZD) and MD [1]. The paper, moreover, has an inhomogeneous structure due to the formation effect. This inhomogeneity is caused, on the one hand, by material variations in the properties of the fibers created in natural processes or by differences in the amount of added fillers and additives. On the other hand, it is due to mechanical effects such as differences in the sheet formation due to filtration and thickening during the production of the pulp in the manufacturing process, as well as to the different tensile forces when passing through the paper machine. The effects of inhomogeneity become noticeable mainly in the form of unstable mechanical properties in the longitudinal and transverse directions or differences in the thickness and density of the material, as well as in surface roughness [2].

Since paper is by nature a heterogeneous material, it exhibits diffuse unstable strain under uniaxial tension, which can be well observed by the Digital Image Correlation (DIC) method. The DIC method only requires a simple experimental setup and has a wide range of measurement sensitivity and resolution, and it provides local strain information by comparing before and after images during the test [3]. In a previous work by Hagman and Nygards [4] speckle analysis was used to observe the localization in the paper when it is plastically stretched. However, a few studies have reproduced the non-uniform strain distribution in numerical simulations.

The Finite Element Method (FEM) is a widely used numerical method for solving scientific and engineering problems. It can be applied to complex geometries and boundary conditions, as well as to time-dependent problems and non-linear material behavior. However, FEM is inherently deterministic and therefore limited to describing the characteristics of a system under prescribed model assumptions. In particular, it cannot directly and reliably represent a system with some degree of uncertainty. The combination of microstructure models with finite element simulations is a well-established approach for investigating the relationship between microstructural geometry and the macroscopic properties of the material. An established way to deal with fiber networks is the micromechanical approach, which can include fiber properties, bonding between fibers, and fiber arrangement in 2D or 3D [5]. However, the complexity of micromechanical models, the difficulty of characterizing fiber properties at the microscopic level, and the excessive measurement and computational costs limit the scope of application of direct simulation of fiber networks [6].

For product and process development, a modeling tool that captures the material behavior with low effort is extremely important. Therefore, the continuum modeling approach with high computational efficiency is widely used to predict mechanical response. When using this approach, it is usually assumed that the materials are uniform, and local inhomogeneities are neglected. However, the stochastic finite element method (SFEM) is an extension of the FEM to include random parameters. The SFEM can represent randomness in one or more of the main components of classical FEM, namely geometry, external forces, and material properties [7]. Despite the inhomogeneity of properties between different regions of the actual material, most finite element analyses of paperboard forming processes use the assumption of an ideal and homogeneous continuum model. An inhomogeneous material model can be built at the microscopic level using various methods, such as cellular automata, Monte Carlo, Voronoi polygons, and crystal plasticity finite element methods [8]. At the macroscopic level, the conventional finite element software also provides the possibility to assume parameter distributions.

For paperboard, inhomogeneity is mainly reflected in three aspects: distribution of thickness, density, and fiber orientation. These three local paper structures can be related to local strains and local material failure using regression models for unbleached cork kraft paper with low grammage according to Lahti et al. [9]. Since all studies (e.g., [4, 9]) are based solely on experimental methods, the results are valid only for the material studied, and the degree of influence of these three major factors is not investigated. In this work, the effect of the aforementioned distributions on the material properties is investigated using numerical methods with the uniaxial tensile test, and the simulation results are validated with experimental data based on the DIC method. A sensitivity study of the influence is also conducted to find out which parameter of inhomogeneity has the greatest influence on the mechanical properties of paperboard.

2 Material and Methods

2.1 Material

The material used in this work is made of 100 % recycled fibers and is usually utilized as printing paperboard. The thickness is 0.3 mm and the grammage, defined as mass per unit area, is $250 \, g/mm^2$. It is in the transition zone between paper and paperboard materials and has typical characteristics of paper material, i.e., better tensile strength in the MD direction and better ductility in the CD direction.

2.2 Experimental Tensile Tests

A Zwick/Roell Z100 material testing machine with a video extensometer was used to perform the tensile tests. As for the DIC system, GOM Aramis 5M, an optical system for 3D deformation analysis, provided the information on the strain distribution during the uniaxial tensile test, and Zeiss Quality Suite software was used to analyze the captured images. All the experiments were performed under laboratory conditions (50 % relative humidity and 23 °C).

The specimen geometry is a simple square with a gripping length of 90 mm and width of 30 mm, and the test speed is 20 mm/min. The tensile specimens are first sprayed with black random spots and then clamped on the tensile test machine (see

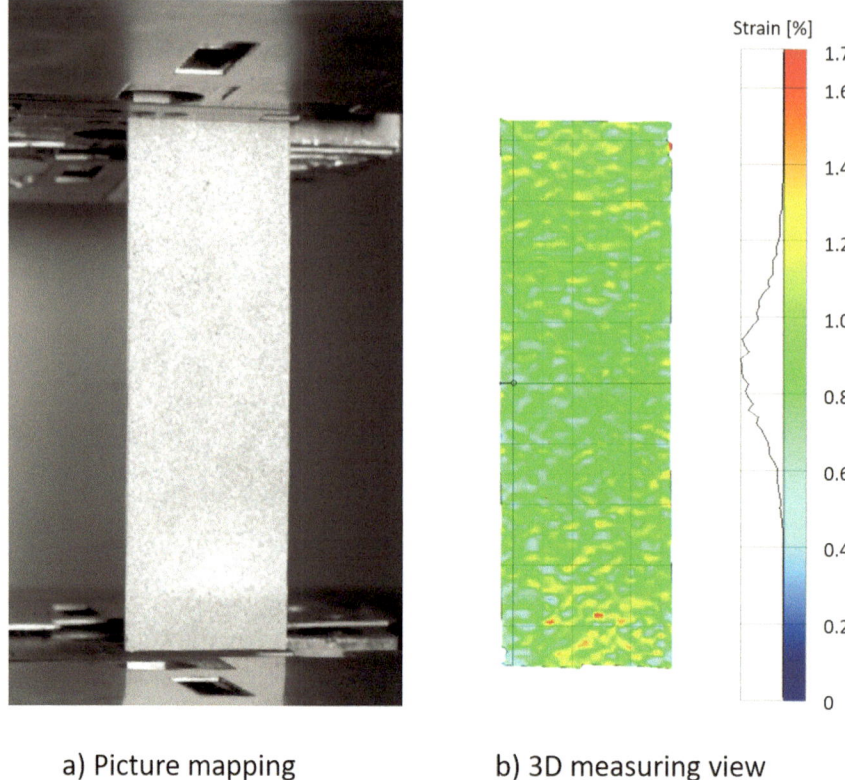

a) Picture mapping   b) 3D measuring view

Fig. 1 Tensile specimens sprayed with black random spots for strain measurement and strain distribution during experiment due to inhomogeneity of paperboard

Fig. 1). From the strain analysis of the example given, it is clear that the distribution is inhomogeneous, which is caused by the inherent inhomogeneity of the material.

2.3 Numerical Simulation

The numerical simulation for the uniaxial tensile test was conducted by ANSYS 2021. As aforementioned, the inhomogeneity of paperboard is mainly due to the disorderly arrangement of the fibers, which macroscopically leads to a local thickness and density heterogeneity of the paperboard. Since the paperboard has strong anisotropy, directional inhomogeneity is also an important factor to be considered in the simulation. These three factors are first taken into consideration separately to study the influence on the mechanical properties of two different paper materials. Then, the simulation including all three factors (label: all inhomo) is also performed to compare the force response and mechanical properties. At the same time, the homogeneous material model (label: homo), which has the average value of the inhomogeneous material model, is always used as a reference.

3 Model, Results, and Discussion

3.1 Material Modeling and Validation

In this work, an elastic-plastic orthotropic model with Hill's yield criterion [10] is used as the material model. The elastic model is defined by three elastic moduli E_1, E_2, and E_3; three Poisson's ratios ν_{12}, ν_{13}, and ν_{23}; and three shear moduli G_{12}, G_{13}, and G_{23}. It is assumed that Poisson's ratios in thickness direction are zero, i.e.,

$$\nu_{13} = \nu_{23} = 0 \tag{1}$$

and

$$G_{xz} = G_{yz} = G_{xy} \tag{2}$$

Then the required parameters of the orthotropic material model are reduced to

$$
\begin{bmatrix} \epsilon_{xx} \\ \epsilon_{yy} \\ \epsilon_{xy} \end{bmatrix} =
\begin{bmatrix} \frac{1}{E_x} & \frac{-\nu_{yx}}{E_y} & 0 \\ -\frac{\nu_{xy}}{E_x} & \frac{1}{E_y} & 0 \\ 0 & 0 & \frac{1}{2G_{xy}} \end{bmatrix}
\begin{bmatrix} \sigma_{xx} \\ \sigma_{yy} \\ \sigma_{xy} \end{bmatrix} \tag{3}
$$

The in-plane shear modulus G_{xy} is given by the empirical expression [11]:

$$G_{xy} = \frac{1}{\frac{4}{E_{45}} - \frac{1}{E_x} + 2\frac{\nu_{xy}}{E_x} - \frac{1}{E_y}} \tag{4}$$

and Poisson's ratio ν_{xy} is then calculated by the following equation: [12]:

$$\nu_{xy} = 0.293\sqrt{\frac{E_x}{E_y}} \tag{5}$$

For the plastic part, Hill's quadratic yield criterion is used. Since only the in-plane properties of the paperboard are discussed here, the stress value related to the thickness direction is considered to be 0, i.e., $\sigma_{i3} = 0$. For the material model in this work, the y-axis is chosen as the reference direction, i.e., $R_{yy} = 1$, so 2D-Hill's criterion is given by

$$f(\sigma, \sigma_y) = \sqrt{\sigma^T M \sigma} - \sigma_y = 0 \tag{6}$$

Table 1 Material constants and evaluated values

E_x [MPa]	E_y [MPa]	E_z [MPa]	G_{xy} [MPa]	σ_0 [MPa]	a	b	c	d	R_{xx}	R_{xy}
5349	2350	65	1221	7.64	-80.52	18.33	464.28	1.564	2.208	1.781

where **M** is Hill's orthotropic coefficient matrix given by

$$\mathbf{M} = \begin{bmatrix} \frac{1}{R_{xx}^2} & -\frac{1}{2R_{xx}^2} & 0 \\ -\frac{1}{2R_{xx}^2} & 1 & 0 \\ 0 & 0 & \frac{3}{R_{xy}^2} \end{bmatrix} \tag{7}$$

and σ_y is the yield stress that can, in general, evolve as a function of some material internal variables.

$$\sigma_y = a(1 - e^{-b\hat{\epsilon}_{pl}}) + c\hat{\epsilon}_{pl}^{1/d} + \sigma_0 \tag{8}$$

where a, b, c, and d are plastic parameter constants; σ_0 is the initial yield stress; and $\hat{\epsilon}_{pl}$ is the equivalent plastic strain. Note that Eq. (8) is fitted to the response in the y-direction since $R_{yy} = 1$, the shear parameter in the planer direction is determined with the behavior in the 45° tensile direction, and the plastic behavior in the thickness direction is assumed to have the same value as in the y-direction $R_{zz} = 1$.

Tensile tests at MD, 45°, and CD are used to calculate Hill's parameters in order to establish the material model, and experiments at 22.5° and 67.5° are used for validation. The following material parameters were obtained for the material from the reverse engineering method, as shown in Table 1. The comparison of the experimental and simulation data shows that the material model has a good agreement and can represent the properties very well (see Fig. 2).

3.2 Analysis of Simulation Results

In the simulation, the three distributions (density, fiber orientation, and thickness) shown in Fig. 3 for MD and CD are assigned to the sample of size $90 \times 30 \, mm^2$. The resolution is four elements per mm^2, which means that the sample consists of 10800 elements in total. According to the requirements, each element is assigned a different value of density, orientation, thickness, or both of them. For a better comparison of the degree of influence, they all follow a Weibull distribution with a positive deviation of 20 %, and the mean values (density: 1000 kg/m^{-3}; fiber orientation: 90° for MD and 0° for CD; thickness: 0.3 mm) are used in the simulation of the homogeneous material model as reference.

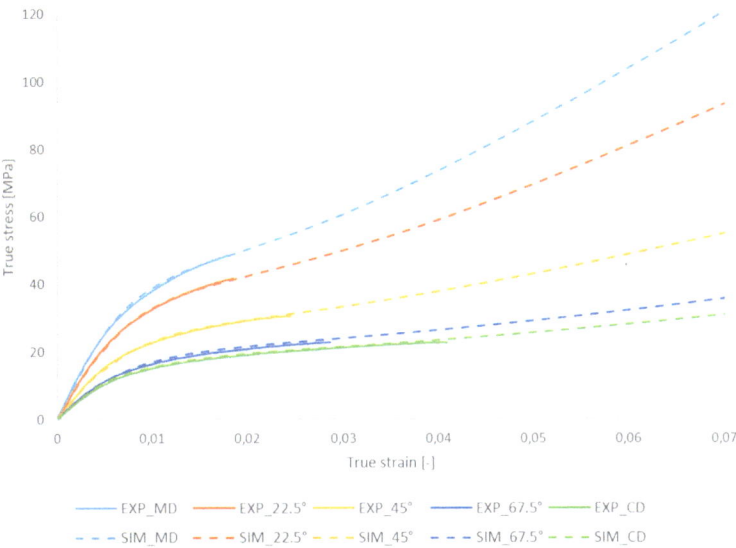

Fig. 2 Validation of the numerical material model (label: SIM) with the experimental data (label: EXP)

3.2.1 Force Response Due to Structural Parameters

Figure 4 shows the force response of the three structural distributions in terms of tensile test in MD and CD, respectively. It is shown that the influences of thickness and density inhomogeneities are in both tensile directions smaller than in the homogeneous material used as a reference for the same displacement. The offset of the curve due to these two factors is small, especially for density, where the effect is almost unnoticeable. This is because, in the simulation, density only affects the gravity of the sample, which can be negligible for the tensile test (see Fig. 5 and derived equation below).

The total force in the z-direction F_z is the sum of the tensile force F and the gravity of the specimen.

$$F_z = F + m \cdot g = F + \rho \cdot V \cdot g \qquad (9)$$

where the volume V is the multiplication of length L, width w, and thickness t.

$$V = L \cdot w \cdot t = L \cdot A \qquad (10)$$

where A is the cross-section area, and the stress σ is

$$\sigma = \frac{F_z}{A} = \frac{F + \rho \cdot V \cdot g}{A} = \frac{F + \rho \cdot L \cdot w \cdot t \cdot g}{w \cdot t} = \frac{F}{w \cdot t} + \rho \cdot L \cdot g \qquad (11)$$

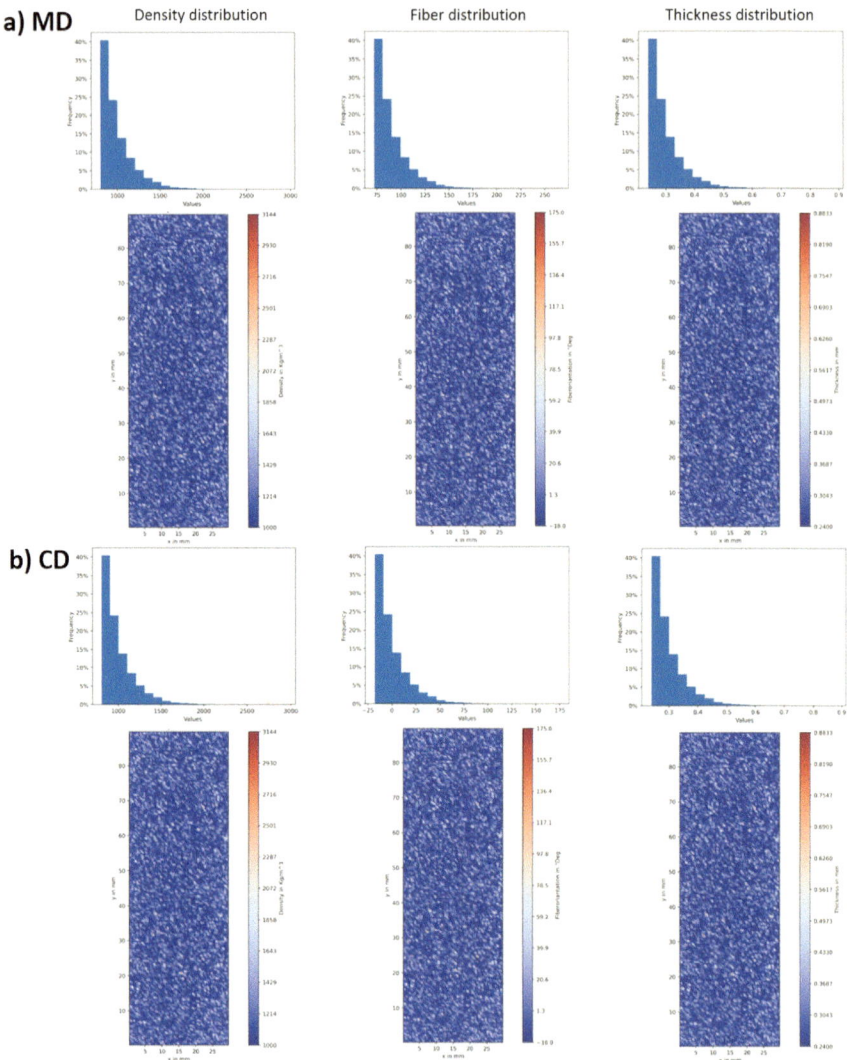

Fig. 3 Three distributions and the mapping on the a) MD and b) CD tensile samples: **a** Density $[kg/m^{-3}]$; **b** Fiber orientation $[°]$; **c** Thickness $[mm]$

Since F is at least two orders of magnitude larger than gravity, i.e., $\rho \cdot V \cdot g$, it is clear that thickness t has a much greater effect here than density ρ. The authors would like to clarify that this is only a derivation of the influencing factors, so that before and after deformation it has no effect on the equation. Since volume constancy is assumed, a 10% increase in thickness, for example, will obviously have a greater effect on the results than a 10% increase in density.

The effect of fiber orientation is more complex. Since the homogeneous material

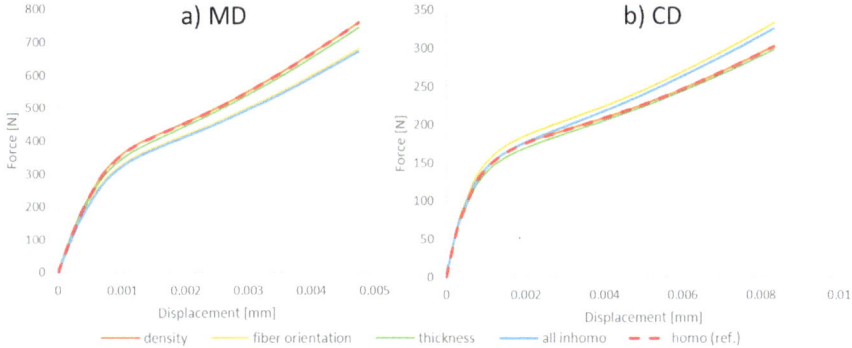

Fig. 4 Force response in MD and CD

Fig. 5 Force response in
MD and CD

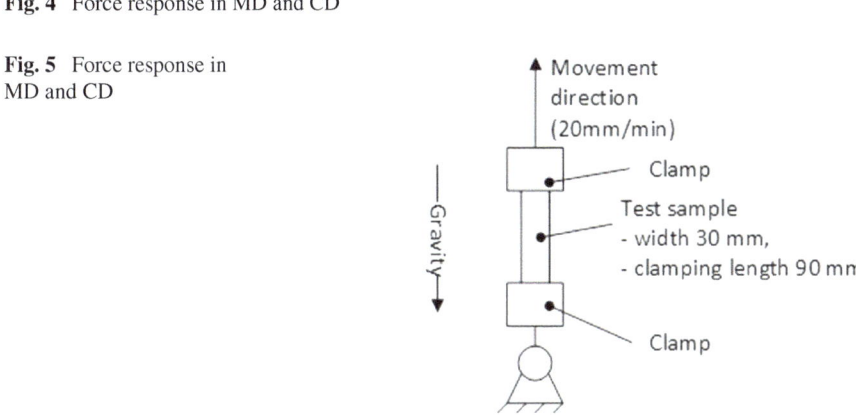

in the MD direction (all fibers oriented at 90°) has the highest strength, the sample
with inhomogeneous fiber distribution results in lower reaction force; conversely,
the homogeneous material in the CD direction (all fibers oriented at 0°) has the
lowest strength, so the sample with inhomogeneous fiber distribution requires more
reaction force. The results of considering all three inhomogeneities at the same time
are consistent with the results of the fiber orientation, further demonstrating the
absolute dominance of the influence of fiber orientation in the three inhomogeneous
parameters.

3.2.2 Influence of Structural Parameters on Material Properties

Figure 6 shows the stress-strain curve of the three structural distributions in MD and
CD. The conclusions remain consistent with the force-displacement curves, except
for the two curves affected by material thickness (thickness and all inhomo). This
is due to the relationship between strain, force, and thickness, so that the effect of
thickness inhomogeneity on the stress-strain curve is largely eliminated, and the

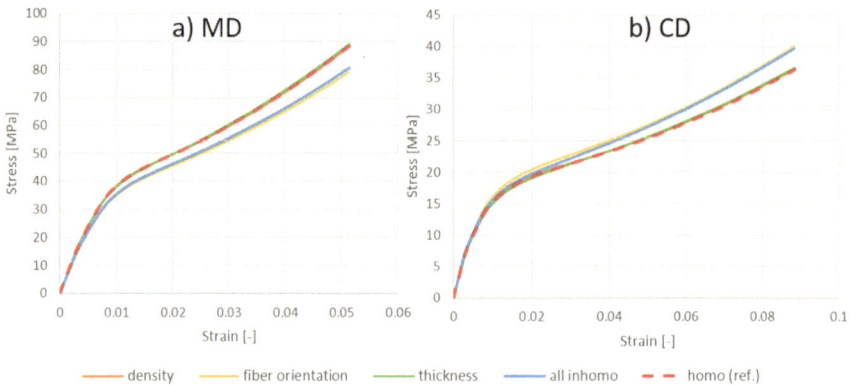

Fig. 6 Stress-strain curve in MD and CD

deviation from the homogeneous material curve is reduced compared to the force-displacement curve.

In order to analyze the elastic properties more specifically, E-modulus and Poisson's ratio are compared according to different distributions, as shown in Fig. 7. As can be seen from the diagram, the influences of density and thickness distributions can be neglected, while the fiber orientation distribution decreases these two parameters in MD, but increases them in CD.

As for the plastic part, it is known from the stress-strain curve that the plastic part continues the difference in the elastic part in a consistent manner, affecting in the same direction and to the same extent as the elastic part.

3.2.3 Material Failure Due to Inhomogeneity

Since these simulations use an implicit analysis method, there is no direct visualization of material failure. However, the failure of the paper material can be described with stress-based or strain-based failure criteria. Since the strain-based failure surface, i.e., the forming limit diagram, is more useful in the case of a 3D forming process, the maximum strain value of the entire sample is compared to effectively predict the fracture of the material, as shown in Fig. 8. It is obvious that the inhomogeneous samples have larger maximum strain values both in MD and CD, especially the fiber orientation. Despite some differences in the force-displacement and stress-strain curves in both directions and in the three inhomogeneous parameters in the previous subsection, the conclusions for fracture are consistent. Therefore, it can be concluded that the inhomogeneity leads to an increase in the maximum strain, which in turn results in an earlier failure of the material.

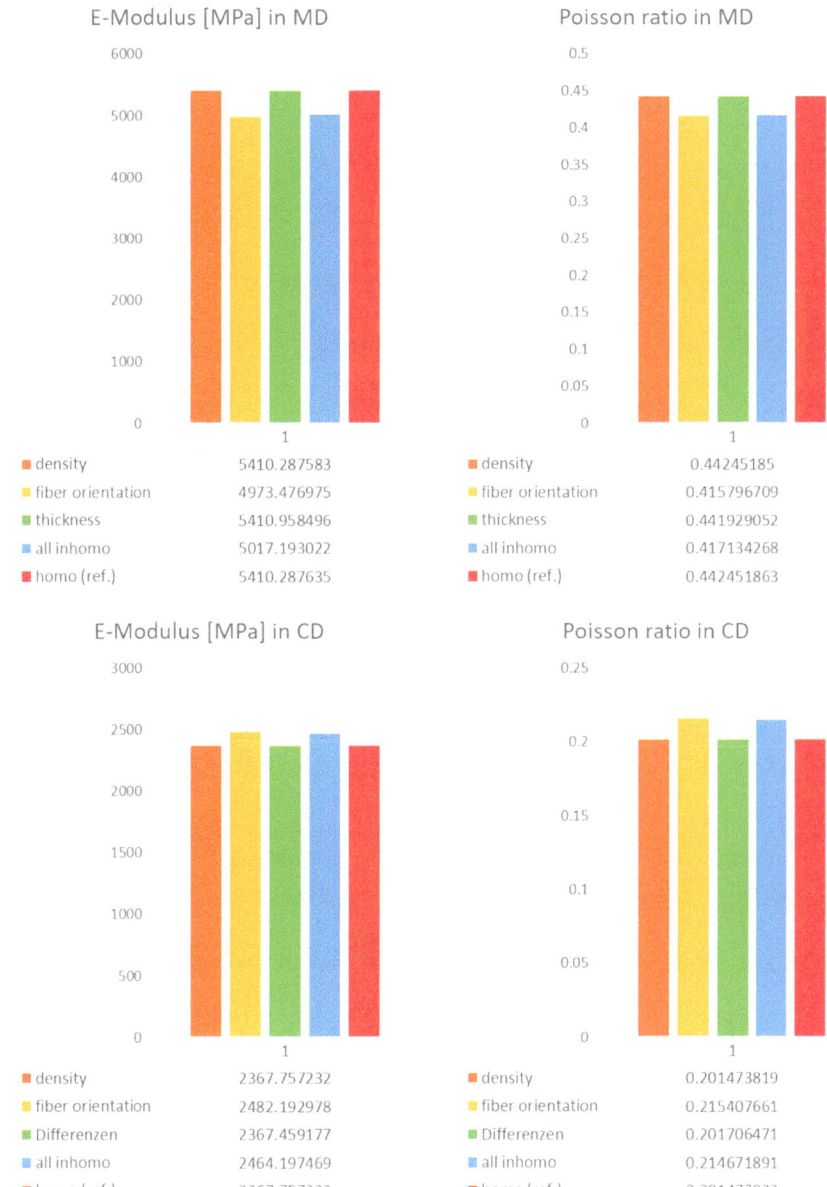

Fig. 7 Comparison of E-modulus and Poisson's ratio in MD and CD

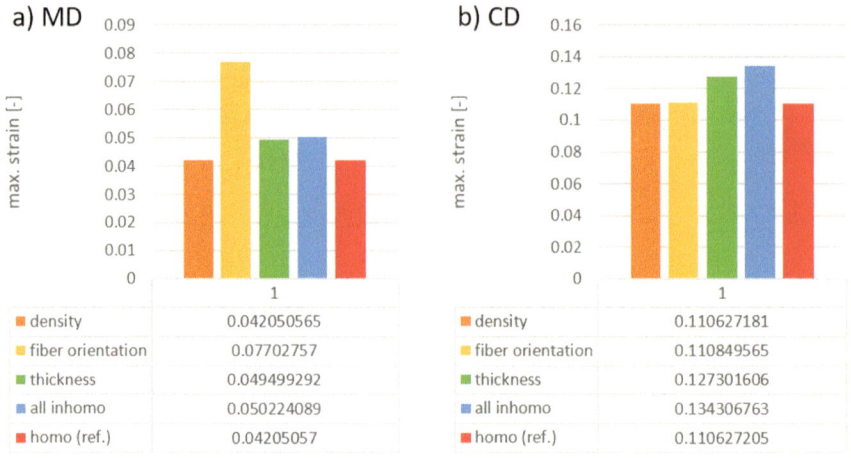

Fig. 8 The maximum strain value on the sample at the same moment

3.3 Discussion

The analysis of the presented simulation results shows that the inhomogeneous nature of the paper material can be well represented numerically. By assigning different local densities, local thicknesses, or local directional tensors to the material, the material response resulting from the inhomogeneity can be obtained. To the best of the authors' knowledge, most established finite element software can implement this feature. Therefore, this allowed the investigation of materials with inhomogeneities other than paperboard, such as composites with fiber substrates. In other words, as long as the specific inhomogeneous property parameters of the material are obtained by measurement, its mechanical properties can be explored by numerical analysis. Conversely, if the desired mechanical properties are available, it is also possible to invert the desired inhomogeneous parameters and thus make appropriate adjustments in the manufacturing process.

Returning to the paper material, the measurements on the microstructure of the paper and the identification and study of weak spots in the previous literature [9] have shown that the inhomogeneity of the paperboard properties is due to the randomness of the fibers and their arrangement during the production process. Therefore, inhomogeneity has an effect on the fracture stress and the maximum elongation strain. Moreover, the degree of influence of different structural parameters varies for different paperboard materials. For most paperboards, i.e., those with highly anisotropic properties, fiber orientation has the greatest effect. Due to the composition of the paper and the production process, it is not difficult to understand that the fibers themselves and the fiber-to-fiber bond have the most direct impact on the performance of the paperboard. For paperboard with special manufacturing processes or special fibers, this conclusion may not apply and requires a separate investigation.

Numerical analysis shows that the inhomogeneity of the paper structure affects the material properties and failure behavior, since the structural inhomogeneity leads to localized strain overload and premature fracture. In terms of fracture mechanism, it is always the weakest point that causes the specimen to fail. Inhomogeneous materials, although having exactly the same average properties as homogeneous materials, will also fail earlier due to a larger number of weak spots. This is advantageous for predicting the failure of inhomogeneous materials.

4   Conclusion

In this work, the inhomogeneous properties of paper, as a fiber-based inhomogeneous material, are analyzed using numerical simulations. Similar distributions in terms of density, fiber orientation, and thickness are assigned to the elemental points of the sample, and their response to mechanical properties is investigated individually and together. Through this study, it was found that inhomogeneities can be well represented numerically. For paperboard with strong anisotropy, fiber orientation has a dominant effect, followed by thickness, while density has only a relatively minimal effect. In addition, the inhomogeneity leads to element points of increased strain on the specimen, resulting in earlier failure of the specimen compared to a homogeneous material of the same character. This method is of great importance for predicting the failure of inhomogeneous materials and, conversely, for determining production parameters.

References

1. Stenberg N (2003) A model for the through-thickness elastic-plastic behaviour of paper. Int J Solids Struct 40(26):7483–7498. https://doi.org/10.1016/j.ijsolstr.2003.09.003
2. Hagman A, Nygårds M (2017) Thermographical analysis of paper during tensile testing and comparison to digital image correlation. Exp Mech 57:325–339. https://doi.org/10.1007/s11340-016-0240-4
3. Pan B, Qian K, Xie H et al (2017) Two-dimensional digital image correlation for in-plane displacement and strain measurement: a review. Measur Sci Technol 20(6):062001. https://doi.org/10.1088/0957-0233/20/6/062001
4. Hagman A, Nygårds M (2012) Investigation of sample-size effects on in-plane tensile testing of paperboard. Nordic Pulp Paper Res J 27:295–304. https://doi.org/10.3183/NPPRJ-2012-27-02-p295-304
5. Simon JW (2021) A review of recent trends and challenges in computational modeling of paper and paperboard at different scales. Arch Comput Methods Eng 284:2409–2428. https://doi.org/10.1007/s11831-020-09460-y

6. Alzweighi M, Mansour R, Tryding J et al (2022) Evaluation of Hoffman and Xia plasticity models against bi-axial tension experiments of planar fiber network materials. Int J Solids Struct 238(111358). https://doi.org/10.1016/j.ijsolstr.2021.111358

7. Stefanou G (2009) The stochastic finite element method: past, present and future. Comput Methods Appl Mech Eng 198(9–12):1031–1051. https://doi.org/10.1016/j.cma.2008.11.007

8. Tang D, Zhou K, Tang W et al (2022) On the inhomogeneous deformation behavior of magnesium alloy beam subjected to bending. Int J Plasticity 150:103180. https://doi.org/10.1016/j.ijplas.2021.103180

9. Lahti J, Dauer M, Keller DS et al (2020) Identifying the weak spots in packaging paper: local variations in grammage, fiber orientation and density and the resulting local strain and failure under load. Cellulose 27:10327–10343. https://doi.org/10.1007/s10570-020-03493-z

10. Hill R (1948) A theory of the yielding and plastic flow of anisotropic metals. Proc Royal Soc Lond Ser A Math Phys Sci 193(1033):281–297. https://doi.org/10.1098/rspa.1948.0045

11. Nygårds M (2008) Experimental techniques for characterization of elasticplastic material properties in paperboard. Nordic Pulp Paper Res J 23(4):432–437. https://doi.org/10.3183/npprj-2008-23-04-p432-437

12. Baum GA, Habeger CC, Fleischman EH (1981) Measurement of the orthotropic elastic constants of paper. The institute of paper chemistry, Appleton, Wisconsin 26

Numerical Simulations of Friction Stir Spot Welding Process Using Coupled Eulerian–Lagrangian Approach

Tuan Anh Do, Minh Chien Nguyen, Yang-Jiu Wu, Pai-Chen Lin, and Xuan Van Tran

Abstract The friction stir spot welding process involves many complex problems with thermal–mechanical coupling, boundary contact, and large deformation. Among these problems, large deformation is the most difficult to deal with conventional finite element method. In this study, 3D thermal–mechanical coupling finite element analysis based on the Coupled Eulerian–Lagrangian (CEL) method is employed to simulate the friction stir spot welding process of two aluminum sheets. The computational temperature histories at different locations and the weld thermal heat-affected zones are compared with those measured from experiments. The computational results are in relatively good agreement with the experiments. The CEL approach shows no mesh distortion and can be applied for extreme deformation problems; temperature diffusion, material flows, and mixture zone results can be obtained to evaluate the quality of the joint; CEL approach is computation cost where the mesh need to be refined.

Keywords Friction Stir Spot Welding · Eulerian–Larangian approach · Process modeling · Welding · Lightweight materials

1 Introduction

Friction Stir Spot Welding (FSSW) is an advanced joining technique for lightweight materials such as aluminum alloys. The FSSW process is a variant of friction stir welding (FSW), proposed to make spot welding. Like friction stir welding (FSW),

T. A. Do · M. C. Nguyen · X. Van Tran (✉)
Institute of Southeast Vietnamese Studies, Thu Dau Mot University, Binh Duong, Vietnam
e-mail: xuantv@tdmu.edu.vn

Y.-J. Wu · P.-C. Lin
National Chung-Cheng University, Chiayi 62102, Taiwan

T. A. Do
Viettel Aerospace Institute, Hanoi, Vietnam

© The Rightsholder, under exclusive licence to [Springer Nature Switzerland AG], part of Springer Nature 2024
J. Kusiak et al. (eds.), *Numerical Methods in Industrial Forming Processes*, Lecture Notes in Mechanical Engineering, https://doi.org/10.1007/978-3-031-58006-2_18

this process creates a softened region near the tool by heat generated from friction between the rotating tool and the sheets. Instead of moving the tool in FSW, the tool is fixed in the longitudinal direction of the workpiece while rotating. FSSW can make joints without melting the base metal, which is an advantage compared with the other spot welding processes (resistance spot welding) [1, 2]. In FSSW, the voids and defects from rapid solidification of fused aluminum alloy can be easily avoided.

As shown in Fig. 1, the FSSW process consists of three stages: plunging, stirring, and retracting [3]. During the plunging step, the tool with a probe pin penetrates the workpieces with high rotational speed and constant indentation rate. An anvil beneath the lower sheet is used to support the tool's downward force. At stirring, after an indentation depth, the tool is stopped but continuously rotating. The materials adjacent to the tool are softened and mixed to form a solid-state joint [4]. Finally, the tool is drawn out of the sheets as shown in Fig. 1c. The quality of weld joint depends on the tool parameters, the rotation speed, the stirring time, and the characteristics of material being joined.

In [6, 7], Sundaram et al. performed FSSW experiments to optimize welding joints of similar and dissimilar material sheets. Because experiments are expensive and time-consuming, numerical simulation needs to be performed. Many studies were conducted to investigate various aspects of the FSSW process and the effects of the parameters on the weld geometries [8, 9]. However, it remains challenging to understand the complex FSSW process, Finite Element Analysis (FEA) is the most powerful to understand the complexities of thermal–mechanical problem during FSSW. The FEA of FSSW involves many complex problems such as thermal–mechanical coupling, boundary contact, and large deformation. Among the challenges, large deformation is the most difficult to deal with conventional FEM.

In [10], Awang and Mucino used a conventional Lagrangian approach to analyze energy generation of FSSW. The result showed a good agreement of the peak temperature and energy dissipation with the experiment [11]. However, the simulation only reached the beginning plugging stage of the process. Using the Lagrangian approach, due to extreme material deformation, severe mesh distortions are observed, and the calculation cannot run until the end [12].

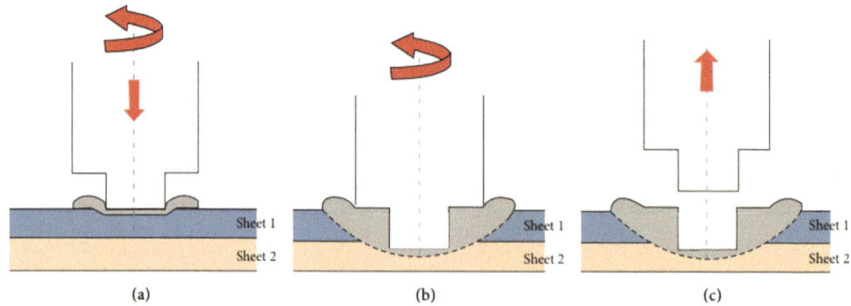

Fig. 1 Illustration of the FSSW process: **a** plunging, **b** stirring, and **c** retracting [5]

In order to deal with large deformation problem, the coupled Eulerian–Lagrangian (CEL) method is one of the most advanced technologies [13–15]. This analysis technique combines two mesh approaches—Lagrangian and Eulerian—in the same analysis. The purpose of this technique is to avoid non-converging problems when performing simulations that involve high/extreme deformations. In contrast to the Lagrangian formulation, in Eulerian mesh, the node is fixed while the material can flow in the mesh. In [16], the CEL technique is used for FSSW simulation for one metal sheet to study the temperature evolution, mainly to investigate the feasibility of the method.

In this paper, we use the CEL approach to simulate the complex FSSW process of two overlap AL6061-T6 aluminum sheets. In Sect. 2, the modeling CEL technique will be introduced. The material model, friction contact treatment, heat transfer, and mass scaling problems are discussed. In Sect. 3, the numerical results are presented in comparison with experiment. Conclusion and future work are presented in Sect. 4.

2 Numerical Simulation

2.1 Experimental Setup

Su et al. [17] performed FSSW to produce joints on similar aluminum AL6061-T6 sheets (101.6 × 25.4 × 2 mm) with (25.4 × 25.4 mm) overlap area. The sheets were welded using a 5 horsepower milling machine under displacement-controlled conditions. The plunge motion was produced at a constant rate of 2 mm/s. The welding temperatures were measured by K-type thermocouples (having a response time of 0.5 s) at locations A, B, C, and D of 0.4 mm underneath the bottom surface of the lower sheet. Location A is at the center and B, C, and D are located at 4, 5, and 6 mm in the radial direction from A, respectively, as indicated in Fig. 2.

Three welding tools, named T1, T2, and T3, are used to perform the FSSW with tool radii of 4, 5, and 6 mm, respectively, see Fig. 4. These tools are made of high-speed steel SKD 11 thermal-treated. We consider tool geometry with a flat shoulder and flat pin. The tool rotation speed is set from 400 to 2000 rpm.

A weld joint made by FSSW in a lap-shear specimen is shown in Fig. 3a and a cross section of the weld is presented in Fig. 3b. To optimize the welded joint, welding

Fig. 2 Position of thermocouples

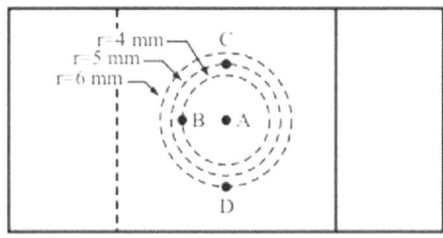

process parameters (tool pin radius, shoulder radius, stirring time, indentation depth, and rotation speed) have been changed. Due to the limit of material and equipment, several configurations have been considered as in Fig. 4. Welding parameters are summarized in the Table 1.

To consider a larger number of sets of parameters, numerical simulations are needed. In the next section, we will reproduce these experiments by numerical simulations. CEL analysis developed in Abaqus/Explicit has been employed to study this problem. The model consists of one Eulerian domain for workpiece and Lagrangian domains for the tool and the anvil. The material temperature can be very high and

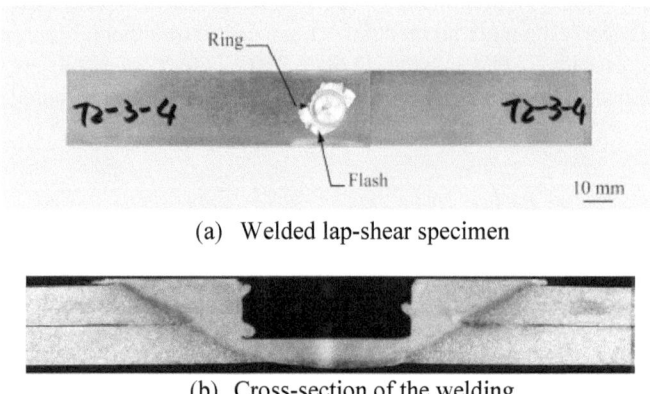

(a) Welded lap-shear specimen

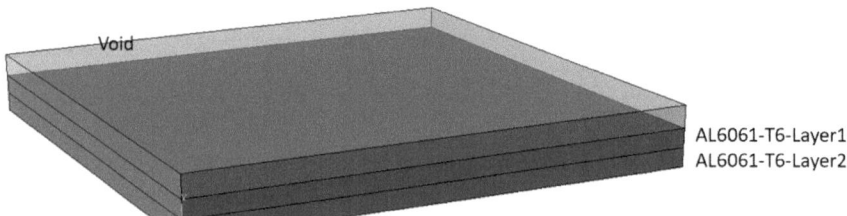

(b) Cross-section of the welding

Fig. 3 **a** A weld joint with a 6061-T6 FSSW in a lap-shear specimen and **b** a cross section of the weld

Fig. 4 Geometry and material assignment in the Eulerian domain

Symbol	Welding parameter	Value
r1	Tool pin radius (mm)	1.65, 2.0, 2.35
r2	Tool shoulder radius (mm)	4.0, 5.0, 6.0
VR2	Rotation speed (rpm)	700
d	Indentation depth (mm)	1.9
T2	Stirring time (s)	From 1.0 to 15.0

Table 1 Welding parameters and values

Fig. 5 Tool configuration

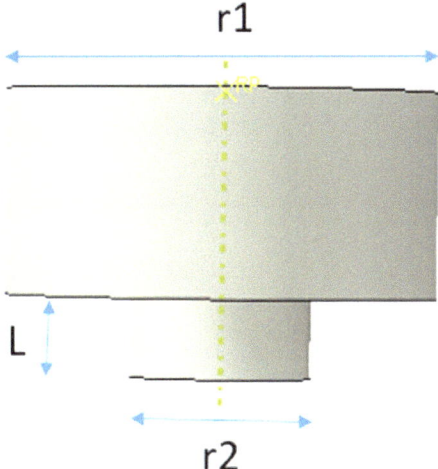

significantly influences the mechanical response. Therefore, fully coupled thermal stress analysis is chosen in this study to account for heat generation by tool–workpiece friction and plastic deformation.

2.2 Geometry Configuration

To reduce the computational cost, only the overlap part of the two sheets is considered in simulation. We consider a Eulerian domain of 25.4 mm × 25.4 mm × 3.2 mm which is divided into three parts vertically as in Fig. 4. Two aluminum sheets of 1 mm each in the lower part and 1.5 mm of air in the upper part so that the material can flow in the domain. We assume that the material is not allowed to enter or leave the Eulerian domain.

The tool (Fig. 5) and the anvil (Fig. 6) have been modeled using the Lagrangian formulation with thermally coupled elements. The tool is modeled as a rigid body. The anvil is fixed at the bottom.

2.3 Material Model

In this study, it is assumed that the workpiece material is isotropic, plastic hardening as a function of strain and strain rate. FSSW induces high strain rate deformationand temperature softening. We employed strain, temperature-dependent viscoplastic Johnson–Cook model for aluminum workpiece. The flow stress is expressed as [18, 19]

Fig. 6 Model boundary conditions

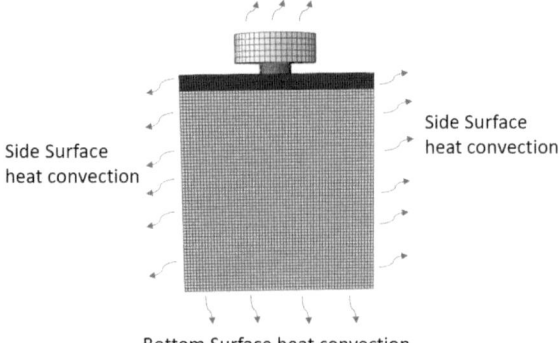

Side Surface heat convection

Side Surface heat convection

Bottom Surface heat convection

$$\overline{\sigma} = \left[A + B\overline{\varepsilon}^n \right]\left[1 + C ln\left(\frac{\dot{\overline{\varepsilon}}}{\dot{\varepsilon_0}} \right)\right]\left[1 - \left(\frac{T - T_{room}}{T_{melt} - T_{room}} \right)^m \right] \tag{1}$$

where $\overline{\sigma}, \overline{\varepsilon}, \dot{\overline{\varepsilon}}, \dot{\varepsilon_0}, T_{room}, T_{melt}$ are flow stress, plastic strain, effective strain rate, reference strain rate, room temperature, and melting temperature, respectively.

In the Johnson–Cook damage model, the damage parameter is defined as

$$w = \Sigma \left(\frac{\Delta\overline{\varepsilon}^{pl}}{\overline{\varepsilon}_f^{pl}} \right) \tag{2}$$

where $\Delta\overline{\varepsilon}^{pl}$ is a variation of equivalent strain rate, and $\overline{\varepsilon}_f^{pl}$ is the strain rate at failure and determined by

$$\overline{\varepsilon}_f^{pl} = \left[d_1 + d_2 exp\left(\frac{d_3 p}{q} \right)\right]\left[1 + d_4 ln\left(\frac{\dot{\overline{\varepsilon}}}{\dot{\varepsilon_0}} \right)\right]\left(1 + d_5\hat{\theta} \right) \tag{3}$$

where $d_1 - d_5$, are failure parameters, $\hat{\theta}, p, q$ are transition temperature, pressure stress, and von Mises stress, respectively.

The AL6061-T6 parameters are chosen as below by Lesuer et al. [20] (Tables 2 and 3).

Table 2 Material parameter for AL6061-T6

Density (kg/ m³)	Young's modulus (GPa)	Thermal expansion (1/K)	Specific heat (J/kg.K)	Thermal conductivity (W/mK)	Poisson's ratio
2690	69.0	2.4e-5	880	180	0.29

Table 3 Johnson–Cook parameter for AL6061-T6

A(MPa)	B(MPa)	n	m	C	$\theta_{melting}$(K)
2690	69.0	2.4e-5	880	180	0.29
d_1	d_2	d_3	d_4	d_5	
0.77	1.45	0.47	0.0	1.60	

2.4   Contact and Boundary Conditions

In the models, the bottom surface of the anvil is fixed in all directions, and material is not allowed to enter or exit the domain as described in Fig. 4. The displacement of the tool is described by applying the boundary conditions at the reference points (center of top surface of the tool). During the plunging stage, the tool penetrates at a velocity of 2 mm/s and a rotation speed of 70 rad/s. During stirring, the tool is stopped at indentation depth, while continuously rotating; therefore, the displacement is removed while the rotation speed is maintained.

The contact interactions between the tool and the workpiece are included as general contact. Heat sources are from friction at tool–workpiece interface and plastic deformations. The general contact algorithm implements the penalty method with a friction coefficient. Following Backar et al. [21], the friction coefficient depends on the temperature. For simplification, we consider a constant value of 0.5 which corresponds to the friction coefficient at ambient temperature. It is assumed that the energy generated by friction is converted to heat and 90% of energy in plastic deformation is dissipated as heat [22].

The heat dissipation to the ambient should be also considered. For simplification, the heat removal from the bottom surface is modeled to be due to heat convection with a film coefficient of 3 W/m^2 (air convection). The heat transfer from the side face and tool surface is simplified to be due to the convection heat transfer with a coefficient of 3000 W/m^2 (replacing heat conduction in the anvil and specimen). Heat transfers between material layers are considered as thermal heat conductance with pressure dependence as derived from [23].

3   Results and Discussions

The history output is extracted for nodal temperature of 2 points at 0.4 mm from the anvil top surface (Point A and B as in Fig. 2).

3.1 Temperature Evolution

The temperature evolution is plotted in Fig. 7 for two points A and B. Experimental data is available in [17] and used as a reference in this paper. Three tool configurations have been considered as follows (Table 4).

As can be observed from Fig. 7, the temperature increases when the tool pin contact with the workpiece. It is the result of heat generation from the friction between pin surface and aluminum surface. At 0.8 s, when the tool shoulder meets the surface, contact area is larger, more heat is generated, and temperature rises faster. At stirring, the temperature gradually increases to the maximum value and is then stable. For T1, the average errors at points A and B are 4.7% and 3.6%, respectively. For T2, the average errors are higher, 9.1% at point A and 7.8% at point B. For T3, we get 8.7% and 6.0% of average errors at points A and B, respectively. The temperature evolution at points A and B are in good agreement with the experiment.

3.2 Thermal Heat-Affected Zone

Figure 8 displays the plots of temperature at the 1 s stirring stage for three tools. The softening zone with the a close to 925 K, as well as TMAZ and HAZ, can be observed by visualization module. For the welds made by the three weld tools under identical processing conditions, the TMAZ sizes are generally proportional to tool size and the TMAZ profiles are quite similar. The TMAZ sizes are calculated equal to 4.4 mm, 6.2 mm, and 9.1 mm for simulation results of T1, T2, and T3, respectively. To compare, the TMAZ sizes estimated for experiments of T1, T2, and T3 are about 4 mm, 5.8 mm, and 8.2 mm, respectively. The computational results of the heat-affected zones are in relatively good agreement with the experiments.

4 Conclusion

In this paper, a novel 3D thermal–mechanical coupling finite element analysis based on the CEL method has been used to simulate the friction stir spot welding process for joining two aluminum sheets. Details on material model, boundary conditions, and contacts have been presented. The computational results are in relatively good agreement with the experiments. Some conclusions are as follows: the CEL approach shows no mesh distortion and can be applied for extreme deformation problem; temperature diffusion, material flows, and mixture zone results can be obtained to evaluate the quality of the joint; CEL approach is computation cost where the mesh need to be refined. In the future work, this model can be used to investigate the influence of friction coefficient, heat convection and combined with machine learning to find the best parameters for the FSSW process.

Fig. 7 Comparison of temperature history between numerical result and experiment **a** T1, **b** T2, and **c** T3

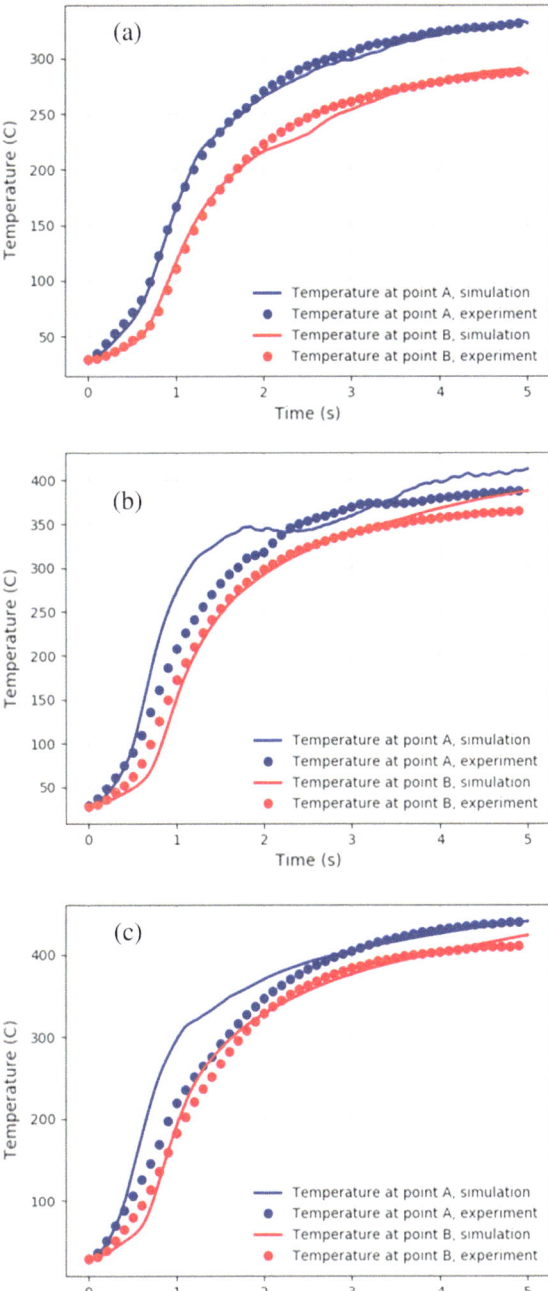

Table 4 Welding parameters and values

Case	Tool	Pin radius (m)	Shoulder radius (m)
(a)	T1	1.65	4.0
(b)	T2	2.0	5.0
(c)	T3	2.35	6.0

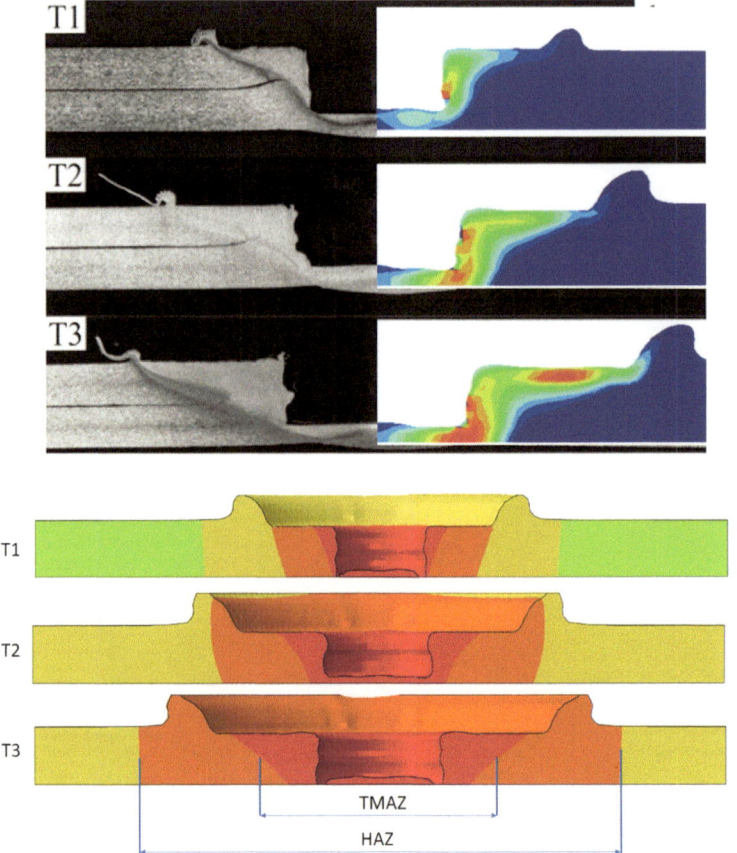

Fig. 8 Comparison of TMAZ between experimental and numerical results of T1, T2, and T3

Acknowledgements This research is funded by Thu Dau Mot University, Binh Duong Province, Vietnam, under grant number NNC.21.2.012.

References

1. Shen Z, Ding Y, Gerlich AP (2019) Advances in friction stir spot welding. Critical Rev Solid State Mater Sci
2. Tran V-X, Pan J, Pan T (2010) Fatigue behavior of spot friction welds in lap-shear and cross-tension specimens of dissimilar aluminum sheets. Int J Fatigue
3. Yang XW, Fu T, Li WY (2014) Friction stir spot welding: a review on joint macro- and microstructure, property, and process modelling. Adv Mater Sci Eng
4. Gerlich A, Su P, North TH (2005) Tool penetration during friction stir spot welding of Al and Mg alloys. J Mater Sci
5. Nguyen N-T, Kim D-Y, Kim HY (2011) Assessment of the failure load for an AA6061-T6 friction stir spot welding joint. Proc Inst Mech Eng B J Eng Manuf 225:1746–1756
6. Manickam S, Balasubramanian V (2016) Optimizing the friction stir spot welding parameters to attain maximum strength in Al/Mg dissimilar joints. J Welding Joining
7. S. S. Y. C. Ozdemir U (2012) Effect of pin penetration depth on the mechanical properties of friction stir spot weldedaluminum and copper. Mater Testing, 233–239
8. Kim DEA (2010) Numerical simulation of friction stir spot welding process for aluminum alloys. Metals Mater Int
9. Constantin MA, Boşneag A, Iordache MD, Bădulescu C, Nitu EL (2016) Numerical simulation of friction stir spot welding. Appl Mech Mater. Trans Tech Publications Ltd
10. Awang M, Mucino VH (2010) Energy generation during friction stir spot welding (FSSW) of Al 6061–T6 plates. Mater Manuf Process 209:167–174
11. Su P, Gerlich A, North TH, Bendzsak GJ (2006) Energy utilisation and generation during friction stir spot welding. Sci Technol Welding Joining
12. Cox CD (2014) Friction stir spot welding: engineering analysis and design. Nashville
13. Chauhana P, Jainb R, Pala SK, Singhc SB (2018) Modeling of defects in friction stir welding using coupled Eulerian and Lagrangian method. J Manuf Processes
14. Coupled Eulerian–Lagrangian prediction of thermal and residual stress environments in dissimilar friction stir welding of aluminum alloys. J Adv Joining Process
15. Ansari MA, Samanta A, Behnagh RA, Ding H (2019) An efficient coupled Eulerian-Lagrangian finite element model for friction stir processing. Int J Adv Manuf Technol
16. Vartak A, Raut NN (2019) Numerical Analysis of Friction Stir Spot Welding (FSSW) using Abaqus. Int J Res Analyt Rev
17. Su Z-M, Qiu Q-H, Lin P-C (2016) Design of friction stir spot welding tools by using a novel thermal-mechanical approach. Materials
18. GR J, WH C (1983) A constitutive model and data for metals subjected to large strains, high strain rates, and high temperatures. Proceedings of the 7th international symposium on ballistics, pp 541–547
19. GR J, WH C (1985) Fracture characteristic of three metals subjected to various strains, strain rates, temperatures and pressures. Eng Fract Mech, 31–48
20. Lesuer DR, Kay GJ, LeBlanc MM (2001) Modeling large-strain, high-rate deformation in metals. In: Third Biennial Tri-laboratory engineering conference modeling and simulation, Pleasanton, CA
21. Backar A, Elhofy M, Nassef G (2020) Finite elements modelling of friction stir welding. Int J Adv Sci Technol 29(3):29–43
22. Knysh P, Korkolis YP (2015) Determination of the fraction of plastic work converted into heat in metals. Mech Mater
23. Haynes WM, Lide DR, Bruno TJ (2015) CRC handbook of chemistry and physics

Modeling of Viscoelasticity of Thermoplastic Polymers Employed in the Hot Embossing Process

F. Rabhi, G. Cheng, and T. Barriere

Abstract The manufacturing of micro-scale components requires mastery of shaping processes ranging from micromechanics to electronic microfabrication. The hot embossing (HE) process is widely developed in various fields, since it allows to emboss complex structures at the micro/nanoscale such as optical sensors, diffractive lenses, microfluidic channels, and so on. The development of micro-structured parts via this process requires an in-depth analysis of the surface quality obtained and the mold filling rate. It is essential to analyze the influence of polymer properties to optimize the final mold filling to reduce cycle time and obtain defect-free replicated components. In this research, compression tests were carried out with poly(methyl methacrylate) (PMMA) and polycarbonate (PC), at different forming temperatures to determine their behavior law properties. Numerical simulation of the polymer forming processing was carried out by using Abaqus finite element software, taking into account the mechanical properties of both polymers and the characteristics of microchannels. The aim was to analyze the effect of the elastic–viscoplastic properties of the materials on the mold filling rate at different temperatures. Numerical simulation of the HE process with PMMA shows that the mold cavity is completely filled with elastic-viscoplastic behaviors, and the filling rate increases as a function of mold displacement. On the other hand, for PC, the embossed temperature has an influence on the filling ratio of the mold.

Keywords Hot embossing · Viscoplastic model · Numerical simulation

F. Rabhi (✉) · T. Barriere
Université de Franche-Comté, CNRS, Institut FEMTO-ST, 25000 Besançon, France
e-mail: faleh.rabhi@femto-st.fr

G. Cheng
INSA CVL, Université Tours, Université Orléans, LaMé, 3 Rue de La Chocolaterie, CS 23410, 41034 Blois Cedex, France

© The Rightsholder, under exclusive licence to [Springer Nature Switzerland AG], part of Springer Nature 2024
J. Kusiak et al. (eds.), *Numerical Methods in Industrial Forming Processes*, Lecture Notes in Mechanical Engineering, https://doi.org/10.1007/978-3-031-58006-2_19

1   Introduction

The demand for microstructures continues to grow. It needs to improve the manufacturing process for fabricating the microcomponents at lower cost and with short
production time. Hot embossing (HE) is a conventional and mature process for manufacturing small-scale devices on a large scale. Poly(methyl methacrylate) (PMMA)
and polycarbonate (PC) are mainly used as processing materials in HE [1].

Cheng et al. [2] have developed a micro-scale HE process using an innovative
injection-molding press and an instrumented mold for a PMMA plate to produce
microfluidic systems slightly above the glass transition temperature (T_g). Wang et al.
[3] optimized various parameters in HE processing and obtained a filling accurately
and efficiently. Kasztelanic et al. [4] employed the HE process to develop the optical
devices. Their work was aimed at optimizing the process to eliminate product defects.
Deshmukh et al. [5] employed the simulation to analyze the replication accuracy of
micro-structured components by optimizing HE processing parameters. Worgull et al.
[6] employed Moldflow to simulate mold filling and applied ANSYS to modeling
demolding to reduce defect in the replicated micro-structured components obtained
by the HE process.

A large number of experimental tests have been carried out to analyze the behavior
of amorphous thermoplastic polymers and to identify their properties within the
forming range applied. Cheng et al. [7] employed thermo-mechanical compression tests to identify by inverse strategy the properties of polymers with various
physical constitutive behavior models. The behavior law described by the two-
layer viscoplastic (TLVP) model dedicated to viscoelastic behavior is composed
of elastoplastic and viscoelastic branches and was employed for the forming simulation process by Charkaluk et al. [8]. Abdel-Wahab et al. [9] investigated the plastic,
elastic, and viscous properties with the same TLVP model. The PMMA material
parameters were identified by an inverse method based on various thermo-mechanical
experiment tests.

In this paper, the HE process for producing microchannels in PMMA and PC was
investigated numerically with the TLVP model. Thermo-mechanical compression
tests were carried out to identify material behavior parameters for the selected polymers. HE simulation was carried out to determine the influence of material parameters
on mold filling efficiency.

2   Description of HE Process and TLVP Model

2.1   Description of HE Process

HE process is one of the polymer replication processes for elaborating the microstructures for the production runs of microcomponents in small and medium series [10].
Figure 1 shows the steps of the HE process to elaborate micro-scale components

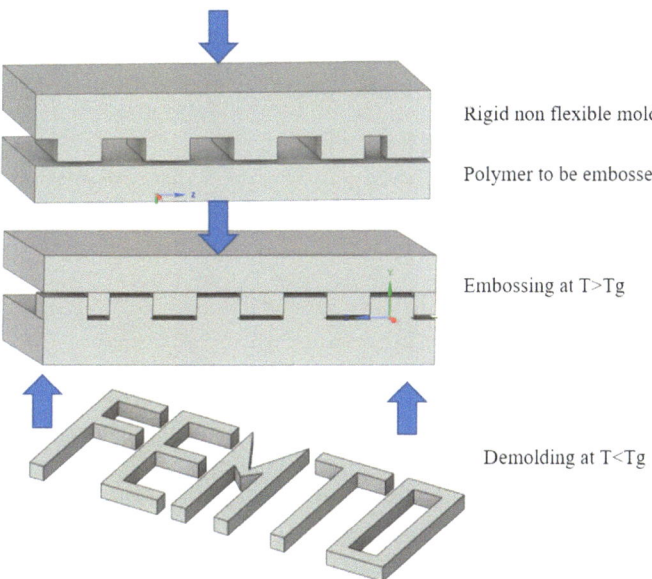

Rigid non flexible mold

Polymer to be embossed

Embossing at T>Tg

Demolding at T<Tg

Fig. 1 Schematic picture of the HE technique

(molding, cooling, demolding). The process involves pressing a polymer plate into a structural mold, the tooling of which is regulated to a temperature higher than T_g of the material used. Holding time and pressure were required to allow the polymer to fill all the structures in the mold. Once the mold and plate have been cooled, the replicated structural plate is removed from the mold, as it has sufficient rigidity to be recovered [11].

The main advantage of HE is the possibility to manufacture devices with various patterns thanks to its easy operation, high accuracy, mass production, short cycle time, and cost-effective. The HE is employed to produce different geometries in microscales such as circular patterns; threadlike lines and hexagonal shapes [12]. HE process can be used to replicate the geometries from simple (triangular, rectangular with shape repetition...) to complex shapes (microfluidic, MEMS...).

2.2 Description of TLVP Model

It was used to describe the behaviors of elastic viscoplastic in the range of forming temperatures used in the embossing process. It considers the elastic, plastic, and viscous deformations of the polymeric material, as shown in Fig. 2. It has already been implemented in the Abaqus® finite element software. All the parameters have been identified by the inverse method based on the experimental database of thermocompression tests, to provide the behavior of PMMA and PC.

Fig. 2 One-dimensional idealization of the TLVP model [13]

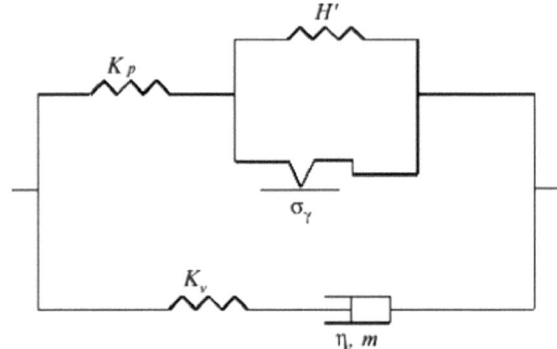

Fig. 2 One-dimensional idealization of the TLVP model [13]

The total modulus is calculated as $K = Kp + Kv$, where K_p and K_v are the elastic modulus of the elastoplastic and the elastic modulus of the viscoelastic, respectively. The proportion of elasticity in the viscoelastic relative to total elasticity f is expressed by the following relationship:

$$f = \frac{K - K_p}{K} \tag{1}$$

The total stress σ is obtained by the addition of the viscous stress σ_v and stress σ_p expressed by the following relationship:

$$\sigma = \sigma_p + \sigma_v \tag{2}$$

The elastic strain ε^{el} is divided into a viscoelastic part ε_v^{el} and elastoplastic part ε_p^{el}:

$$\varepsilon^{el} = f\varepsilon_v^{el} + (1 - f)\varepsilon_p^{el} \tag{3}$$

The total strain ε contains the elastic ε^{el}, plastic strains ε^{pl} and the viscous strain ε^v which is expressed by the following relationship:

$$\varepsilon = \varepsilon^{el} + f\varepsilon^v + (1 - f)\varepsilon^{pl} \tag{4}$$

3 Results of Experimental Characterization Tests

Uniaxial thermo-mechanical compression tests were carried out to identify the material parameters of the viscoplastic constitutive behavior by using the inverse method and the method of least squares. The plastic parameters of the polymers are identified

Table 1 Set of values for the parameters identified for the two materials studied

	Tg + 20 °C		Tg + 30 °C	
	σ^p (MPa)	ε^p	σ^p (MPa)	ε^p
PMMA	0	0	0	0
	0.51	0.01	0.10	0.01
	0.90	0.04	0.30	0.05
	1.49	0.11	0.5	0.11
PC	0	0	0	0
	0.39	0.40	0.14	0.40
	0.62	0.50	0.21	0.50
	0.90	0.56		

Table 2 TLVP model parameters at Tg + 20 °C

Parameters	PMMA	PC
K_p (MPa)	3.61	2.17
K (MPa)	33.38	31.73
f	0.89	0.93
A (Pa)	6.63×10^{-6}	1.71×10^{-4}
n	0.88	0.70
m	0	0

from stress–strain curves at two testing temperatures for PMMA and PC polymers. All the obtained plastic parameters are summarized in Table 1.

The value of the parameter f is determined using Eq. (1). The parameters identified for the materials studied in the elastic–viscoplastic model are summarized in Table 2.

4 Results of Numerical Simulation

The aim of the study is to optimize the efficiency of cavity filling during the HE process. Two-dimensional axisymmetric geometric representing the compression of the polymer plate in the structural mold is employed in the simulation of forming process, see Fig. 3. Elastic, elastoplastic, and elastic–viscoplastic behaviors are considered with the TVLP model in the simulation. It was carried out with different parameters identified in Tables 1 and 2 at two temperatures with different imposed displacements (0.07, 0.12, and 0.17 mm). Numerical simulation was carried out for both polymers (PMMA, PC) in order to determine the von Mises stresses as well as to study the influences of the material properties on the mold filling rate.

Fig. 3 Description of the 2D
model studied

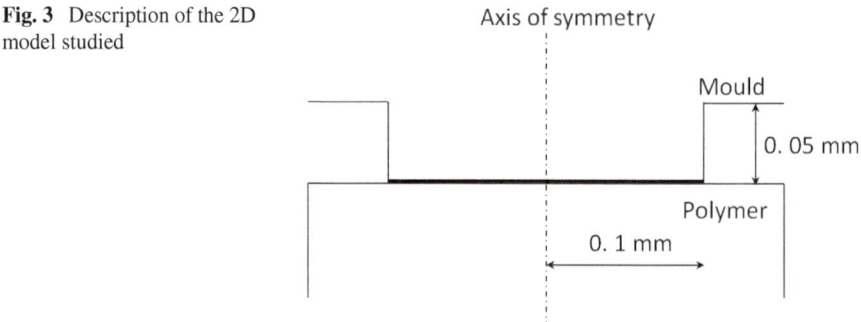

4.1   Results and Discussions of Mold Filling Ratio for PC and PMMA

The simulation results for PC of the HE process at Tg + 20 °C in terms of mold filling rate are shown in Fig. 4. The filling rate increases with the imposed displacement of the mold. The mold is 95% filled with the elastic–viscoplastic behavior applied on the polymer substrate. This indicates that viscous behavior plays a very important role in filling the micro-cavity during shaping.

Filling of the mold with the combination of PC's elastic, plastic, and viscous properties at Tg + 20 °C is studied by numerical simulation during the HE process, see Fig. 5. The value of von Mises stress is homogeneous with different imposed displacements. As a conclusion, the cavity-filling ratio increased with the imposed displacement. When the maximal imposed displacement of 0.17 mm is performed, the simulation result shows that the micro-cavity was almost completely filled. Based on the filling simulation results obtained at the same imposed displacement (0.17 mm),

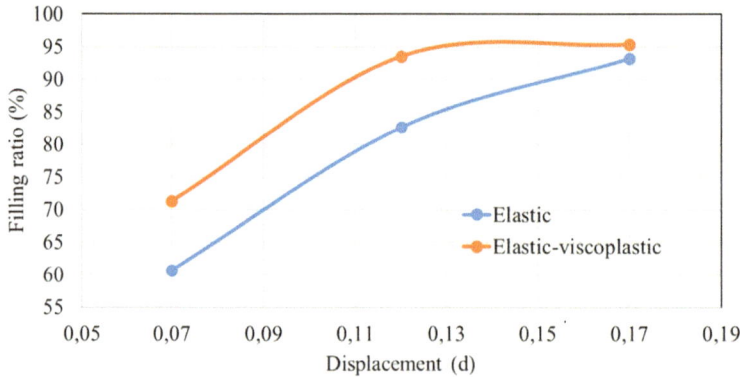

Fig. 4 Evolution of cavity filling ratio during simulation as a function of imposed displacements and constitutive behaviors for PC

the case of elastic–viscoplastic behavior generates the higher filling ratio than that of elastic behavior.

This result demonstrates that the micro-cavity was almost completely filled when the elastic, plastic, and viscous behaviors were considered with the TVLP model during the simulation of the HE process in the micro-scale.

The simulation results for PMMA with the elastic and elastic–viscoplastic behaviors of the HE process at Tg + 20 °C in terms of mold filling rate are shown in Table 3. The cavity filling ratio increased with the imposed displacement and the elastic–viscoplastic behavior generated an increase in the filling ratio compared to the elastic proprieties.

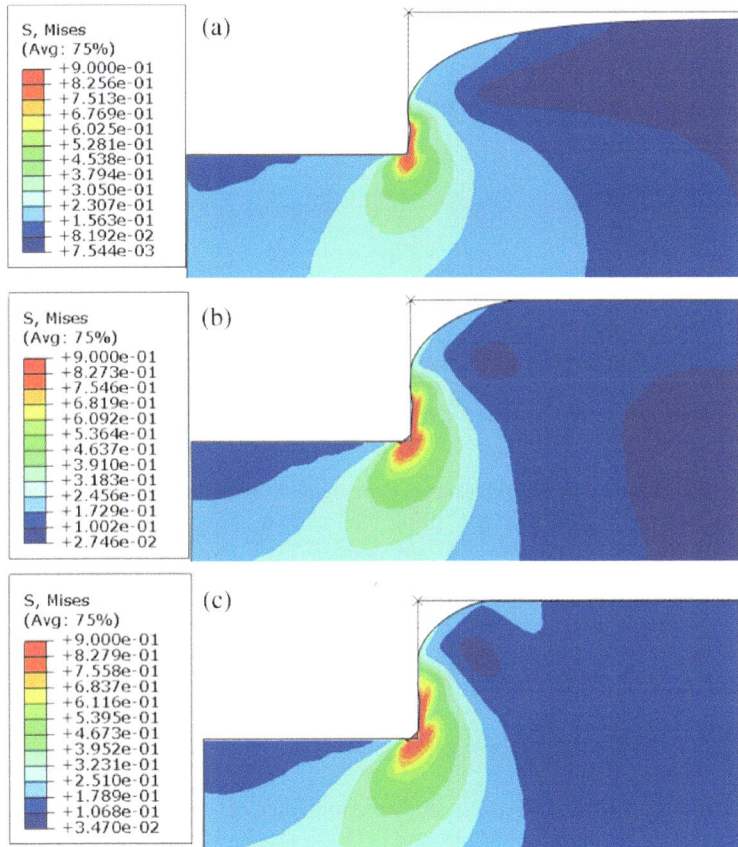

Fig. 5 Von Mises stress for PC during HE process with TLVP model at Tg + 20 °C versus imposed displacements: **a** 0.07, **b** 0.12, and **c** 0.17

Table 3 Evolution of cavity filling ratio during simulation with different imposed displacements and constitutive behavior laws for PMMA

Displacement (mm)	Elastic (%)	Elastic–viscoplastic (%)
0.07	65.00	98.50
0.12	89.60	99.00
0.17	93.30	99.36

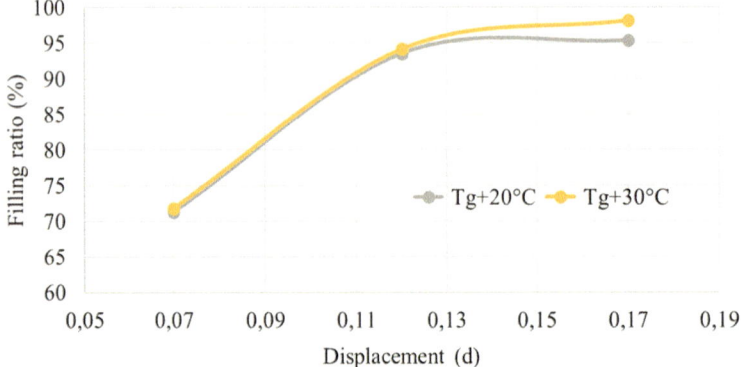

Fig. 6 Effect of temperature on cavity filling ratio versus different imposed displacements for PC with elastic–viscoplastic behavior

4.2 *Effect of Temperature on Mold Filling Ratio for PC*

The cavity filling ratio of PC substrate is investigated with various embossing temperatures to analyze its effect on the filling rate, shown in Fig. 6. It shows a comparison of filling rate as a function of temperature and imposed displacement, taking into account elastic–viscoplastic properties. The filling ratio increases slightly with temperature during the shaping of micro-structured component.

5 Conclusions

The aim of this current work was to study the influence of behavior laws on the filling efficiency for the shaping of PMMA and PC by using the HE process at micro-scale. The TLVP model was employed to model the elasto-viscoplastic property.

The elastic, plastic, and viscous parameters were identified using the inverse method, based on the evolution of the true stress–strain curves obtained from the results of uniaxial thermal mechanical compression at different temperatures.

A filling rate of 99% was obtained taking into account the elasto-viscoplastic property, which means that the viscous property of PMMA influences the filling rate. For the PC, the polymer substrate is better filled with higher temperatures in the HE process.

References

1. Cheng G, Sahli M, Gelin JC, Barriere T (2016) Physical modelling, numerical simulation and experimental investigation of microfluidic devices with amorphous thermoplastic polymers using a hot embossing process. J Mater Process Technol 229:36–53. https://doi.org/10.1016/j.jmatprotec.2015.08.027
2. Cheng G, Sahli M, Gelin JC, Barrière T (2014) Process parameter effects on dimensional accuracy of a hot embossing process for polymer-based micro-fluidic device manufacturing. Int J Adv Manuf Technol 75:225–235. https://doi.org/10.1007/s00170-014-6135-6
3. Wang J, Yi P, Deng Y, Peng L, Lai X, Ni J (2017) Recovery behavior of thermoplastic polymers in micro hot embossing process. J Mater Process Technol 243:205–216. https://doi.org/10.1016/j.jmatprotec.2016.12.024
4. Kasztelanic R, Cimek J, Kujawa I, Golebiewski P, Filipkowski A, Stepien R, Sobczak G, Krzyzak K, Pierscinski K, Buczynski R (2023) Mid-infrared ZBLAN glass optical components made by hot embossing technique. Opt Laser Technol 157:108655. https://doi.org/10.1016/j.optlastec.2022.108655
5. Deshmukh SS, Kar T, Som S, Goswami A (2022) Investigation of replication accuracy of embossed micro-channel through hot embossing using laser patterned copper mold 60:2222–2229. https://doi.org/10.1016/j.matpr.2022.03.128.
6. Worgull M, Heckele M (2004) New aspects of simulation in hot embossing. Microsyst Technol 10:432–437. https://doi.org/10.1007/s00542-004-0418-z
7. Cheng G, Sahli M, Gelin JC, Barrière T (2016) Physical modelling, numerical simulation and experimental investigation of microfluidic devices with amorphous thermoplastic polymers using a hot embossing process. J Mater Process Technol 229:36–53. https://doi.org/10.1016/j.jmatprotec.2015.08.027
8. Charkaluk E, Bignonnet A, Constantinescu A, Dang Van K ()2002 Fatigue design of structures under thermomechanical loadings. Fatigue Fract Eng Mater Struct 25:1199–1206. https://doi.org/10.1046/j.1460-2695.2002.00612.x
9. Abdel-Wahab AA, Ataya S, Silberschmidt VV (2017) Temperature-dependent mechanical behaviour of PMMA: experimental analysis and modelling. Polym Test 58:86–95. https://doi.org/10.1016/j.polymertesting.2016.12.016
10. Worgull M, Hétu JF, Kabanemi KK, Heckele M (2006) Modeling and optimization of the hot embossing process for micro- and nanocomponent fabrication. Microsyst Technol 12:947–952. https://doi.org/10.1007/s00542-006-0124-0
11. Ng SH, Wang ZF, Tjeung RT, de Rooij NF (2007) Development of a multi-layer microelectrofluidic platform. Microsyst Technol 13:1509–1515. https://doi.org/10.1007/s00542-006-0341-6.
12. Jaishree, Bhandari A, Khatri N, Mishra YK, Goyat MS (2023) Superhydrophobic coatings by the hot embossing approach: recent developments and state-of-art applications. Mater Today Chem 30:101553. https://doi.org/10.1016/j.mtchem.2023.101553
13. Rabhi F, Cheng G, Barriere T, Hocine NA (2020) Influence of elastic-viscoplastic behaviour on the filling efficiency of amorphous thermoplastic polymer during the micro hot embossing process. J Manuf Process 59:487–499. https://doi.org/10.1016/j.jmapro.2020.09.032

Effect of Welding Processes on Microstructure and Mechanical Properties of Laser-Welded Al-Si-Coated 22MnB5 Steel

Qi He, Zeran Hou, Jie Kong, Lulu Deng, Xiaolong Ma, and Junying Min

Abstract Tailor welded blank (TWB) has been developed in the automotive industry to improve structural safety and material efficiency. Al-Si-coated 22MnB5 steel is the main material for laser tailor welding due to its resistance to oxidation and decarburization during hot stamping, which can otherwise deteriorate the performance of welded joints. This study investigates the use of partial ablation welding (PAW) and filler wire welding (FWW) as compared to traditional self-fusion welded (SFW) to enhance the mechanical performances of Al-Si-coated 22MnB5 steel welded joints. The effects of welding processes on the microstructure of welded joints as well as the mechanism of performance enhancement were investigated. Tensile testing results showed that PAW and FWW joints consistently fractured at the base material, whereas the SFW joints fractured at the weld seam. The maximum tensile loads of PAW and FWW joints were, respectively, 12.75 and 12.93% higher than that of SFW joints. Additionally, the microhardness distribution of PAW and FWW joints is uniform with no softening zone. Finally, the study discusses the improved mechanical properties of PAW and FWW joints in terms of microstructural evolution and diffusion of Al elements.

Keywords 22MnB5 steel · Laser welding · Microstructure · Al-Si coating · Mechanical properties

Q. He · Z. Hou · J. Kong · X. Ma · J. Min (✉)
School of Mechanical Engineering, Tongji University, Shanghai 201804, China
e-mail: junying.min@tongji.edu.cn

L. Deng
Pan Asia Technical Automotive Center Company Limited, Shanghai 201201, China

© The Rightsholder, under exclusive licence to [Springer Nature Switzerland AG], part of Springer Nature 2024
J. Kusiak et al. (eds.), *Numerical Methods in Industrial Forming Processes*, Lecture Notes in Mechanical Engineering, https://doi.org/10.1007/978-3-031-58006-2_20

1   Introduction

Ultra-high strength steel (UHSS) is increasingly used in automotive manufacturing for reducing vehicle weight and improving safety and crashworthiness [1–3]. Press hardened steel (PHS), as a type of UHSS, is typically used for automotive crash-resistant structural parts such as A-pillars, B-pillars, and roof rails [4, 5]. The most widely used steel grade in PHS is 22MnB5, which has a martensitic microstructure after hot stamping and a tensile strength of approximately 1500 MPa [1]. Tailor welded blanks (TWBs) are used to produce parts with the required geometry, shape, and mechanical properties while reducing weight and improving crash behavior due to the absence of reinforcing blanks and fewer joining elements necessary [6, 7]. For the TWBs of PHSs, the most common method of joining sheets with varying thicknesses or different materials is laser welding to create a laser welded blank (LWB), which is then subjected to hot stamping [8].

To avoid surface oxidation as well as decarburization of sheet metal during hot stamping, the surface is typically pre-coated with an additional protective layer [1, 9]. Windmann et al. [10] discovered that the Al-Si layer has high-temperature oxidation resistance and excellent corrosion resistance. At present, Al-Si coating is commonly used as a protective coating for PHSs. However, many study results showed that the Al-Si layer deteriorates the final mechanical performances of the joints. Sun et al. [11] revealed that Al-10 wt.% Si coating led to approximately 15% ferrite and 85% martensite generated in the weld seam of the coated joints, whereas martensite is predominantly formed in the weld seam of the uncoated joints. Saha et al. [12] examined the microstructures of FZ at different conditions of pre- and post-press hardening. Results showed that the pre-press-hardened FZ microstructures mainly consisted of δ-ferrite, martensite, and a minimal quantity of lower bainite. On the other hand, the post-press-hardened FZ microstructures were primarily composed of martensite and α-ferrite. Chen et al. [13] investigated the microstructure characteristic of laser-welded Al-Si-coated boron steel joints. They discovered that the Al-Si coating fused during the welding process, forming a Fe-Al phase, which was determined to be a solid solution of α-Fe and Al. This study discovered that the Al-Si layer had a substantial effect on the microstructures and mechanical performances of the welded joints. Vierstraete et al. [14] indicated that laser ablation before welding is a useful approach for forming Al-Si-coated tailor-welded blanks. The approach has been patented by ArcelorMittal. Lin et al. [8] found that the use of filled wire during the welding process can dilute the content of aluminum in the weld, thereby reducing the amount of δ-ferrite phase.

Therefore, partial ablation welding (PAW) and filler wire welding (FWW), in comparison with the traditional self-fusion welded (SFW), were used to enhance the mechanical performances of Al-Si-coated 22MnB5 steel welded joints in this work. The mechanical properties of the PAW, FWW, and SFW joints were compared. Meanwhile, the Al content of the various phases formed in the weld was investigated. Based on these analyses, the influences of welding processes on the microstructure

Table 1 Chemical compositions of 22MnB5 steel (wt.%)

Element	C	Mn	Si	Cr	Ti	Ni	Fe
Wt.%	0.23	1.27	0.27	0.41	0.058	0.027	Bal

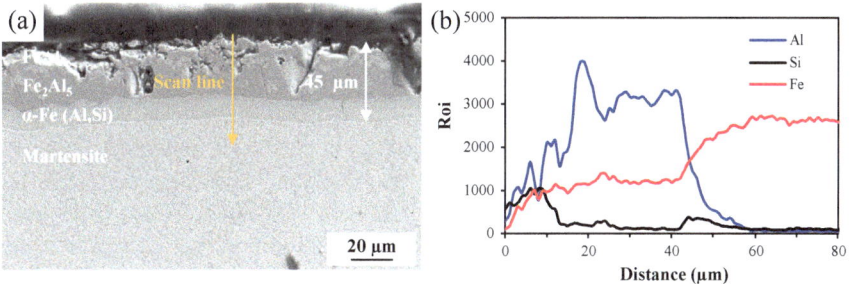

Fig. 1 Microstructure of the Al-Si coating: **a** microstructure, **b** EDS line scanning result across the Al-Si coating

evolution of welded joints as well as the mechanism of performance enhancement were investigated.

2 Experimental Procedure

2.1 Materials

Laser-welded Al-Si-coated 22MnB5 steel sheets with a full martensite microstructure were used in this work. The chemical composition (wt.%) of the base material (BM) is shown in Table 1. The ultimate tensile strength of BM was 1446 MPa, and the elongation was about 6.7%. Figure 1 shows the microstructure and the energy-dispersive X-ray spectroscopy (EDS) line scanning result across the Al-Si layer. The overall thickness of the Al-Si layer was approximately 45 μm and consisted of FeAl, Fe_2Al_5, and α-Fe layers. As shown in Fig. 1b, two platforms were distinguished by the element variation and were determined as FeAl and Fe_2Al_5 phases. The crossing position represented the elemental change between α-Fe and BM.

2.2 Laser Welding Processes

Compared with traditional self-fusion welding (SFW), partial ablation welding (PAW) and filler wire welding (FWW) were used to enhance the mechanical performances of welded joints of Al-Si-coated 22MnB5 steel. The process parameters of

Table 2 Process parameters of three laser welding processes

Laser welding process	Laser power (kW)	Welding speed (m/min)	Welding gap (mm)	Defocus distance (mm)	Feeding speed (m/min)
PAW	3	6.5	0	–	–
FWW	4	2.8	0.35	3	3.2
SFW	4	2.8	0	3	–

the three laser welding processes are listed in Table 2. The laser power and welding speed applied in the PAW process were 3 kW and 6.5 m/min, respectively. 20 μm of Al-Si layer on one side was removed by laser ablation before laser welding. The experimental laser power and welding speed for the FWW and SFW processes were 4 kW and 2.8 m/min, respectively, and the defocus distance was 3 mm above the upper surface of the specimen.

2.3 Mechanical and Microstructural Characterization

Uniaxial tensile tests were conducted at a tensile speed of 3 mm/min using a universal testing machine (MTS E45.105-ATBC). The full-field strain distribution on the specimen's surface was recorded using digital image correlation (DIC) technology. Three specimens were tested for each laser welding process to ensure the repeatability of the experiment. Figure 2 depicts the geometry of a tensile test sample in accordance with the ASTM: E8/E8M standard. Vickers microhardness was determined using a force of 200 g and a dwell duration of 10 s.

Metallography specimens were cut by wire cutting in the direction perpendicular to the weld. The microstructure of the welded joint was observed by optical microscope (OM) and scanning electron microscopy (SEM) after the specimens were inlaid, ground, polished, and etched. The content of aluminum was determined by EDS.

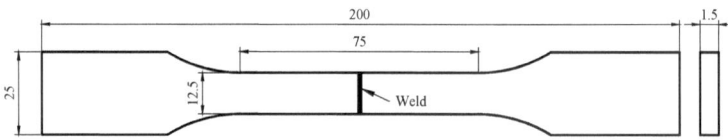

Fig. 2 Schematic diagram of tensile specimen (unit: mm)

3 Results and Discussion

3.1 Tensile Properties

Figure 3 presents the engineering stress–strain curves of the BM and PAW, FWW, and SFW joints. The corresponding mechanical properties are shown in Table 3. The PAW and FWW joints fractured at the base metal, while the SFW joints fractured at the fusion zone. The ultimate tensile strength (UTS) of the PAW and FWW joints was 1427 MPa and 1431 MPa respectively, slightly lower than the BM (1442 MPa). The total elongation of the PAW and FWW joints was also slightly lower than that of the BM. However, the UTS and elongation of SFW joints reduced to 1276 MPa and 1.4%. The UTS of PAW and FWW joints was 12.75% and 12.93% higher than that of SFW joints, respectively, and the elongation was 77.8% and 76.7% higher. Therefore, compared with the SFW process, PAW and FWW processes can significantly increase the tensile strength and elongation of welded joints.

The local ductility of a material is determined by its resistance to crack initiation and propagation, which is responsible for bending, stretch flanging, and crash resistance. In order to determine local ductility, fracture surfaces need to be evaluated [15, 16]. Figure 4 shows the fracture locations and fracture macro-morphology of PAW, FWW, and SFW joints. The three welded joints did not show any significant necking near their failure locations. As shown in Fig. 5, the width and thickness of

Fig. 3 Engineering stress–strain curves of BM and PAW, FWW, and SFW joints

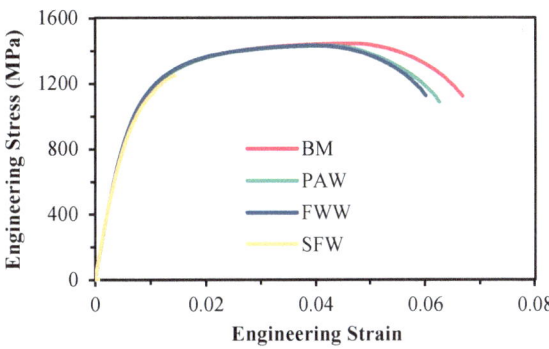

Table 3 Mechanical properties of BM and PAW, FWW, and SFW joints

	Yield strength (MPa)	Ultimate tensile strength (MPa)	Elongation (%)	Fracture location
BM	1035	1442	6.7	BM
PAW	1021	1427	6.3	BM
FWW	1040	1431	6.0	BM
SFW	974	1245	1.4	FZ

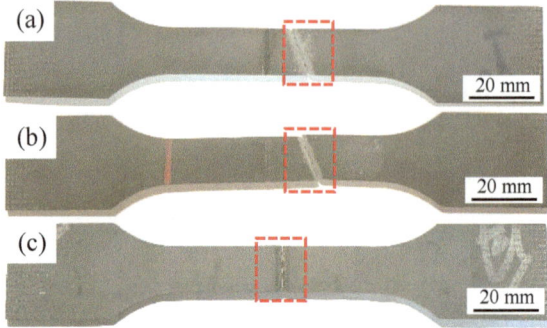

Fig. 4 Failure macro-morphology in different tensile specimens: **a** PAW, **b** FWW, **c** SFW

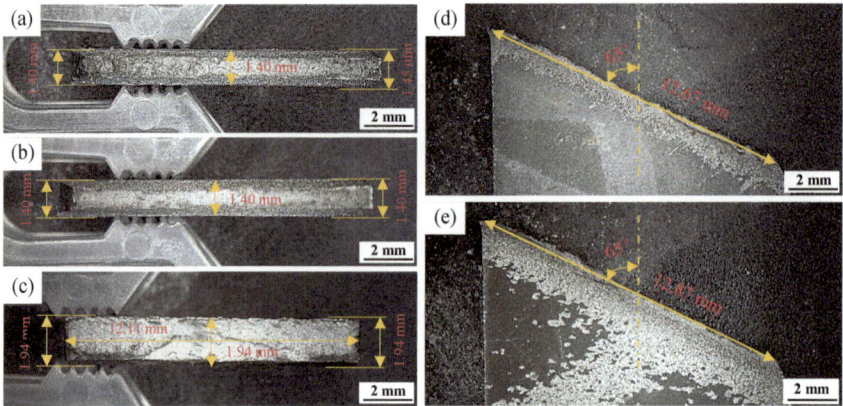

Fig. 5 Measurement of width and thickness of fracture surfaces: **a, d** PAW, **b, e** FWW, **c** SFW

the fracture surface were obtained by an optical microscope equipped with a digital microscope system, and the fracture surface was perpendicular to the microscope's observation direction. ASTM E8 was used to measure the fracture surfaces' width and thickness. According to the corresponding equations, the reduction of area at fracture and local fracture strain of PAW, FWW, and SFW joints was about 0.50%, 0.61%, and 0.08%, respectively. SFW joints fractured at the FZ, and the reduction of area at fracture was significantly lower than that of PAW and FWW joints. All three joints showed less than 5% reduction in area at fracture, indicating brittle fractures.

3.2 Microhardness

The microhardness test indentation points were spaced at 0.2 mm, and 10 points were tested on the left and right sides of the weld seam. Figure 6 presents the microhardness

Fig. 6 Microhardness profile of PAW, FWW, and SFW joints

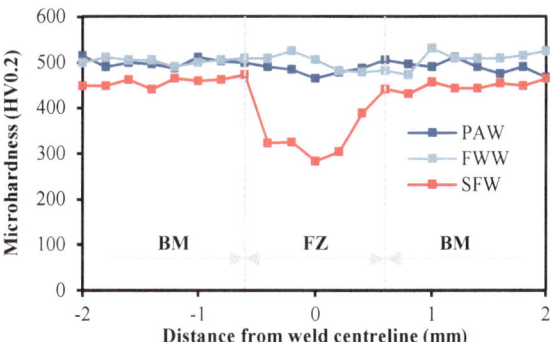

profile of PAW, FWW, and SFW joints. The microhardness distribution of PAW and FWW joints was uniform without a softening zone, and the average microhardness values were 491 and 503 HV, respectively. However, in the FZ of the SFW joint, the microhardness declined significantly due to the presence of ferrite structures. The average microhardness of the FZ of the SFW joint decreased to 308 HV, which was about 68% of the BM, corresponding to the phenomenon of the joint fracturing at the fusion zone.

3.3 Microstructure

Figure 7 presents the morphology of PAW, FWW, and SFW joint cross-sections, respectively. The welds were all hourglass-shaped without obvious defects, such as cracks and porosity. The FZ of PAW and FWW joints was uniform and had a less white phase. In contrast, a light-colored phase was produced in the SFW joint because the entire Al-Si layer was incorporated into the FZ from both sides of the weld. Figure 8 shows the microstructures of the upper and lower regions of the PAW, FWW, and SFW joints. As shown in Fig. 8a–d, the lath martensite microstructures were clearly demonstrated in the FZ of the PAW and FWW joints. In contrast, α-ferrite was generated in the upper and lower areas of the SFW joint, while martensite and a substantial amount of α-ferrite phase existed in the FZ, as shown in Fig. 8e, f. EDS spot scanning was conducted on five spots in each phase to determine the cause of α-ferrite formation and to calculate the average. As presented in Table 4, the content of Al element in α-ferrite (1.73–1.82 wt.%) was considerably higher compared to that in martensite (0.42–0.54 wt.%).

Saha et al. [17] revealed that ferrite was responsible for the decrease in tensile properties in welded joints. Al contents above 1.2 wt.% have been reported to prevent the transformation of ferrite to austenite, resulting in the stabilization and maintenance of the ferrite phase throughout the welding process [17, 18]. Dippenaar et al. [19] discovered that the presence of Al components reduces free energy change, accelerates ferrite phase transition kinetics, and suppresses austenite phase transition

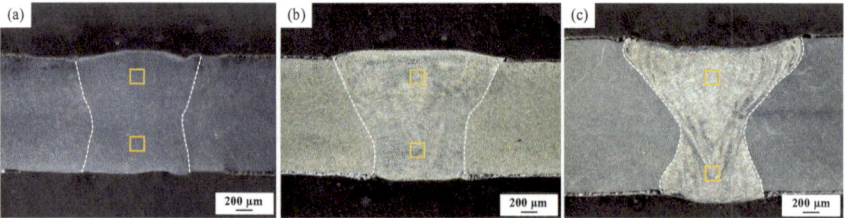

Fig. 7 Cross-sectioned morphology of the welded joints: **a** PAW, **b** FWW, **c** SFW

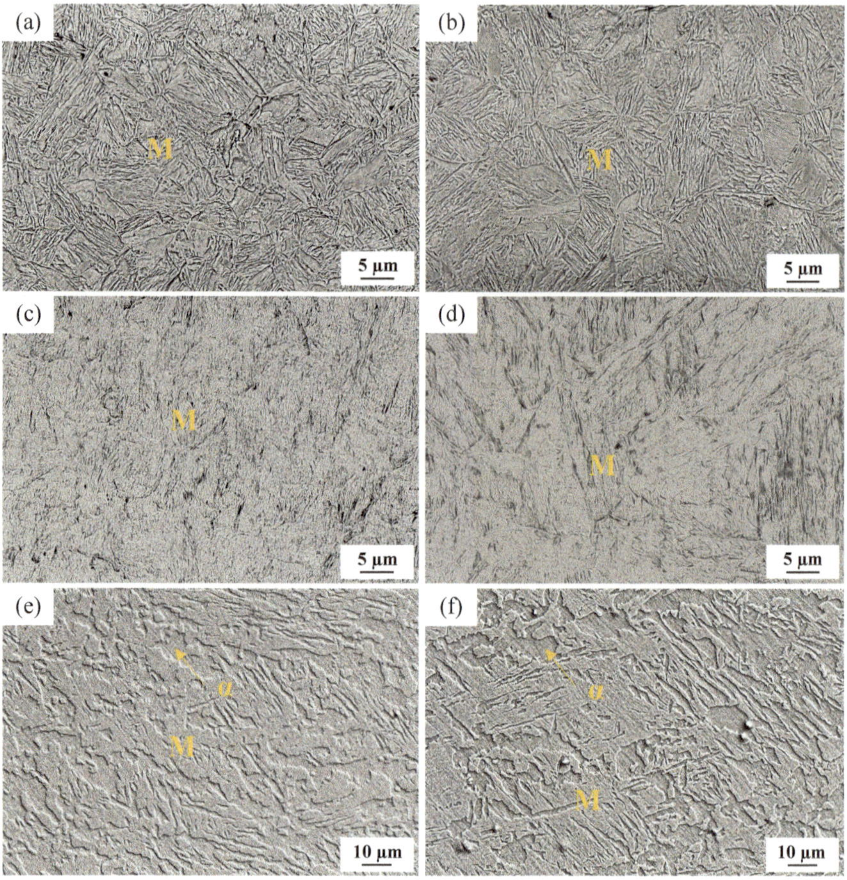

Fig. 8 Microstructure of the upper and lower regions of the FZ: **a, b** PAW, **c, d** FWW, **e, f** SFW

Table 4 EDS spot scanning results of the Al content (wt.%) for different phases in the FZ

Laser welding process	Microstructure	Al content (wt.%)
PAW	Martensitic	0.12
FWW	Martensitic	0.54
SFW	α-ferrite	1.62
	Martensitic	0.48

kinetics. The high Al element content is the primary reason for δ-ferrite formation, which recrystallizes to form α-ferrite after the hot stamping process [20]. Wang et al. [21] identified that the existence of δ-ferrite reduces the initial energy and propagation energy of cracks, which leads to easy ductile cracking in δ-ferrite and initiates brittle cracking in the tempered martensitic matrix.

For the SFW joints, the Al-Si layer melts and spreads into the weld seam during the laser welding process, causing the Al content in the weld to increase. The strong stabilizing effect of Al elements on ferrite prevents the transformation of the primary δ-ferrite phase [11, 22], thus producing α-ferrite phases in the SFW joints after hot stamping. The presence of α-ferrite is where failure cracks can easily propagate, reducing the strength and hardness of the SFW joints.

For the PAW joints, the martensite has a low Al content of less than 0.20 wt.%, as shown in Table 4. The PAW process removed the 20 μm Al-Si layer on one side by laser ablation, and the remaining α-Fe (Al, Si) layer was melted into the PAW joints throughout the welding process. However, no ferrite was found in the PAW joints due to the whole martensite structure, which explains the high strength and hardness of the weld, as observed in the mechanical properties.

Furthermore, the Al content in the FWW joints was approximately 0.50 wt.%, as shown in Table 4. Lin et al. [8] revealed that the use of carbon steel filler wire during laser welding can reduce the probability of δ-ferrite generation. As a result, the initial cracks were difficult to interconnect to form long cracks. Thus, the addition of carbon steel filler wire improved the strength and elongation of the FWW joints.

4 Conclusions

In this study, the effect of different welding processes on the mechanical properties, microstructural evolution, and Al content of laser-welded Al-Si-coated 22MnB5 steel was studied. The welded joints were produced using three different welding processes, namely, PAW, FWW, and SFW, and their mechanical properties were evaluated. The main conclusions of this study were summarized as follows:

1. The results presented that the SFW joint had significantly lower strength and elongation compared to the PAW and FWW joints. Moreover, a brittle fracture occurred at the weld seam of the SFW joint. The ultimate tensile strength of the SFW joint decreased from 1442 to 1245 MPa, and the elongation decreased from

6.7 to 1.4%. In contrast, the ultimate tensile strength of PAW and FWW joints showed an improvement of approximately 12%. The fracture locations of the PAW and FWW joints were consistently at the BM.

2. The formation of α-ferrite in the weld caused softening of the SFW joints. The average microhardness of the SFW joints decreased to 308 HV, which was about 68% of the BM. No softening occurred in the FZ of the PAW and FWW joints, and the microhardness distribution was uniform. The average microhardness of PAW and FWW joints was 491 and 503 HV, respectively.

3. The PAW process removes one side of the Al-Si layer by laser ablation, which can effectively control the content of the Al element. The FWW process dilutes the Al content and inhibits the formation of α-ferrite in the weld seam through the use of carbon-steel filler wire. Al-Si coating provided the Al element during the SFW process, leading to a large amount of α-ferrite produced in the weld. In comparison, the predominant martensite was formed in the PAW and FWW joints.

References

1. Karbasian H, Tekkaya AE (2010) A review on hot stamping. J Mater Process Technol 210(15):2103–2118
2. Kim HY, Park JK, Lee M-G (2013) Phase transformation-based finite element modeling to predict strength and deformation of press-hardened tubular automotive part. Int J Adv Manuf Technol 70(9–12):1787–1801
3. Golem L, Cho L, Speer JG et al (2019) Influence of austenitizing parameters on microstructure and mechanical properties of Al-Si coated press hardened steel. Mater Des 172
4. Wang J, Enloe C, Singh J et al (2016) Effect of prior austenite grain size on impact toughness of press hardened steel. SAE Int J Mater Manuf 9(2):488–493
5. Merklein M, Wieland M, Lechner M et al (2016) Hot stamping of boron steel sheets with tailored properties: a review. J Mater Process Technol 228:11–24
6. Khan MS, Razmpoosh MH, Biro E et al (2020) A review on the laser welding of coated 22MnB5 press-hardened steel and its impact on the production of tailor-welded blanks. Sci Technol Weld Join 25(6):447–467
7. Merklein M, Johannes M, Lechner M et al (2014) A review on tailored blanks—production, applications and evaluation. J Mater Process Technol 214(2):151–164
8. Lin W, Li F, Hua X et al (2018) Effect of filler wire on laser welded blanks of Al-Si-coated 22MnB5 steel. J Mater Process Technol 259:195–205
9. Chen R, Zhang C, Lou M et al (2020) Effect of Al-Si coating on weldability of press-hardened steels. J Mater Eng Perform 29(1):626–636
10. Windmann M, Röttger A, Theisen W (2013) Phase formation at the interface between a boron alloyed steel substrate and an Al-rich coating. Surf Coat Technol 226:130–139
11. Sun Y, Wu L, Tan C et al (2019) Influence of Al-Si coating on microstructure and mechanical properties of fiber laser welded 22MnB5 steel. Opt Laser Technol 116:117–127
12. Saha DC, Biro E, Gerlich AP et al (2016) Fusion zone microstructure evolution of fiber laser welded press-hardened steels. Scripta Mater 121:18–22
13. Chen X, Lei Z, Chen Y et al (2019) Microstructure and mechanical properties of laser welded Al-Si coated hot-press-forming steel joints. Materials (Basel) 12(20)
14. Taylor T, Clough A (2018) Critical review of automotive hot-stamped sheet steel from an industrial perspective. Mater Sci Technol 34(7):809–861

15. Larour P, Freudenthaler J, Weissböck T (2017) Reduction of cross section area at fracture in tensile test: measurement and applications for flat sheet steels. J Phys 896(1):012073
16. Wagner L, Larour P (2018) Influence of specimen geometry on measures of local fracture strain obtained from uniaxial tensile tests of AHSS sheets. Mater Sci Eng 418(1):012074
17. Saha D, Biro E, Gerlich A et al (2016) Fiber laser welding of Al-Si coated press hardened steel. Weld J 95(95):147–156
18. Di HS, Sun Q, Nie XK et al (2017) Mirostructure and properties of laser welded joints of dual phase and press-hardened steels. Procedia Eng 207:1665–1670
19. Dippenaar RJ, Phelan DJ (2003) Delta-ferrite recovery structures in low-carbon steels. Metall and Mater Trans B 34:495–501
20. Xu W, Yang S, Tao W et al (2022) Effect of Al-Si coating removal state on microstructure and mechanical properties of laser welded 22MnB5 steel. J Mater Eng Perform
21. Wang P, Lu S, Xiao N et al (2010) Effect of delta ferrite on impact properties of low carbon 13Cr–4Ni martensitic stainless steel. Mater Sci Eng A 527(13–14):3210–3216
22. Wang X, Zhang Z, Hu Z et al (2020) Effect of Ni foil thickness on the microstructure of fusion zone during PHS laser welding. Opt Laser Technol 125:106014

Determining Critical Wall Angle in Micro-incremental Sheet Forming of SS316L Foils for Formability Assessment

Mainak Palⓘ**, Vandit Pandya, Chandrakant K. Nirala**ⓘ**, and Anupam Agrawal**ⓘ

Abstract Micro-incremental sheet forming (μISF) has advantages over existing micro-forming processes (due to its die-less nature of material deformation) and vast applications in sophisticated industries. In μISF, a flexible ultra-thin sheet (foil) is plastically deformed into a complex 3D geometrical shaped component. It is precisely governed by the user-defined toolpath of the forming tool on the surface of the foil. This study investigated the deformation behavior of 100-μm-thick SS316L foils; the foils were deformed into small conical shapes with a tool-tip with a radius of 500 μm. The μISF process set-up consisted of a rigid fixture to avoid any unwanted distortions during the forming. To test the formability of the foils at different working parameters, experimental and numerical examinations were conducted by varying the wall angle (α = 45°, 60°, 75°) of the forming with three different step depths (Δz = 10, 20, and 30 μm). A finite element analysis of the process was carried out using ABAQUS® software; the results revealed that the forming angle in μISF had a direct relationship with the formability of the material. Higher values of the step depth assisted in increasing the formability of the SS316L foils. To examine the critical wall angle (α_{cr}) and critical foil thickness (t_{cr}) at the fracture location, a higher range of α (70°–74°) was selected in order to establish a correlation with the depth of the forming.

Keywords μISF · Foils · SS316L · Formability · Critical wall angle · Critical foil thickness

M. Pal · V. Pandya · C. K. Nirala · A. Agrawal (✉)
Department of Mechanical Engineering, Indian Institute of Technology Ropar, Rupnagar, Punjab 140001, India
e-mail: anupam@iitrpr.ac.in

© The Rightsholder, under exclusive licence to [Springer Nature Switzerland AG], part of Springer Nature 2024
J. Kusiak et al. (eds.), *Numerical Methods in Industrial Forming Processes*, Lecture Notes in Mechanical Engineering, https://doi.org/10.1007/978-3-031-58006-2_21

1 Introduction

Manufacturing industries have recently started focusing on developing advanced micro-fabrication facilities in order to meet the rising demand for miniaturization and the fast-growing use of custom micro-products that are fabricated from ultra-thin sheets (foils). Scaling down conventional forming techniques for real-world applications is challenging due to the difficulty of reducing the size of the machinery and supporting equipment [1]. The major challenges/issues that have been encountered (understanding the micro-deformation behavior [size-effect], material characterization, product handling, forming limit qualification, process modeling, interfacial considerations, etc.) are the critical elements that determine the industrial acceptability of this process [2, 3]. Forming foils into complex 3D geometries requires great precision to serve high-end applications, like in micro-electronics, avionics, and the biomedical and defense industries [4]. Incremental sheet forming (ISF) technology was first patented in 1967 by Leszak [5]. In 1978, Mason [6] started a small-batch production of metallic components by forming metal sheets. Jeswiet et al. [7] proposed a detailed analysis of the fundamental aspects of the incremental forming process. The process mechanics of the ISF process were nicely defined by Jackson and Allwood [8] and Duflou et al. [9]. Singh and Agrawal [10, 11] studied the combination of structure thinning and ISF to generate complex shapes of thin monolithic structures. Nirala and Agrawal [12] developed a fractal geometry-based toolpath to produce incremental deformations in sheets to achieve the better uniformity of the thickness distribution in ISF.

Micro-incremental sheet forming (μISF) is a new die-less method for the production of miniature parts in which a forming tool navigates through the surface of a foil to deform it to a complex 3D part. Saotome and Okamoto [13] first implemented the μISF experiment to fabricate miniature components from thin Al foils. Obikawa et al. [14] developed a micro-milling machine to produce a collection of tiny dots and pyramids on thin sheets. Bansal et al. [15] produced geometrical profiles of different shapes using μISF for studying the formability of foils with different working parameters. Beltran et al. [16] also executed micro-forming tests on SS304 foils for different geometrical shapes. Later, Song et al. [17] carried out a numerical and experimental investigation to understand the micro-scale deformation mechanism of the μISF process.

In this paper, an attempt has been made to study the formability of SS316L foils at different forming angles (α) for developing conically shaped geometries. Within the selected range of α, the critical values of the wall angle and sheet thickness are calculated at the locations of fractures.

2　Work Methodology

2.1　Material and Experimental Details

A series of experiments were performed on a rectangular 30×25 mm SS316L foil with a thickness of 100 μm in order to validate the repeatability of the μISF process. A forming tool (Tungsten Carbide) using a hemispherical end with a radius of 500 μm was developed in-house using the μ-turning method. Reverse-μEDM is also an established technique for electrode dressing [18], which was studied by Pal et al. [19] for the production of a precise μISF tool. A micro-forming fixture was designed and fabricated to avoid any unwanted distortion of a workpiece at the primary stage of the process. The experiments were performed on a numerically controlled hybrid-μEDM machine (Mikrotools Pte Ltd., DT-110i) following a spiral incremental toolpath that was developed in GUI in MATLAB® R2021a (as shown in Fig. 1a, b). Initially, three different forming angles (α) of 45°, 60°, and 75° were chosen to measure the formability of SS foils at a step depth (Δz) of 30 μm. The tool was given a clockwise rotation of 500 rpm and a feed rate of 20 mm/min. Furthermore, critical wall angle (α_{cr}) of failure was calculated by considering a higher wall angle range (from 70° to 74°). To investigate the effect of Δz on the failure depth of the component, forming tests were conducted at three different levels of Δz (10, 20, and 30 μm) for the 75° wall angle. Based on the obtained results, finite element analysis (FEA) was carried out on ABAQUS® to validate this with the experimental outcome.

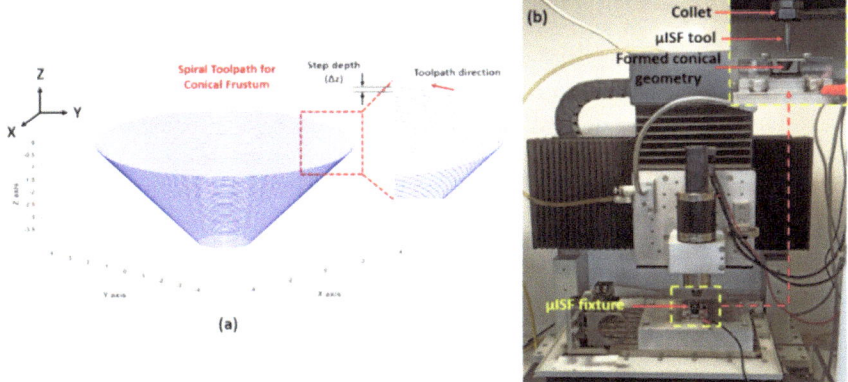

Fig. 1　**a** Incremental spiral toolpath for conically shaped geometry; **b** experimental set-up for μISF

2.2   Numerical Simulation

A 3D elastic–plastic isotropic hardening model based on the von Mises yielding criterion was used for the FEA of the μISF process using ABAQUS®/Explicit. The thickness of the foil (100 μm) was comparatively smaller than the workpiece dimensions and the tool-head radius (500 μm). The tool was considered to be an analytic rigid body, and the SS316L foil was taken to be deformable. For the meshing part, S4R shell elements with a size of 0.4 × 0.4 mm with five integration points along the thickness direction were used with reduced integration. The simulated kinematics of these shell elements provided high accuracy and a low CPU computational run time of 31–32 h. for the current study. The CPU was configured with a clock speed of 2.00 GHz and 128 GB of RAM. The mechanical properties of the material were calculated through uniaxial tensile test (Fig. 2a) and are tabulated in Table 1. The master–slave contact algorithm was used for the tool/foil surface interactions, with a friction coefficient (μ) of 0.34 [20]. Ductile damage fracture criteria based on the void growth model (VGM) were used for predicting the failure during the μISF of the SS316L. This model calculates the onset of the failure of the material based on the growth, nucleation, and coalescence of the voids during the deformation stage. Through this model, the initial fracture of the material is predicted if the damage accumulates up to a certain value. During the plastic deformation, material fragmentation occurs; this results in the growth of voids and nucleation, which cause fractures in the material due to the different plastic strains. The FE model assumes that the material is isotropic and homogenous. The damage initiation and evolution can be introduced through the approach of strain energy, stress triaxiality, fracture strains, etc. in the numerical simulation. Here, the model assumes that the equivalent plastic strain $\mathcal{E}_{pl}$ (1) at which the failure of the material occurs is the function of stress triaxiality η (2) and the equivalent plastic strain rate ($\acute{\varepsilon}_{pl}$) [21]. The stress triaxiality value that was incorporated in the simulation was calculated from the initial simulation runs. For the ductile materials, the damage-initiation criteria work when Situation (3) achieves a satisfactory response in which $\mathcal{E}_f$ is the equivalent plastic strain of the fracture and W_D represents the damage-initiation variable. $W_D \geq 0$ is taken for each increment during the finite element analysis.

$$\mathcal{E}_{pl} = \phi(\eta, \acute{\varepsilon}_{pl}) \tag{1}$$

$$\eta = \text{Hydrostatic stress } (p)/\text{von Mises stress } (q) \tag{2}$$

$$W_D = \int_0^{\mathcal{E}f} \frac{\mathrm{d\mathcal{E}pl}}{\phi(\eta,\acute{\varepsilon}pl)} = 1 \tag{3}$$

For calculating the value of the step time and step distance of the tool movement in the x, y, and z coordinates (toolpath amplitude), Eqs. (4) and (5) were used. The FEA of the deformed conical geometry is shown in Fig. 2b.

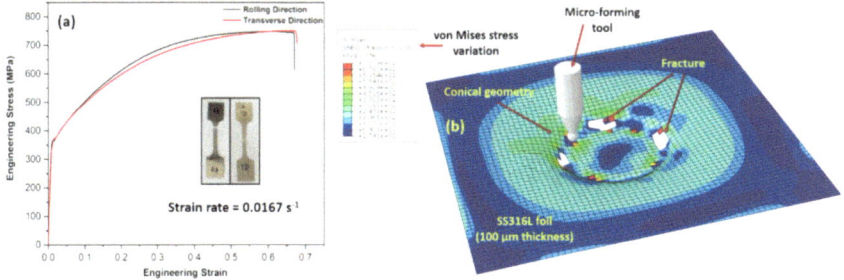

Fig. 2 **a** Engineering stress–strain curves in rolling direction (RD) and transverse direction (TD) for SS316L; **b** FEA result of deformed conical geometry

Table 1 Properties of SS316L foil used in numerical simulation	Mechanical properties	Magnitude
	Density	8.05 gm/cm^3
	Poisson's ratio	0.3
	Ultimate tensile strength (UTS)	754 MPa
	Modulus of elasticity	189 GPa
	Elongation	38%

$$\text{Step distance } (d) = \left[(x_2 - x_1)^2 + (y_2 - y_1)^2 + (z_2 - z_1)^2 \right]^{1/2} \qquad (4)$$

$$\text{Step time } (t) = \text{Step distance } (d) / \text{Tool velocity } (v) \qquad (5)$$

3 Results and Discussion

3.1 Formability Assessment of Formed Micro-components

To study the formability of the micro-components, the μISF experiment was initially carried out at three different forming angles (α) (45°, 60°, and 75°) at a constant step depth of 30 μm. The wall angle in μISF is a major parameter in determining the formability of the sheet metal [7]. The opening diameter of the conical geometry was taken as 10 mm. From the experimental results, it could be observed that the part that was formed with angles α of 45° and 60° achieved the target average depths of 4.8 and 8.32 mm, respectively, without any fractures. However, early failure of the part could be observed with an angle of 75° at an average depth of 1.74 mm (Fig. 3a). The higher wall angle resulted in high stress at the tool-foil interface due to the smaller contact area at the higher wall angles. This led to the excessive thinning of the sheet

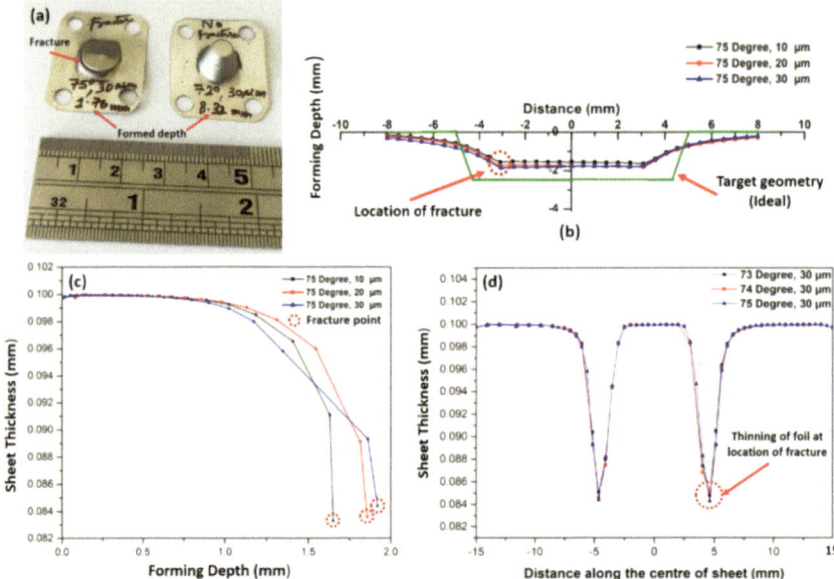

Fig. 3 **a** Formed components at wall angles of 72° and 75°; FEA results: **b** cross-sectional profile at different step depths for 75°; **c** sheet thickness variation with forming depth until fracture; **d** cross-sectional thickness distribution at constant step depth (30 μm)

as per the law of sines [7], which resulted in the premature failure of the micro-part. Furthermore, three different values of Δz were considered for investigating the effect of changing the step depth (Δz) on the forming limit of the part. The results revealed that a higher Δz assisted in increasing the formability of the component. The forming height increased from 1.32 to 1.74 mm with the increase in Δz. The FEA results also showed reasonable agreement with the experimental results as per the formability that was achieved in both cases. The μISF results of the experimental and simulation are shown in Table 2. Figure 3b, c display the simulation outcome of the cross-sectional profile and thickness distribution, respectively, for 75°. In comparison with the 10 and 20 μm levels of Δz, the greatest forming depth (1.92 mm) and lowest foil thickness reduction at the fracture point (84.2 μm) could be observed with the 30 μm step depth. For the 10 μm step depth, the fracture thickness of the foil was close to 83.1 μm (max. thinning) at the lower forming depth of 1.67 mm.

3.2 Determination of Critical Wall Angle (α_{cr}) and Critical Foil Thickness (t_{cr})

The results demonstrated that, at higher wall angles, the fracture of a component is more likely to occur; therefore, the value of α was varied from 70° to 74° in order to

Table 2 Experimental and simulation results of μISF process

Formed geometry	Wall angle (α)	Step depth (Δz) (μm)	Fracture occurred	Average formed depth (experimental) (mm)	Average formed depth (simulation)
Cone A	75°	10	Yes	1.32	1.67 mm
Cone B	75°	20	Yes	1.55	1.85 mm
'Cone C	75°	30	Yes	1.74	1.92 mm
Cone 1	70°	30	No	8.32	–
Cone 2	71°	30	No	8.32	–
Cone 3	72°	30	No	8.32	–
Cone 4	73°	30	Yes	1.86	2.04 mm
Cone 5	74°	30	Yes	1.81	1.97 mm

determine the critical wall angle (α_{cr}) and critical foil thickness (t_{cr}) at the location of the fracture. The experimental and FEA results are shown in Table 2. At an angle of 70°, no failure of the component was initially observed at a forming depth of 8.32 mm.

A similar trend in the formability was witnessed when α was increased to 71° and 72°. However, at an angle of 73°, a sudden failure in the formed geometry could be observed. With an increase in the wall angle from 73° to 75°, a reduction in the forming depth could be observed in both the experimental and simulation results. Therefore, the formability was limited to an α_{cr} of 72°; beyond this angle, the failure of the component took place. The thickness distribution graph in Fig. 3d shows the fracture location where the maximum thinning of the foil was observed. For wall angles of 73°, 74°, and 75°, the thicknesses of the sheets at the times of fractures could be observed as 84.75, 84.62, and 84.2 μm, respectively. Therefore, for designing the μISF process for SS316L foil, it can be concluded that when the foil thickness is reduced below the 85 μm limit (t_{cr}), the failure of the component takes place.

4 Conclusions

This paper presents a new approach for developing miniature-sized products that are made from ultra-thin SS316L foils. μISF experiments were carried out at three different forming angles at a constant step depth. The fracturing of the component could be observed in the high wall angle (α = 75°) as compared to the other two angles. A higher step depth (Δz = 30 μm) helped to increase the formability of the formed geometry. A similar trend could be observed in both the FEA and experimental results. A critical wall angle (α_{cr}) of 72° and critical foil thickness (t_{cr}) of 85 μm could be observed at the location of the fracture. Further work in the direction of the FE validation of sheet thickness distribution, stress variation, etc. with the experimental results will be focused on a detailed study of the process.

References

1. Geiger M, Kleiner M, Eckstein R, Tiesler N, Engel U (2001) Microforming. CIRP Ann Manuf Technol 50:445–462. https://doi.org/10.1016/S0007-8506(07)62991-6
2. Qin Y (2006) Micro-forming and miniature manufacturing systems—development needs and perspectives. J Mater Process Technol 177:8–18. https://doi.org/10.1016/j.jmatprotec.2006.03.212
3. Messner A, Engel U, Kals R, Vollertsen F (1994) Size effect in the FE-simulation of micro-forming processes. J Mater Process Tech 45:371–376. https://doi.org/10.1016/0924-0136(94)90368-9
4. Pradeep Raja C, Ramesh T (2021) Influence of size effects and its key issues during micro-forming and its associated processes—a review. Eng Sci Technol Int J 24:556–570. https://doi.org/10.1016/j.jestch.2020.08.007
5. Leszak E (196) Apparatus and process for incremental dieless forming. United States Patent Office, pp 1964–1966
6. Mason B (1978) Sheet metal forming for small batches. B.Sc thesis, University of Nottingham
7. Jeswiet J, Micari F, Hirt G, Bramley A, Duflou J, Allwood J (2005) Asymmetric single point incremental forming of sheet metal. CIRP Ann Manuf Technol 54:88–114. https://doi.org/10.1016/s0007-8506(07)60021-3
8. Jackson K, Allwood J (2009) The mechanics of incremental sheet forming. J Mater Process Technol 9:1158–1174. https://doi.org/10.1016/j.jmatprotec.2008.03.025
9. Duflou JR, Habraken AM, Cao J, Malhotra R, Bambach M, Adams D et al (2018) Single point incremental forming: state-of-the-art and prospects. Int J Mater Form 11:743–773. https://doi.org/10.1007/s12289-017-1387-y
10. Singh A, Agrawal A (2016) Investigations on structural thinning and compensation stratagem in deformation machining stretching mode. Manuf Lett 9:1–6. https://doi.org/10.1016/j.mfglet.2016.06.001
11. Singh A, Agrawal A (2017) Experimental and numerical investigations on structural thinning, thinning evolution and compensation stratagem in deformation machining stretching mode. J Manuf Process 26:216–225. https://doi.org/10.1016/j.jmapro.2017.02.013
12. Nirala HK, Agrawal A (2018) Fractal geometry rooted incremental toolpath for incremental sheet forming. J Manuf Sci Eng 140:1–9. https://doi.org/10.1115/1.4037237
13. Saotome Y, Okamoto T (2001) An in-situ incremental microforming system for three-dimensional shell structures of foil materials. J Mater Process Technol 113:636–640. https://doi.org/10.1016/S0924-0136(01)00651-3
14. Obikawa T, Satou S, Hakutani T (2009) Dieless incremental micro-forming of miniature shell objects of aluminum foils. Int J Mach Tools Manuf 49:906–915. https://doi.org/10.1016/j.ijmachtools.2009.07.001
15. Bansal A, Jiang B, Ni J (2019) Die-less fabrication of miniaturized parts through single point incremental micro-forming. J Manuf Process 43:20–25. https://doi.org/10.1016/j.jmapro.2019.03.046
16. Beltran M, Malhotra R, Nelson AJ, Bhattacharya A, Reddy NV, Cao J (2013) Experimental study of failure modes and scaling effects in micro-incremental forming. J Micro Nano-Manuf 1:1–15. https://doi.org/10.1115/1.4025098
17. Song X, Zhang J, Zhai W, Taureza M, Castagne S, Danno A (2018) Numerical and experimental investigation on the deformation mechanism of micro single point incremental forming process. J Manuf Process 36:248–254. https://doi.org/10.1016/j.jmapro.2018.10.035

18. Nirala CK, Saha P (2016) Evaluation of μEDM-drilling and μEDM-dressing performances based on online monitoring of discharge gap conditions. Int J Adv Manuf Technol 85:1995–2012. https://doi.org/10.1007/s00170-015-7934-0

19. Pal M, Kishore H, Agrawal A, Nirala CK (2022) Fabrication of precise hemispherical end tool for micro incremental sheet forming using reverse-μEDM. Procedia CIRP 107:1600–1605. https://doi.org/10.1016/j.procir.2022.06.001

20. Sridhar R, Shanmugasundaram D, Rajenthirakumar D (2023) Investigation on single-point incremental forming process of SS316 sheets at elevated temperatures. Lecture notes in mechanical engineering. Pp 85–96. https://doi.org/10.1007/978-981-19-3866-5_8

21. Dassault Simulia ABAQUS/CAE (2018) ABAQUS analysis user manual, version 6.8. 2022 damage initiation for ductile metals, p 2080

Numerical Analysis of Damage and Failure in Anisotropic Sheet Metals During Biaxial Loading

Michael Brünig, Sanjeev Koirala, and Steffen Gerke

Abstract In the paper, the influence of stress state and loading direction with respect to the principal axes of anisotropy on damage and fracture behavior of the anisotropic aluminum alloy EN AW-2017A is discussed. The focus is on numerical calculations on the micro-level considering void-containing representative volume elements revealing information on damage mechanisms. Using experimental data taken from uniaxial and biaxial tests, material parameters are identified. Based on numerical studies on the micro-scale with differently loaded void-containing cubes, it is shown that the stress state, the load ratio and the loading direction with respect to the principal axes of anisotropy have an influence on evolution of damage processes on the micro-scale and on the corresponding damage strains.

Keywords Ductile damage · Anisotropic metals · Micro-mechanical studies

1 Introduction

Numerical analysis of deformation and failure behavior of complex structures requires accurate modeling of inelastic behavior of materials. In this context, various constitutive theories and corresponding robust and efficient numerical techniques have been discussed during the last decades. For example, failure in ductile metals is mainly caused by nucleation, growth and coalescence of micro-defects leading to the evolution of macro-cracks. Therefore, a straight-forward way for the formulation of appropriate constitutive models should be based on the analysis of the behavior of individual micro-defects in elastic-plastic materials [5, 7]. The results of these calculations on the micro-scale can then be used to develop macroscopic phenomenological approaches which, for example, can be used to analyze the deformation and failure behavior of ductile metals during forming operations.

M. Brünig (✉) · S. Koirala · S. Gerke
Institut für Mechanik und Statik, Universität der Bundeswehr München, 85577 Neubiberg,
Germany
e-mail: michael.bruenig@unibw.de

J. Kusiak et al. (eds.), *Numerical Methods in Industrial Forming Processes*, Lecture
Notes in Mechanical Engineering, https://doi.org/10.1007/978-3-031-58006-2_22

To get insight into damage and failure mechanisms in ductile metals and to examine the behavior of micro-defects caused by various loading conditions, three-dimensional finite element simulations of microscopic cell models have been performed by different research groups; see, for example, [1, 5, 10, 11, 16–18]. These investigations based on the assumption of isotropic elastic-plastic material behavior showed that the current stress state remarkably affects the damage and failure processes on the micro-level as well as the corresponding macroscopic behavior. The numerical results of the unit cell calculations can be used to propose and validate damage evolution equations in phenomenological continuum models and to determine micro-mechanically motivated constitutive parameters [5, 7].

Manufacturing processes such as rolling, deep drawing or extrusion cause anisotropies in ductile metals resulting from internal changes in the crystallographic structure. These deformation-induced anisotropies must be taken into consideration in material models used to simulate the deformation and failure behavior of thin metal sheets. Different anisotropic yield criteria for hydrostatic-stress-independent material behavior have been presented in the literature based on quadratic [13, 19], non-quadratic [2, 12, 15] or spline functions [20]. In addition, the Hoffman yield condition [14] has been developed to take into account the strength-differential effect in anisotropic materials.

In the present paper, micro-mechanical numerical simulations with spherical void-containing representative volume elements are discussed. The plastic anisotropy of the investigated ductile metal is modeled by the Hoffman yield criterion. Different load combinations are taken into account with respect to the principal axes of anisotropy. Numerical results of these micro-mechanical calculations are used to reveal the effect of stress state, of load ratio and of loading direction on damage mechanisms as well as on corresponding damage strains.

2   Constitutive Model

The numerical analysis is based on the continuum damage model presented by [3–5] which has been enhanced for anisotropic plasticity by [6, 8, 9]. The framework uses the introduction of the damage strain tensor, $\mathbf{A}^{da}$, characterizing the formation of macroscopic strains caused by damage processes on the micro-scale. In addition, the kinematics take into account the additive decomposition of the strain rate tensor into elastic, $\dot{\mathbf{H}}^{el}$, effective plastic, $\overset{\ast}{\dot{\mathbf{H}}}{}^{pl}$, and damage parts, $\dot{\mathbf{H}}^{da}$ [3].

Anisotropic plastic behavior of the investigated aluminum alloy EN AW-2017A is modeled by the Hoffman yield condition [14]

$$f^{pl} = \mathbf{C} \cdot \bar{\mathbf{T}} + \sqrt{\frac{1}{2}\bar{\mathbf{T}} \cdot \mathcal{D}\,\bar{\mathbf{T}}} - c = 0, \tag{1}$$

where $\bar{\mathbf{T}}$ denotes the effective Kirchhoff stress tensor defined in the fictitious undamaged configuration and the tensor of coefficients

$$\mathbf{C} = C^i_{.j} \, \mathbf{g}_i \otimes \mathbf{g}^j = C_{(i)} \, \mathbf{g}_i \otimes \mathbf{g}^i \tag{2}$$

with the components (in Voigt notation)

$$\left[C^i_{.j} \right] = [C_1 \ C_2 \ C_3 \ 0 \ 0 \ 0]^T \tag{3}$$

has been used. In addition, further material parameters describing the plastic anisotropy are given by the tensor

$$\mathcal{D} = D^{i.k}_{.j.l} \, \mathbf{g}_i \otimes \mathbf{g}^j \otimes \mathbf{g}_k \otimes \mathbf{g}^l \tag{4}$$

with

$$\left[D^{i.k}_{.j.l} \right] = \begin{bmatrix} C_4 + C_5 & -C_4 & -C_5 & 0 & 0 & 0 \\ -C_4 & C_4 + C_6 & -C_6 & 0 & 0 & 0 \\ -C_5 & -C_6 & C_5 + C_6 & 0 & 0 & 0 \\ 0 & 0 & 0 & C_7 & 0 & 0 \\ 0 & 0 & 0 & 0 & C_8 & 0 \\ 0 & 0 & 0 & 0 & 0 & C_9 \end{bmatrix} \tag{5}$$

and

$$c = c_o + R_o \epsilon^{pl} + R_\infty \left(1 - e^{-b \epsilon^{pl}} \right) \tag{6}$$

represents the equivalent yield stress of the undamaged metal. For the investigated ductile anisotropic aluminum alloy EN AW-2017A the parameters are listed in Table 1.

They have been identified by a uniaxial tension test of a flat specimen cut in the rolling direction of the aluminum alloy sheet. The anisotropy parameters C_i in Eqs. (3) and (5) are determined considering the stress-strain behavior of uniaxially loaded specimens cut in different directions with respect to the rolling direction [9]. These parameters are listed in Table 2.

Based on the yield criterion (1) generalized invariants of the effective Kirchhoff stress tensor $\bar{\mathbf{T}}$ are defined [9]: the first Hoffman stress invariant is given by

$$\bar{I}^H_1 = \frac{1}{a} \mathbf{C} \cdot \bar{\mathbf{T}} \quad \text{with} \quad a = \frac{1}{3} \, \mathrm{tr} \mathbf{C} \tag{7}$$

Table 1 Plastic material parameters

	c_o [MPa]	R_o [MPa]	R_∞ [MPa]	b
RD	333	488	142	19

Table 2 Anisotropy parameters

C_1	C_2	C_3	C_4	C_5	C_6	C_7	C_8	C_9
−0.0424	−0.0102	0.0000	0.8123	1.3607	1.3103	3.7580	3.0000	3.0000

whereas the second and third deviatoric stress invariants are defined as

$$\bar{J}_2^H = \frac{1}{2}\bar{\mathbf{T}} \cdot \mathcal{D}\bar{\mathbf{T}} \tag{8}$$

and

$$\bar{J}_3^H = \det\left(\mathcal{D}\,\bar{\mathbf{T}}\right). \tag{9}$$

Based on these definitions the generalized Hoffman stress triaxiality

$$\bar{\eta}^H = \frac{\bar{I}_1^H}{3\sqrt{3\bar{J}_2^H}} \tag{10}$$

and the generalized Hoffman Lode parameter

$$\bar{L}^H = \frac{-3\sqrt{3}\,\bar{J}_3^H}{2\,(\bar{J}_2^H)^{(3/2)}} \tag{11}$$

are defined to characterize the dependence of anisotropic metals on the current stress state. Formation of plastic strains is governed by the flow rule

$$\dot{\bar{\mathbf{H}}}^{pl} = \dot{\gamma}\bar{\mathbf{N}} \tag{12}$$

with the equivalent plastic strain rate $\dot{\gamma}$ and the normalized deviatoric effective stress tensor

$$\bar{\mathbf{N}} = \frac{\mathcal{D}\,\bar{\mathbf{T}}}{\|\mathcal{D}\bar{\mathbf{T}}\|}. \tag{13}$$

Furthermore, the onset and evolution of damage in plastically anisotropic ductile materials are modeled by the damage criterion

$$f^{da} = \alpha I_1^H + \beta\sqrt{J_2^H} - \sigma = 0 \tag{14}$$

where I_1^H and J_2^H are the generalized first and second deviatoric Hoffman invariants of the Kirchhoff stress tensor formulated with respect to the damaged configurations, and σ is the equivalent damage stress measure. The parameters α and β depend on stress state and loading direction and have been identified by a series of experiments performed with different biaxially loaded specimens; see [9] for further details. In addition, the evolution of macroscopic irreversible strains caused by damage mechanisms on the micro-level is characterized by the damage rule

$$\dot{\mathbf{H}}^{da} = \dot{\mu} \left(\frac{1}{\sqrt{3}} \tilde{\alpha} \mathbf{1} + \tilde{\beta} \mathbf{N} \right) \tag{15}$$

where

$$\mathbf{N} = \frac{\mathrm{dev}\mathbf{T}}{\|\mathrm{dev}\mathbf{T}\|} \tag{16}$$

is the normalized deviatoric part of the Kirchhoff stress tensor and the parameters $\tilde{\alpha}$ and $\tilde{\beta}$ represent the stress and loading direction dependence of the damage strain rate tensor (15).

3 Numerical Analysis

In the proposed continuum framework, the formation of damage is characterized by the evolution of macroscopic damage strains corresponding to different damage and failure processes on the micro-level. To detect the stress state and loading direction dependence of the parameters $\tilde{\alpha}$ and $\tilde{\beta}$ in the damage rule (15), numerical simulations with a void-containing representative volume element with initial porosity of 3% have been performed undergoing various three-dimensional loading conditions. One eighth of the unit cell model is shown in Fig. 1. The numerical simulations are performed using the finite element program ANSYS enhanced by a user-defined subroutine taking into account the proposed anisotropic continuum model. Eight-node elements of type SOLID185 are used. With symmetry boundary conditions the unit cell can be seen as a part of a pre-damaged structural element. The solid elements are elastically and plastically deformed whereas the changes in size and shape of the initially spherical void are related to the damage strains. Based on the proposed kinematic approach the components of the macroscopic strain rate tensor in the principal directions (i) are decomposed

$$\dot{H}_{(i)}^{\text{unit-cell}} = \dot{H}_{(i)}^{el} + \dot{H}_{(i)}^{pl} + \dot{H}_{(i)}^{da} \tag{17}$$

into elastic, plastic and damage strain rates. Elastic and plastic strain rates in the solid elements on the micro-level, $\dot{\mathbf{h}}^{el}$ and $\dot{\mathbf{h}}^{pl}$, lead to the elastic-plastic macroscopic strain rates

$$\dot{\mathbf{H}}^{ep} = \dot{\mathbf{H}}^{el} + \dot{\mathbf{H}}^{pl} = \frac{1}{V} \int_{V_{\text{matrix}}} \left(\dot{\mathbf{h}}^{el} + \dot{\mathbf{h}}^{pl} \right) dv \tag{18}$$

where V represents the current volume of the representative volume element and V_{matrix} is the current volume of the matrix material (solid elements). With Eqs. (17) and (18) the macroscopic damage strain rate tensor can be written in the form

Fig. 1 Finite element mesh
of one eighth of the unit cell

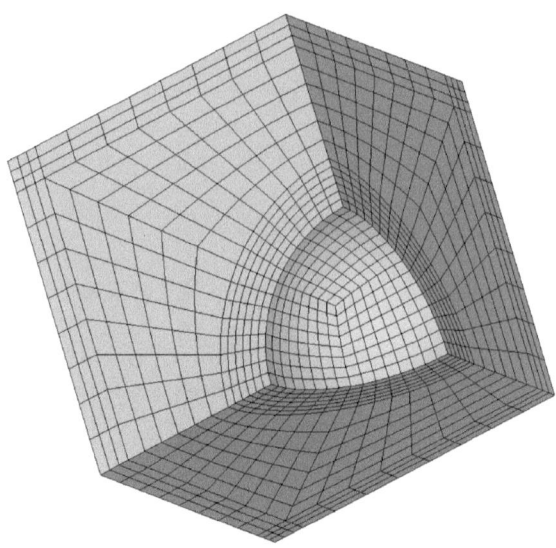

$$\dot{H}_{(i)}^{da} = \dot{H}_{(i)}^{\text{unit-cell}} - \dot{H}_{(i)}^{ep} \tag{19}$$

leading to the principal components of the damage strain tensor

$$A_{(i)}^{da} = \int \dot{H}_{(i)}^{da} \mathrm{d}t. \tag{20}$$

In addition, the void volume fraction f of the unit cell is determined using the volumetric part of the damage strain rate tensor leading to

$$\dot{f} = (1 - f)\,\mathrm{tr}\dot{\mathbf{H}}^{da} \tag{21}$$

and

$$f = \int \dot{f}\mathrm{d}t, \tag{22}$$

see [3] for further details. In the numerical analysis, the amount of strains and their rates are taken to be described by corresponding scalar-valued measures, the equivalent strain rate

$$\dot{\epsilon}_{eq} = \sqrt{\frac{2}{3}\dot{\mathbf{H}} \cdot \dot{\mathbf{H}}} \tag{23}$$

and the equivalent strain

$$\epsilon_{eq} = \int \dot{\epsilon}_{eq}\mathrm{d}t. \tag{24}$$

Fig. 2 Formation of
principal components of the
damage strain tensor for
$\eta^H = 0.389$ and
$L^H = -0.731$

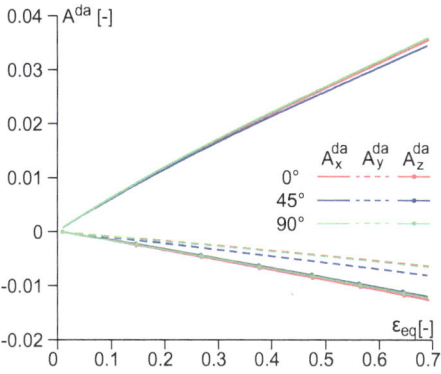

4   Numerical Results

The effect of different load ratios $F_x/F_y/F_z$ on the damage behavior of the unit
cell has been examined. In the present paper, two cases are discussed. The numerical results are compared with experimental observations reported in the literature.
For example, in [6] the X0-specimen had been biaxially loaded with the load ratios
$F_1/F_2 = 1/0$ and $1/{-}1$ leading to tensile and shear dominated stress states, respectively.

For the load ratio $F_1/F_2 = 1/0$, the Hoffman stress triaxiality $\eta^H = 0.389$ and
the Hoffman Lode parameter $L^H = -0.731$ had been predicted in corresponding
numerical simulations [6]. For comparison, the unit cell is loaded by the load ratio
$F_x/F_y/F_z = 1/0/0$ leading to the same stress parameters η^H and L^H. The formation of
the principal values of the damage strain tensor $A^{da}_{(i)}$ (20) versus the equivalent strain
measure (24) is shown in Fig. 2. In particular, A^{da}_x increases up to 0.035 whereas A^{da}_y
and A^{da}_z show a small decrease up to –0.006 and –0.01, respectively. This behavior
is nearly identical for the different loading directions with respect to the rolling
direction. In addition, the formation of the void volume fraction f (22) is shown in
Fig. 3. Nearly identical behavior can be seen for the loading directions 0° and 90°

Fig. 3 Formation of the void
volume fraction for
$\eta^H = 0.389$ and
$L^H = -0.731$

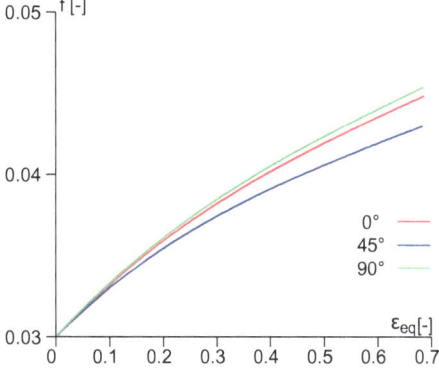

Fig. 4 Formation of principal components of the damage strain tensor for η^H = 0.0 and L^H = 0.0

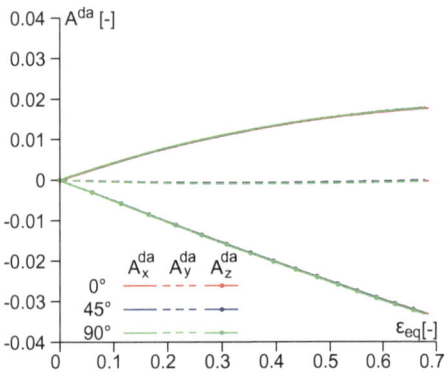

with maximum values of 0.0415 whereas for loading direction 45° the porosity only reaches 0.040. This damage behavior was also visible in the pictures of scanning electron microscopy in [6] where large pores occur for loading in rolling (0°) and in transverse (90°) direction and less and smaller voids were visible after loading in a diagonal direction (45°). These damage mechanisms on the micro-level correspond to the slightly smaller macroscopic damage strain components and the smaller void volume fraction f numerically predicted for the diagonal direction.

For the biaxial experiments with the load ratio $F_1/F_2 = 1/-1$, the Hoffman stress triaxiality $\eta^H = 0.0$ and the Hoffman Lode parameter $L^H = 0.0$ had been predicted in corresponding numerical simulations [6]. These stress parameters have also been achieved in the unit cell calculations with the load ratio $F_x/F_y/F_z = 1/0/-1$. The evolution of the principal values of the damage strain tensor $A_{(i)}^{da}$ (20) versus the equivalent strain measure (24) is shown in Fig. 4. In this shear loading case, the damage strain component A_x^{da} increases up to 0.017 and A_z^{da} shows a decrease up to -0.032 whereas the component A_y^{da} remains 0.0. This means that during shear loading the initially spherical void is deformed into an ellipsoid. The void volume fraction f shown in Fig. 5 only shows a slight decrease of the initial void. In both

Fig. 5 Formation of the void volume fraction for η^H = 0.0 and L^H = 0.0

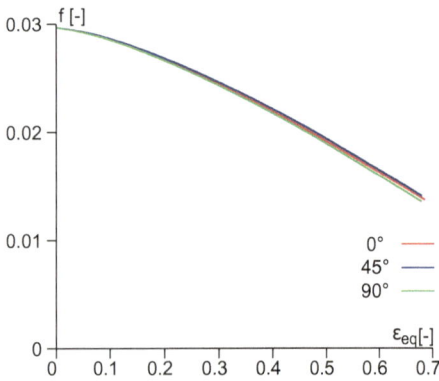

figures, nearly no effect of the loading direction on the macroscopic strain behavior can be seen. This damage behavior had also been seen in the pictures of scanning electron microscopy published in [6]. The photos showed predominant shear mechanisms on the micro-scale with only a few initial voids which were remarkably deformed in shear direction. These damage processes on the micro-scale correspond to the numerically predicted macroscopic principal damage strain components and the void volume fraction representing shear mechanisms on the macro-level with slight compression (decrease in the void volume fraction).

5 Conclusions

In the present paper, the effect of the stress state and the loading with respect to the rolling direction on damage and failure of the aluminum alloy EN AW-2017A has been examined. The results of two load cases corresponding to biaxial loading scenarios of the X0-specimen have been discussed. The stress state was characterized by the generalized stress triaxiality and the generalized Lode parameter expressed in terms of stress invariants based on the Hoffman yield criterion for anisotropic materials. Elastic-plastic constitutive parameters for the investigated ductile metal were taken from experiments performed with specimens cut from sheets in different directions with respect to the rolling direction. These parameters were used in numerical calculations considering void-containing representative volume elements. Different three-dimensional load ratios have been taken into account. Numerically predicted formations of the principal components of the damage strain tensor and of the void volume fraction as well as the corresponding damage and failure mechanisms on the micro-level have been discussed. These numerical results have been qualitatively compared with pictures from scanning electron microscopy published in the literature. The numerically predicted formation of damage strains can be seen as quasi-experimental results. They can be used to develop laws for the damage strain rates which will be taken into account in numerical simulation of experiments which will be used to validate the proposed rules for the damage strain rates with stress state and loading direction dependencies. This will be discussed in a forthcoming paper. Furthermore, the validated laws can then be used to numerically predict the deformation and failure behavior of structural components built with anisotropic sheet metals.

Acknowledgements Financial support from the Deutsche Forschungsgemeinschaft (DFG, German Research Foundation) under project number 394286626 is gratefully acknowledged.

References

1. Barsoum I, Faleskog J (2011) Micromechanical analysis on the influence of the Lode parameter on void growth and coalescence. Int J Solids Struct 48:925–938
2. Barlat F, Aretz H, Yoon JW, Karabin ME, Brem JC, Dick RE (2005) Linear transformation-based anisotropic yield functions. Int J Plast 21:1009–1039
3. Brünig M (2003) An anisotropic ductile damage model based on irreversible thermodynamics. Int J Plast 19:1679–1713
4. Brünig M (2016) A thermodynamically consistent continuum damage model taking into account the ideas of CL Chow. Int J Damage Mech 25:1130–1143
5. Brünig M, Gerke S, Hagenbrock V (2013) Micro-mechanical studies on the effect of the stress triaxiality and the Lode parameter on ductile damage. Int J Plast 50:49–65
6. Brünig M, Gerke S, Koirala S (2021) Biaxial experiments and numerical analysis on stress-state-dependent damage and failure behavior of the anisotropic aluminum alloy EN AW-2017A. Metals 11:1214
7. Brünig M, Hagenbrock V, Gerke S (2018) Macroscopic damage laws based on analysis of microscopic unit cells. ZAMM - Zeitschrift für Angewandte Mathematik und Mechanik 98:181–194
8. Brünig M, Koirala S, Gerke S (2022) Analysis of damage and failure in anisotropic ductile metals based on biaxial experiments with the H-specimen. Exp Mech 62:183–197
9. Brünig M, Koirala S, Gerke S (2023) A stress-state-dependent damage criterion for metals with plastic anisotropy. Int J Damage Mech (in press)
10. Gao X, Wang T, Kim J (2005) On ductile fracture initiation toughness: effects of void volume fraction, void shape and void distribution. Int J Solids Struct 42:5097–5117
11. Gao X, Zhang G, Roe C (2010) A study on the effect of the stress state on ductile fracture. Int J Damage Mech 19:75–94
12. Ha J, Baral M, Korkolis Y (2018) Plastic anisotropy and ductile fracture of bake-hardened AA6013 aluminum sheet. Int J Solids Struct 155:123–139
13. Hill R (1948) A theory of the yielding and plastic flow of anisotropic metals. Proc R Soc Lond 193:281–297
14. Hoffman O (1967) The brittle strength of orthotropic materials. J Compos Mater 1:200–206
15. Hu Q, Yoon JW, Manopulo N, Hora P (2021) A coupled yield criterion for anisotropic hardening with analytical description under associated flow rule: modeling and validation. Int J Plast 136:102882
16. Kim J, Gao X, Srivatsan T (2003) Modeling of crack growth in ductile solids: a three-dimensional analysis. Int J Solids Struct 40:7357–7374
17. Kuna M, Sun D (1996) Three-dimensional cell model analyses of void growth in ductile materials. Int J Fract 81:235–258
18. Scheyvaerts F, Onck P, Tekoglu C, Pardoen T (2011) The growth and coalescence of ellipsoidal voids in plane strain under combined shear and tension. J Mech Phys Solids 59:373–397
19. Stoughton TB, Yoon JW (2009) Anisotropic hardening and non-associated flow rule in proportional loading of sheet metals. Int J Plast 25:1777–1817
20. Tsutamori H, Amaishi T, Chorman RR, Eder M, Vitzthum S, Volk W (2020) Evaluation of prediction accuracy for anisotropic yield functions using cruciform hole expansion test. J Manuf Mater Process 4:43

Prediction of Fracture Behavior in a Quenching and Partitioning Steel Under Different Stress States

Fuhui Shen, Yannik Sparrer, Guijia Li, and Sebastian Münstermann

Abstract Increasing research efforts have been devoted to the development of quenching and partitioning (Q&P) steel, which is considered to be a very promising representative of the third generation of advanced high-strength steel (AHSS). The excellent tensile properties achieved by the novel Q&P treatment make it a potential material to manufacture structural components in automotive industries. In addition to tensile strength and ductility, the formability and fracture properties of Q&P steels shall be thoroughly investigated under different stress states. Therefore, the deformation and fracture properties of a laboratory Q&P steel have been investigated by conducting a comprehensive experimental program and the corresponding finite element simulations. Tensile tests have been performed using flat specimens with different notch configurations to achieve a very wide range of loading conditions. In addition, the deformation and fracture behavior during different experiments have been simulated using finite element methods and damage mechanics approaches. After collecting the critical stress and strain variables from simulation results, a stress state-dependent fracture criterion has been calibrated and validated to characterize the fracture resistance of the Q&P steel in this study.

Keywords Q&P steel · Stress state · Fracture

1 Introduction

Driven by the increasing demands on green mobility and structural safety, extensive research efforts have been devoted to developing advanced high-strength steels (AHSS), which play an essential role in lightweight automotive engineering. The quenching and partitioning (Q&P) steel is considered to be a very promising representative of the third generation of AHSS due to its excellent combination of strength and ductility. In the current development of Q&P steels, most studies are focused

F. Shen (✉) · Y. Sparrer · G. Li · S. Münstermann
Steel Institute, RWTH Aachen University, Intzestraße 1, 52072 Aachen, Germany
e-mail: fuhui.shen@iehk.rwth-aachen.de

© The Rightsholder, under exclusive licence to [Springer Nature Switzerland AG], part of Springer Nature 2024
J. Kusiak et al. (eds.), *Numerical Methods in Industrial Forming Processes*, Lecture Notes in Mechanical Engineering, https://doi.org/10.1007/978-3-031-58006-2_23

on the influence of processing parameters on the final tensile properties [1]. The mechanical performance of Q&P steels under other loading conditions needs to be systematically evaluated for the industrial application of this new grade of AHSS.

The local and global formability are both important factors that need to be considered for the application of AHSS [2–5]. It is observed that the global formability, represented by the forming limit curve as the necking resistance, of a commercial Q&P980 steel, is better than a dual-phase DP980 steel, which is attributed to the strong strain hardening rate [1]. There are few studies in the literature on the local formability (fracture resistance) of Q&P steels, which still requires more systematic investigations [6]. The effects of stress states, especially the Lode angle parameter, on ductile fracture have received intensive attention in the recent development of ductile fracture theories [2, 7–9]. To achieve better accuracy in describing the fracture properties of high-strength metallic materials, many uncoupled ductile fracture criteria, microscopic mechanisms motivated or purely phenomenological, have been proposed and applied in the past two decades [10–12]. A common feature of these fracture criteria is that the fracture strain is expressed as a function of stress state parameters. The stress triaxiality and Lode angle parameter are the most often applied parameters to describe the stress states, which is also adopted in this study. The general aim of this study is to investigate the effects of stress states on the fracture properties of a laboratory Q&P steel with excellent tensile properties. The damage mechanics approach is adopted in this study, which is based on a combination of experimental characterization and finite element simulations.

2   Materials and Experiments

The mechanical properties of a laboratory Q&P steel have been systematically investigated in this study. The uniaxial tensile tests at room temperature and quasi-static loading conditions were performed using a smooth dog bone (SDB) specimen to obtain the engineering stress and strain curves, as shown in Fig. 1a. It can be seen that very good uniaxial tensile properties were achieved in the laboratory Q&P steel in this study. To investigate the fracture properties of the material, tensile tests were performed along the rolling direction using different specimen geometries at room temperature. Various notch configurations were used to obtain different stress states in the critical positions of these flat specimens. The overview of all specimens is shown in Fig. 1b. The force and displacement (at a gauge length of 40 mm) results were measured during the tensile tests.

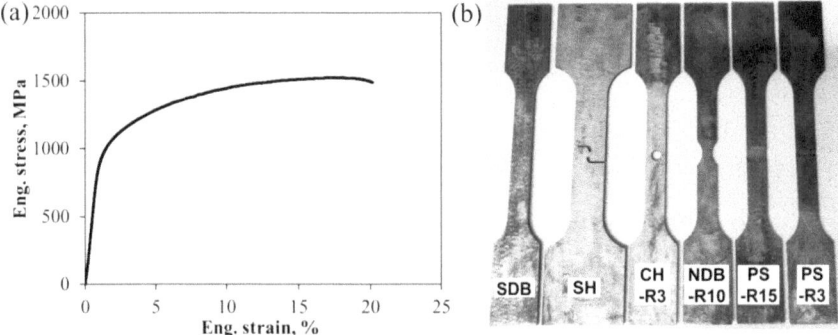

Fig. 1 Engineering stress and strain curve of the investigated Q&P steel (**a**) and overview of all specimens used for tensile tests (**b**). Specimen geometries: shear (SH), central hole (CH-R3), notched dog bone (NDB-R10), and plane strain tension (PS-R15 and PS-R3)

3 Models

The material is assumed to be isotropic in the mechanical properties, the elasto-plastic deformation behavior is thus described by the Mises plasticity model. The Voce hardening law is calibrated based on experimental results of uniaxial tensile tests at room temperature. The fracture properties of the investigated Q&P steel are described using the uncoupled damage mechanics approach. The fracture behavior under different stress states is predicted using the unified fracture criterion that was recently proposed by Shen et al. [13, 14]. The stress state is described using a general expression of stress triaxiality η and Lode angle parameter $\bar{\theta}$. These two variables are calculated from invariants of stress tensors (I_1, J_2, J_3).

$$\eta = I_1/\sqrt{27 \cdot J_2} \tag{1}$$

$$\bar{\theta} = 1 - \frac{6}{\pi}\theta = 1 - \frac{2}{\pi}\cos^{-1}(\sqrt{27/4} \cdot J_3 \cdot J_2^{-3/2}) \tag{2}$$

For proportional loading conditions, the stress state influence on fracture initiation strain is quantified using the Bai-Wierzbicki [10] phenomenological function, where $C_{1\sim4}$ are four material parameters that need to be calibrated. To compensate for the non-proportional loading effects during deformation, the overall stress state of the specimen is represented by the average values of η and $\bar{\theta}$ (η_{avg}, $\bar{\theta}_{\text{avg}}$). The fracture indicator I_f is accumulated over the non-proportional loading paths, and fracture occurs when I_f reaches the unity.

$$\bar{\varepsilon}_f(\eta, \bar{\theta}) = \left(C_1 \exp^{-C_2\eta} - C_3 \exp^{-C_4\eta}\right)\bar{\theta}^2 + C_3 \exp^{-C_4\eta} \tag{3}$$

$$I_{\mathrm{f}} = \int\limits_{0}^{\overline{\varepsilon}^{\mathrm{p}}} \frac{1}{\overline{\varepsilon}_{\mathrm{f}}(\eta, \overline{\theta})} d\overline{\varepsilon}^{\mathrm{p}} \tag{4}$$

4   Finite Element Simulations

Finite element simulations of deformation and fracture behavior of different specimens subjected to remote tensile tests have been carried out in this study using the ABAQUS/Explicit with a user-defined subroutine. The critical region of fracture specimen models with half thickness has been discretized using solid elements (C3D8R) with a fine mesh ($0.1 \times 0.1 \times 0.1$ mm^3). The classical J2 Mises plasticity model is applied in this study to simulate the elastoplastic deformation behavior of the material. The strain hardening behavior of the material is described by the Voce hardening law. The distribution of equivalent plastic strain on different cross-section planes at the fracture displacement is shown in Fig. 2 for NDB-R10 and PS-R15 specimens. In both specimens, the crack initiates at the symmetry center, which is the peak position of equivalent plastic strain. In addition, the localization in the thickness direction is not very pronounced, as shown in Fig. 2. The strong strain hardening behavior of the investigated Q&P steel is beneficial for the global formability as the necking resistance is improved, which is consistent with the observations in a medium-Mn steel with superior tensile properties [2].

In order to calibrate the fracture criterion, the evolution of local stress state variables and equivalent plastic strain has been extracted from the critical positions of different specimens, as shown in Fig. 3. The solid curves represent the evolution history of stress state variables while the dashed lines correspond to the average values of stress triaxiality (black curves) and Lode angle parameter (blue curves)

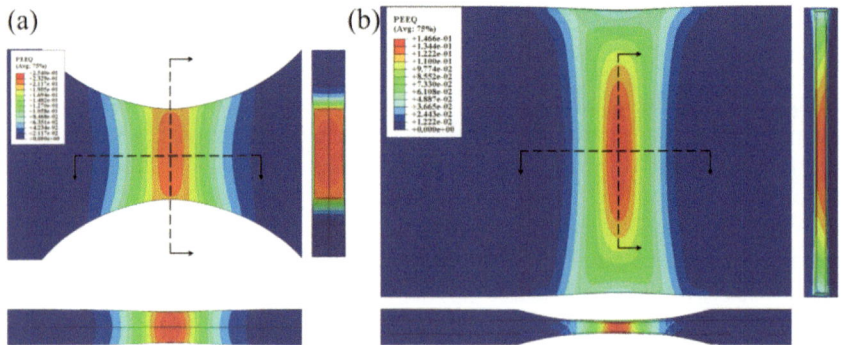

Fig. 2 The equivalent plastic strain contour on the mid-thickness plane and two other cross-section planes at the fracture displacements in the **a** NDB-R10 and **b** PS-R15 specimens

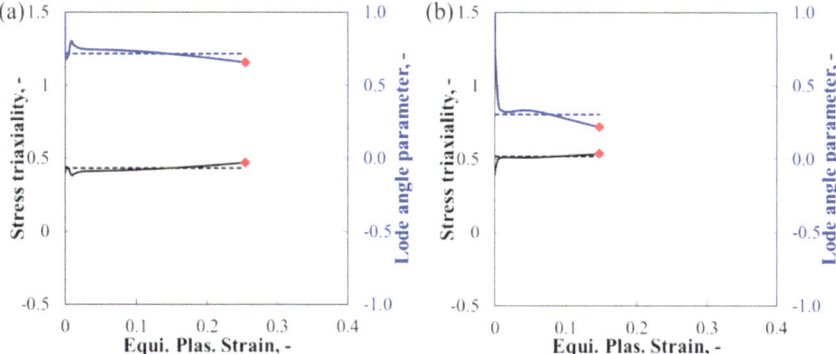

Fig. 3 Evolution of local stress state variables in the critical elements of different tensile specimens during deformation. **a** NDB-R10 and **b** PS-R15

until the fracture point (red symbols). There is a slight increase of stress triaxiality with increasing strain in these specimens. In general, it can be seen that the stress state in the critical elements does not change significantly during plastic deformation. It is observed that the overall stress triaxiality is similar in NDB-R10 and PS-R15 specimens. However, the failure strain of the PS-R15 specimen is much smaller than the NDB-R10 specimen, which proves the significance of the Lode angle parameter on the failure strain of the investigated Q&P steel.

After collecting the critical stress and strain variables from simulation results of different geometries, the four parameters in the fracture criterion have been optimized. In the final step, the fracture criterion has been implemented in the finite element simulations to predict the fracture behavior in different specimens. The numerical simulation results (dashed curves) of force and displacement curves are compared with experimental results (dotted curves) in Fig. 4 for the NDB-R10 and PS-R15 specimens. The fracture behavior of the investigated Q&P steel in tensile tests using different specimens can be accurately predicted by the calibrated failure criterion.

5 Conclusions

Superior tensile properties are obtained in a laboratory Q&P steel, which provides excellent necking resistance. The fracture properties of the Q&P steel have been investigated by performing tensile tests using specimens with optimized sample geometries. The uncoupled damage mechanics approach is adopted to simulate the deformation and fracture behavior of the material. The failure strain of the investigated laboratory high-strength Q&P steel shows a clear dependence on the stress state, in particular, the Lode angle parameter.

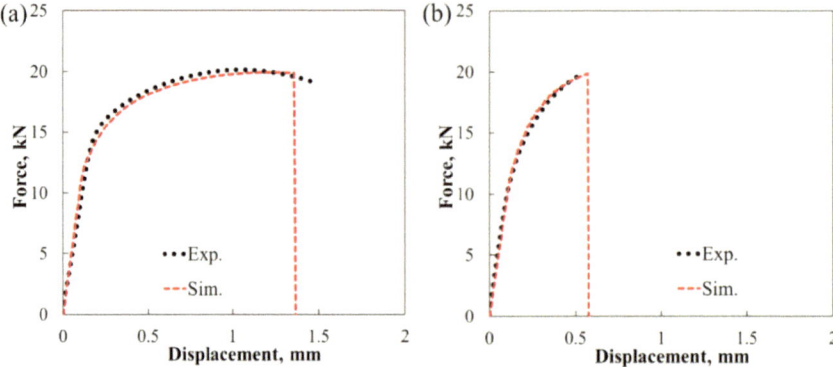

Fig. 4 Comparison between experimental and simulated fracture behavior in different specimens during tensile tests. **a** NDB-R10 and **b** PS-R15

References

1. Chen X, Niu C, Lian C, Lin J (2017) The evaluation of formability of the 3rd generation advanced high strength steels QP980 based on digital image correlation method. Procedia Eng 207:556–561
2. Shen F, Wang H, Liu Z, Liu W, Könemann M, Yuan G, Wang G, Münstermann S, Lian J (2022) Local formability of medium-Mn steel. J Mater Process Technol 299
3. Hance B (2016) Advanced high strength steel: deciphering local and global formability. In: Proceedings of the international automotive body congress, Dearborn, MI
4. Heibel S, Dettinger T, Nester W, Clausmeyer T, Tekkaya AE (2018) Damage mechanisms and mechanical properties of high-strength multiphase steels. Materials (Basel) 11
5. Shen FH, Wang HS, Xu H, Liu WQ, Münstermann S, Lian JH (2022) Local formability of different advanced high strength steels. Key Eng Mater 926:917–925
6. Han S, Chang Y, Wang C, Han Y, Dong H (2022) Experimental and numerical investigations on the damage induced in the shearing process for QP980 steel. Materials (Basel) 15
7. Mu L, Jia Z, Ma Z, Shen F, Sun Y, Zang Y (2020) A theoretical prediction framework for the construction of a fracture forming limit curve accounting for fracture pattern transition. Int J Plast 129
8. Bao Y, Wierzbicki T (2004) On fracture locus in the equivalent strain and stress triaxiality space. Int J Mech Sci 46:81–98
9. Shen F, Münstermann S, Lian J (2020) Investigation on the ductile fracture of high-strength pipeline steels using a partial anisotropic damage mechanics model. Eng Fract Mech 227
10. Bai Y, Wierzbicki T (2008) A new model of metal plasticity and fracture with pressure and Lode dependence. Int J Plast 24:1071–1096
11. Lou Y, Huh H, Lim S, Pack K (2012) New ductile fracture criterion for prediction of fracture forming limit diagrams of sheet metals. Int J Solids Struct 49:3605–3615
12. Mohr D, Marcadet SJ (2015) Micromechanically-motivated phenomenological Hosford-Coulomb model for predicting ductile fracture initiation at low stress triaxialities. Int J Solids Struct 67–68:40–55
13. Shen F, Münstermann S, Lian J (2022) A unified fracture criterion considering stress state dependent transition of failure mechanisms in bcc steels at −196 °C. Int J Plast 156
14. Shen F, Münstermann S, Lian J (2023) Cryogenic ductile and cleavage fracture of bcc metallic structures—influence of anisotropy and stress states. J Mech Phys Solids 105299

Cyclic Bending Behaviors of Extruded AZ31 Magnesium Alloy Beams

Kecheng Zhou, Ding Tang, Dayong Li, and Huamiao Wang

Abstract Bending behavior holds significant importance, as it plays an unavoidable role in a variety of forming processes. Therefore, gaining insight into the cyclic bending mechanical behavior of magnesium (Mg) alloys is crucial for designing, fabricating, and ensuring the performance of end products. This study conducts cyclic bending simulations of an extruded AZ31 Mg alloy beam using crystal-plasticity-based methods. The study encompasses the macroscopic aspects of stress distribution and strain distribution, along with a detailed discussion of microscopic factors such as the relative activities of deformation mechanisms and twin volume fraction during cyclic bending. Different from conventional alloys with high crystal symmetry, the apparent the neutral layer shifting phenomena is observed in the extruded AZ31. The simulated results reveal that this distinct behavior is ascribed to the alternating twinning and detwinning mechanisms.

Keywords Magnesium alloy · Crystal plasticity · Cyclic bending · Twinning-detwinning

1 Introduction

The global interest in magnesium (Mg) alloys has surged owing to their lightweight characteristics, high specific strength, and exceptional recyclability [1–4]. The distinctive hexagonal close-packed (HCP) structure, the unidirectional deformation twinning mechanism, and the developed texture all play a role in the pronounced

K. Zhou · D. Tang · D. Li · H. Wang (✉)
State Key Laboratory of Mechanical System and Vibration, Shanghai Jiao Tong University, Shanghai 200240, China
e-mail: wanghm02@sjtu.edu.cn

D. Li · H. Wang
Materials Genome Initiative Center, Shanghai Jiao Tong University, Shanghai 200240, China

J. Kusiak et al. (eds.), *Numerical Methods in Industrial Forming Processes*, Lecture Notes in Mechanical Engineering, https://doi.org/10.1007/978-3-031-58006-2_24

anisotropic behavior exhibited by wrought Mg alloys [5–8]. In order to accommodate arbitrary deformation in magnesium (Mg) alloys, a minimum of five independent deformation systems are necessary. Experimentally observed major deformation mechanisms in Mg alloys include basal <a> slip ($\{0001\}\langle11\overline{2}0\rangle$) [9], prismatic <a> slip ($\{10\overline{1}0\}\langle11\overline{2}0\rangle$) [10], pyramidal <a> slip ($\{10\overline{1}1\}\langle11\overline{2}0\rangle$) [11], pyramidal <c+a> slip ($\{11\overline{2}2\}\langle11\overline{2}3\rangle$) [12] and extension twin ($\{10\overline{1}2\}\langle10\overline{1}1\rangle$) [13]. The cyclic deformation behaviors of wrought Mg alloys are examined under relatively straightforward paths, including tension–compression cycles and simple shear [14–17]. Extensive documentation supports the crucial roles of twinning and detwinning in the cyclic plasticity of both pure Mg and Mg alloys [18–20]. In common manufacturing processes or in-service conditions, bending and cyclic bending behaviors are frequently encountered as instances of inhomogeneous cases [21–23]. In the pursuit of improved processing technology and elevated product quality, understanding and quantifying the cyclic bending behaviors in Mg alloys are of paramount importance.

The investigation presented in this work focuses on the cyclic bending behaviors of Mg alloys featuring different curvature amplitudes. The study utilizes the EVPSC-BEND, which is designed for bending loads based on the elasto-viscoplastic self-consistent model [24, 25]. The moment–curvature loading curves, relative activity and twin volume fraction (TVF) during cyclic loading, and the distribution of stress and strain components at certain loading processes are discussed.

2 Simulation Methods

A brief overview of the EVPSC-BEND approach for simulating pure bending is provided here, with a more comprehensive description available elsewhere [24, 25]. When the beam is in a state of pure bending, a local coordinate (x) is established along the thickness of the beam, as illustrated in Fig. 1. The beam is discretized into numerous layers along the thickness direction. The brown dashed line denotes the neutral layer and serves as the coordinate origin. The intrados and extrados coordinates are designated as a and b, respectively. The curvature of the neutral layer is $1/\rho$.

The stress increments ($\Delta\Sigma_r^i$, $\Delta\Sigma_\theta^i$) in response to bending ($\Delta\rho$) within the beam can be expressed as follows:

$$
\begin{bmatrix}
\frac{t\overline{M}_{\theta r}^{e1}}{\rho+x_1}+\overline{M}_{\theta\theta}^{ei} & 0 & \cdots & 0 \\
\frac{t\overline{M}_{\theta r}^{e2}}{\rho+x_2} & \frac{t\overline{M}_{\theta r}^{e2}}{\rho+x_2}+\overline{M}_{\theta\theta}^{e2} & \cdots & 0 \\
\vdots & \vdots & \ddots & \vdots \\
\frac{t\overline{M}_{\theta r}^{en}}{\rho+x_n} & \frac{t\overline{M}_{\theta r}^{en}}{\rho+x_n} & \cdots & \frac{t\overline{M}_{\theta r}^{en}}{\rho+x_n}+\overline{M}_{\theta\theta}^{en}
\end{bmatrix}
\begin{bmatrix}
\Delta\Sigma_\theta^1 \\
\Delta\Sigma_\theta^2 \\
\vdots \\
\Delta\Sigma_\theta^n
\end{bmatrix}
=
\begin{bmatrix}
\left(\frac{t\overline{M}_{\theta r}^{e1}}{(\rho+x_1)^2}\Sigma_\theta^1-\frac{x_1}{\rho^2}\right)\Delta\rho-\Delta E_\theta^{p1} \\
\left(\frac{t\overline{M}_{\theta r}^{e2}}{(\rho+x_2)^2}\Sigma_{k=1}^2\Sigma_\theta^k-\frac{x_2}{\rho^2}\right)\Delta\rho-\Delta E_\theta^{p2} \\
\vdots \\
\left(\frac{t\overline{M}_{\theta r}^{en}}{(\rho+x_n)^2}\Sigma_{k=1}^n\Sigma_\theta^n-\frac{x_n}{\rho^2}\right)\Delta\rho-\Delta E_\theta^{pn}
\end{bmatrix}
\tag{1}
$$

Fig. 1 Numerous discretized layers beam under pure bending

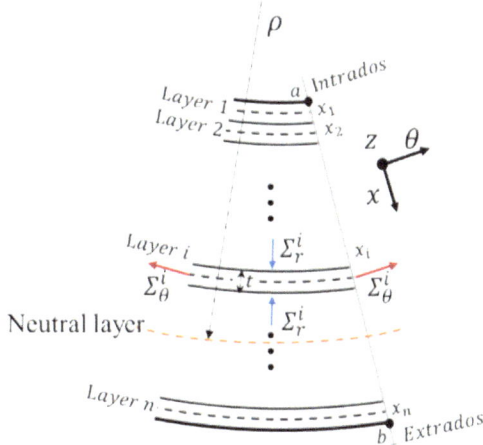

Here, ΔE_θ^{pi} represents the increment of plastic strain of the ith layer, while $\overline{M}_{\theta r}^{ei}$ and $\overline{M}_{\theta\theta}^{ei}$ denote the components of elastic compliance. In the EVPSC-TDT model, every layer is treated as a homogeneous effective medium (HEM) consisting of numerous grains. The relationship between the strain rate $\dot{E}_{ij}$ and the stress of the HEM (Σ_{ij}) is given by:

$$\dot{E}_{ij} = \dot{E}_{ij}^e + \dot{E}_{ij}^p = \overline{M}_{ijkl}^e \check{\Sigma}_{kl} + \dot{E}_{ij}^p \tag{2}$$

where $\overline{M}_{ijkl}^e$ is the elastic compliance, and $\dot{E}_{ij}^e$ and $\dot{E}_{ij}^p$ are the elastic and plastic parts of the strain rate, respectively.

The strain rate of the single crystal $\dot{\varepsilon}_{ij}$ comprises both elastic and plastic components: $\dot{\varepsilon}_{ij}^e$ and $\dot{\varepsilon}_{ij}^p$. The plastic deformation of Mg alloys arises from both slip and twinning:

$$\dot{\varepsilon}_{ij} = \dot{\varepsilon}_{ij}^e + \dot{\varepsilon}_{ij}^p = M_{ijkl}^e \dot{\sigma}_{kl} + \sum_\alpha \dot{\gamma}^\alpha P_{ij}^\alpha \# \tag{3}$$

Here, M_{ijkl}^e represents the elastic compliance, $\dot{\gamma}^\alpha$ denotes the shear rat, and P_{ij}^α is the Schmid tensor for a slip/twinning system α. For both slip and twinning systems, the resolved shear stress (RSS) serves as the driving force to induce the shear rate $\dot{\gamma}^\alpha$. And the RSS on a system α is calculated by the Cauchy stress σ_{ij}: $\tau^\alpha = \sigma_{ij} P_{ij}^\alpha$.

The shear rate for a slip system α can be defined by

$$\dot{\gamma}^\alpha = \dot{\gamma}_0 \left| \frac{\tau^\alpha - \tau_b^\alpha}{\tau_{cr}^\alpha} \right|^{\frac{1}{m}} sgn\left(\tau^\alpha - \tau_b^\alpha\right) \# \tag{4}$$

incorporating the reference shear rate $\dot{\gamma}_0$, the critical resolved shear stress (CRSS) τ_{cr}^α, the back stress τ_b^α, and the strain rate sensitivity m.

For a twin system α, the shear rate can be expressed similarly:

$$\dot{\gamma}_T^\alpha = \begin{cases} \dot{\gamma}_0 \left| \frac{\tau^\alpha - \tau_b^\alpha}{\tau_{cr}^\alpha} \right|^{1/m} & \tau^\alpha - \tau_b^\alpha \text{ suit the situation} \\ 0 & \text{else} \end{cases} ; \dot{f}_T^\alpha = \frac{|\dot{\gamma}_T^\alpha|}{\gamma^{tw}} \tag{5}$$

where $\dot{f}_T^\alpha$ represents the rate of TVF associated with a certain twinning operation, γ^{tw} is the characteristic twinning shear. The reduction of matrix (MR) and propagation of twin (TP) lead to an increase in TVF. Conversely, the propagation of matrix (MP) and reduction of twin (TR) result in a decrease in TVF, indicating the occurrence of de-twinning.

The evolution of the TVF of the αth twinning system, f^α can be expressed as:

$$\dot{f}^\alpha = f^M(\dot{f}_{MR}^\alpha + \dot{f}_{MP}^\alpha) + f^\alpha(\dot{f}_{TP}^\alpha + \dot{f}_{TR}^\alpha)\# \tag{6}$$

where $f^M = 1 - f^{tw} = 1 - \sum_\alpha f^\alpha$ is the volume fraction of the matrix.

If the TVF, represented as f^{tw}, attains a threshold value V^{th}, twinning within a grain is halted. This threshold is determined by the accumulated twin fraction V^{acc} and the effective twinned fraction V^{eff},

$$V^{th} = \min\left(1.0, A_1 + A_2 \cdot \frac{V^{eff}}{V^{acc}}\right)\# \tag{7}$$

where A_1 and A_2 are two governing parameters.

For a slip or twinning system α, the change of the CRSS τ_{cr}^α can be expressed as:

$$\dot{\tau}_{cr}^\alpha = \frac{d\hat{\tau}^\alpha}{d\Gamma} \sum_\beta h^{\alpha\beta} |\dot{\gamma}^\beta|\# \tag{8}$$

in this equation, Γ stands for the accumulated shear strain of all the deformation systems in the grain, expressed as $\Gamma = \sum_\alpha \int \dot{\gamma}^\alpha dt$. The coefficients $h^{\alpha\beta}$ are latent hardening coupling parameters that empirically address obstacles on system α associated with system β. The threshold stress $\hat{\tau}^\alpha$ is defined by an extended Voce law:

$$\hat{\tau}^\alpha = \tau_0^\alpha + (\tau_1^\alpha + h_1^\alpha \Gamma)\left(1 - \exp\left(-\frac{h_0^\alpha \Gamma}{\tau_1^\alpha}\right)\right)\# \tag{9}$$

here, $\tau_0, h_0, h_1,$ and $\tau_0 + \tau_1$ stand for the initial Critical Resolved Shear Stress (CRSS), the initial hardening rate, the asymptotic hardening rate, and the back-extrapolated CRSS, respectively.

The Bauschinger effect, linked to back stress, is induced by the generation of geometrically necessary dislocations. In the EVPSC-TDT model, to approximate the development of a permanent forest dislocation structure and the structure of reversible dislocations, the back stress (τ_b^α) is introduced. The rate of back stress ($\dot{\tau}_b^\alpha$) is expressed as follows:

$$\dot{\tau}_b^\alpha = \xi^\alpha sgn(\dot{\gamma}^\alpha) \sum_\beta h^{\alpha\beta} |\dot{\gamma}^\beta| - \eta^\alpha \tau_b^\alpha \sum_\beta h^{\alpha\beta} |\dot{\gamma}^\beta|\# \tag{10}$$

the effects of linear kinematic hardening and dynamic recovery are represented by the first and second terms in the above equation, respectively. ξ^{α}, η^{α} are two governing material coefficients.

The bending moment on the beam is given by the equation:

$$B = \int_a^b x w \Sigma_\theta(x) dx = t \sum_{k=1}^n w x_k \Sigma_\theta^k \# \tag{11}$$

where w denotes the beam width, t is the layer thickness, and n represents the number of layers. As demonstrated in our prior research, opting for 10 discretized layers has demonstrated satisfactory outcomes and will be applied in the current study.

3 Results and Discussion

Figure 2a illustrates the coordinate system utilized in the cyclic bending simulation, while the initial texture, represented by the {0001} and $\{10\bar{1}0\}$ pole figures (Fig. 2b), has been previously investigated in related research [24, 25]. The parameters associated with slip systems, including basal $\langle a \rangle (\{0001\}\langle11\bar{2}0\rangle)$, prismatic $\langle a \rangle (\{10\bar{1}0\}\langle11\bar{2}0\rangle)$ and pyramidal $\langle c+a \rangle (\{\bar{1}\bar{1}22\}\langle\bar{1}\bar{1}23\rangle)$, and extension twin system $\{10\bar{1}2\}\langle10\bar{1}1\rangle$, in the EVPSC-BEND model, were determined by fitting the stress–strain curves obtained from tensile and compressive experiments along the longitude direction (LD) of the sample. The value of these parameters is listed in Table 1. The simulated results, as depicted in Fig. 2c, exhibit good agreement with the corresponding experimental results.

The simulated moment–curvature curve has been compared with the experimental result, as shown in Fig. 3 (The corresponding four-point bending test has been described in detail in the previous work [24]). The comparison demonstrates a good agreement, confirming the predictability of the EVPSC-BEND model. Based on the credibility of this model, this study further conducted simulations of cyclic bending. As illustrated in Fig. 4, the cyclic Moment–curvature curves depict the curvature amplitudes of Cycle A, B, and C as 0.003, 0.006, and 0.010, respectively. The marker points indicate three instances when the loading reached its maximum amplitude. The asymmetry of the yielding moment becomes increasingly evident as the amplitude increases from Cycle A to Cycle C. The relative activity and evolution of twin volume fraction (TVF) during cyclic bending are presented in Fig. 5. The extension twin remains at a high level throughout the entire cycle, while the apparent TVF fluctuation illustrates the involvement of the detwinning mechanism in the deformation. Basal slip's role diminishes as the curvature amplitude increases from Cycle A to Cycle C. In each monotonous bending process ($O - P_1$, $P_1 - P_2$, $P_2 - P_3$), prismatic slip exhibits a nearly monotonic increase, except for a short but significant drop after P_1 and P_2. The amount of pyramidal slip remains low. The mechanical behavior of magnesium alloys is closely related to the deformation systems they activate, particularly the twinning system. The results of activity analysis indicate that

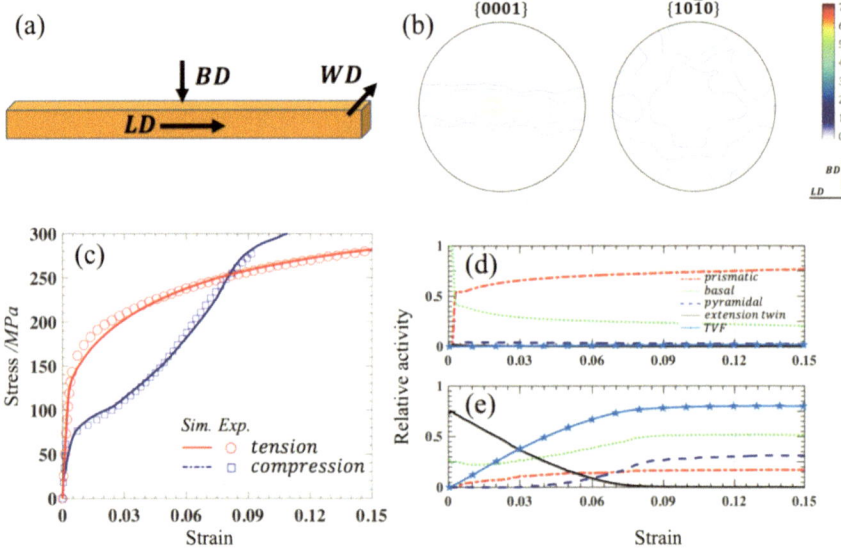

Fig. 2 **a** The layout of the pure-bending simulation. **b** The initial texture of the simulation. **c** Comparison of stress–strain curves for uniaxial tension and uniaxial compression along the LD direction between experimental and simulation results. **d** The relative activity of tension. **e** The relative activity of compression

Table 1 List of the parameters involved in the EVPSC-BEND model. $h^{\alpha\beta}$ denotes the latent hardening parameters associated with slip systems corresponding to the extensive twin systems

Mode	τ_0	τ_1	h_0	h_1	ξ^α	η^α	$h^{\alpha\beta}$	A_1	A_2
Basal	5	25	300	20	200	50	2	–	–
Prismatic	65	35	320	10	400	50	1	–	–
Pyramidal	80	100	800	20	500	50	1	–	–
Extensive twin	12	0	0	0		50	1	0.4	0.65

Elastic constants (GPa)[26]:$C_{11} = 58.0$, $C_{12} = 25.0$, $C_{13} = 20.8$, $C_{33} = 61.2$, $C_{44} = 16.6$

Reference slip/twinning rate:$\dot{\gamma}_0 = 0.001 s^{-1}$

Rate sensitivity: $m = 0.05$

the twinning-detwinning behavior is significant throughout the entire cyclic bending process. The influence of twinning during the cyclic process is worthy of careful discussion.

The stress distribution at P_1, P_2, and P_3 during loading cycles A–C is presented in Fig. 6, while Fig. 7 displays the corresponding strain distribution at the same points. The stress distribution at P_1–P_3 in loading Cycle A–C follows a pattern of alternation between tensile and compressive zones. For a circumferential force balance of zero, it is imperative that the shaded areas in the tensile and compressive regions are equal, prompting the need for the neutral layer shifting. The strain component E_θ displays

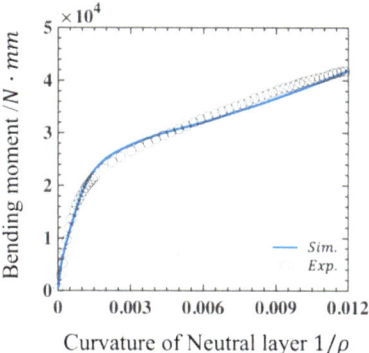

Fig. 3 The simulated and experimental moment–curvature curves

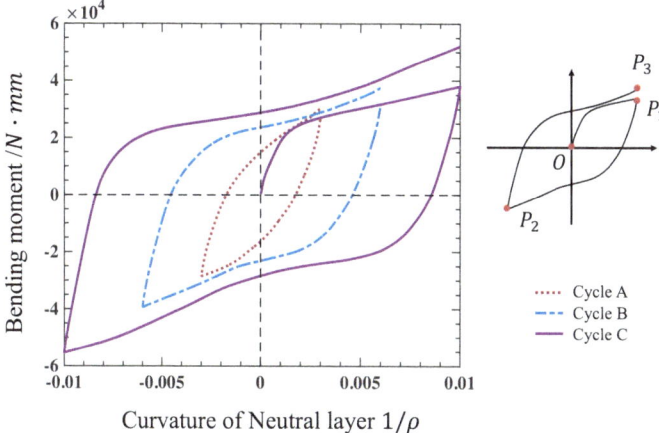

Fig. 4 The cyclic loading curves of three different amplitude cycle

a linear distribution along the thickness, whereas the other two components, E_r and E_z, exhibit significant non-linearity in their distribution. The stress distribution at P_1 reveals that twinning, by influencing the yield and hardening of different layers, creates curves on the bending cross-section resembling uniaxial tension and uniaxial compression. This characteristic distribution is also evident at P_2 and P_3. The distribution in P_2 is reversed relative to P_1 and P_3. Additionally, there is a clear hardening effect from P1 to P3, which weakens the asymmetricity between the tensile and compressive zone.

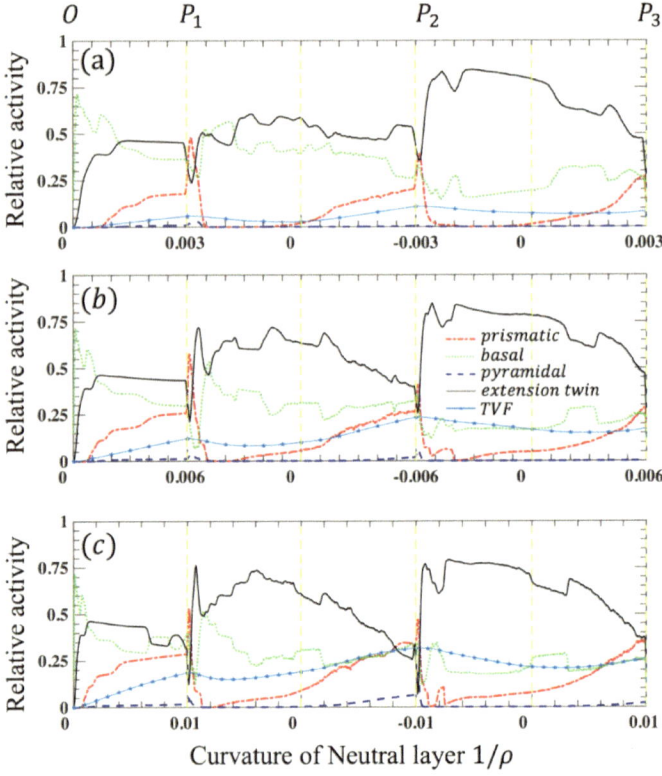

Fig. 5 The relative activity and twin volume fraction under cyclic bending in **a** Cycle A, **b** Cycle B, and **c** Cycle C

In Fig. 8, the distribution of TVF at P_1, P_2, and P_3 during loading cycles A–C is presented. The polar nature of the extension twin and the initial texture lead to the apparent occurrence of twinning and detwinning effects. The wider compressive zone at P_2 (as shown in Fig. 6) results in a higher TVF value at P_2 than at P_1. The TVF distribution at P_3 is similar to that at P_2, but with a higher value. During loading from P_1 to P_2, detwinning occurs in the upper half, and the twin does not completely disappear. From P_2 to P_3, a similar loading condition causes twinning to become more profound again, resulting in a higher TVF at the end of the cycle.

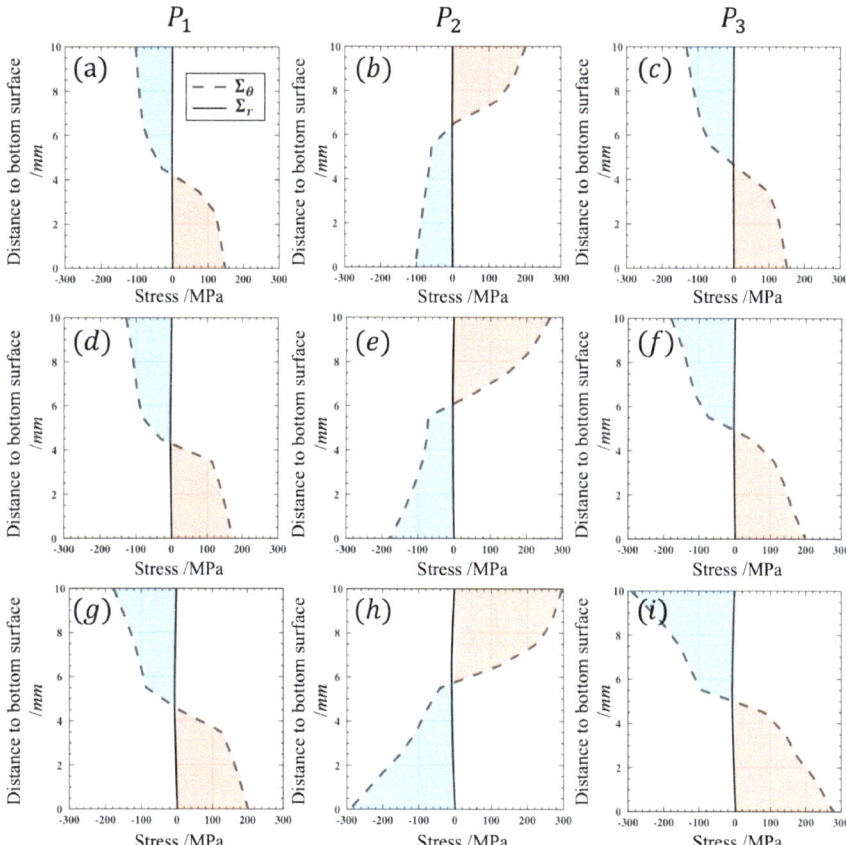

Fig. 6 The stress distribution at different loading stages. **a–c** The distribution at P_{1-3} of Cycle A, respectively; **d–f** The distribution at P_{1-3} of Cycle B, respectively; **g–i** The distribution at P_{1-3} of Cycle C, respectively

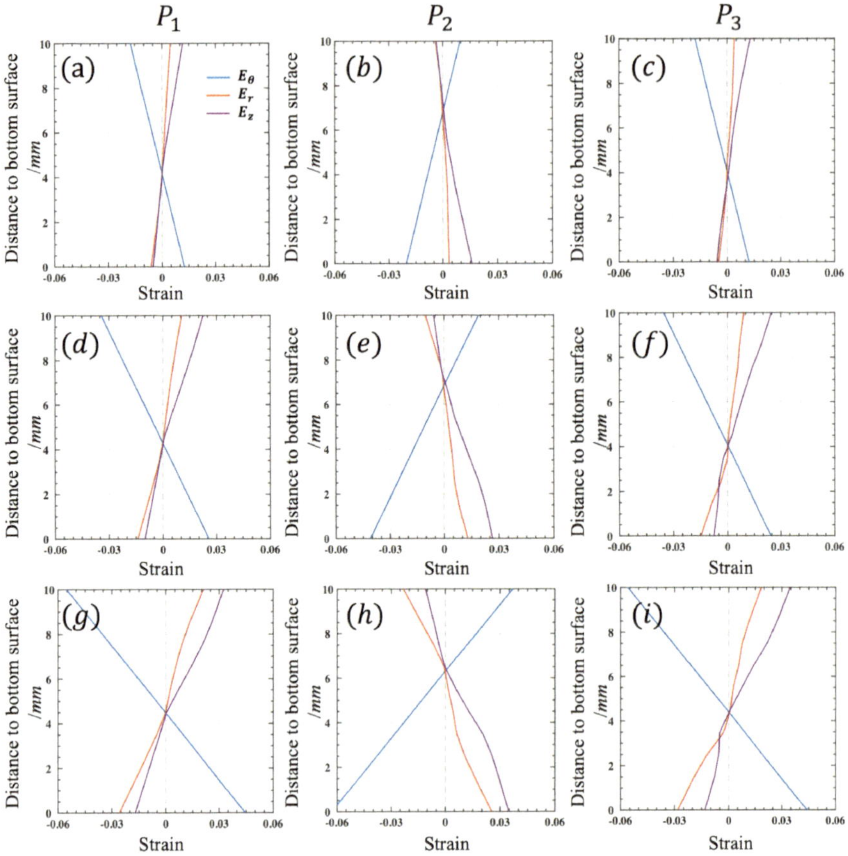

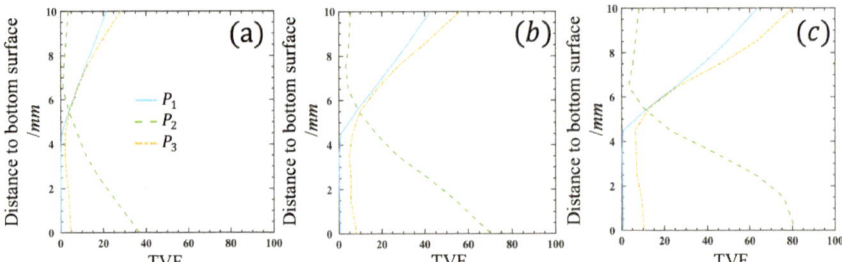

Fig. 7 The strain distribution at different loading stages. **a–c** The distribution at P_{1-3} of Cycle A, respectively; **d–f** The distribution at P_{1-3} of Cycle B, respectively; **g–i** The distribution at P_{1-3} of Cycle C, respectively

Fig. 8 The TVF distribution at different loading stages: **a** Cycle A; **b** Cycle B; **c** Cycle C

4   Conclusion

This work investigated the cyclic bending behavior of Mg alloys with curvature amplitudes of 0.003, 0.006, and 0.010 using the EVPSC-BEND model. The relative activity and twin volume fraction (TVF) during cyclic loading indicate the twinning-detwinning mechanism is actively involved throughout the entire cyclic bending process. The results of stress distribution also demonstrate that the twinning mechanism induces different yield stresses in different layers. The stress distribution results demonstrate that the twinning mechanism induces different yield stresses in different layers. Combined with the distribution of twin volume fraction (TVF), it can be observed that the action of twinning and detwinning reproduces similar stress distributions during different stages of the cyclic process. Our future research will involve conducting cyclic four-point bending experiments to further validate the predictive capability of the model. Additionally, we will perform microscopic characterization, focusing on the changes in texture and twin volume fraction during the cyclic bending process, to further analyze the role of twinning and detwinning mechanism.

Acknowledgements This work was supported by the National Natural Science Foundation of China (Numbers 52075325, 51975365, and 52011540403).

References

1. Eliezer D, Aghion E, (Sam) Froes FH (1998) Magnesium science, technology and applications. Adv Perform Mater 5:201–212. https://doi.org/10.1023/A:1008682415141
2. Agnew SR, Nie JF (2010) Preface to the viewpoint set on: the current state of magnesium alloy science and technology. Scripta Mater 63:671–673. https://doi.org/10.1016/j.scriptamat.2010.06.029
3. Pollock TM (2010) Weight loss with magnesium alloys. Science 328:986–987. https://doi.org/10.1126/science.1182848
4. Yang Y, Xiong X, Chen J, Peng X, Chen D, Pan F (2021) Research advances in magnesium and magnesium alloys worldwide in 2020. J Magnes Alloy 9:705–747. https://doi.org/10.1016/j.jma.2021.04.001
5. Cepeda-Jiménez CM, Pérez-Prado MT (2016) Microplasticity-based rationalization of the room temperature yield asymmetry in conventional polycrystalline Mg alloys. Acta Mater 108:304–316. https://doi.org/10.1016/j.actamat.2016.02.023
6. Chun YB, Davies CHJ (2011) Twinning-induced anomaly in the yield surface of highly textured Mg–3Al–1Zn plate. Scripta Mater 64:958–961. https://doi.org/10.1016/j.scriptamat.2011.01.044
7. Qiao H, Guo XQ, Oppedal AL, El Kadiri H, Wu PD, Agnew SR (2017) Twin-induced hardening in extruded Mg alloy AM30. Mater Sci Eng A 687:17–27. https://doi.org/10.1016/j.msea.2016.12.123
8. Shi B, Yang C, Peng Y, Zhang F, Pan F (2022) Anisotropy of wrought magnesium alloys: a focused overview. J Magnes Alloy 10:1476–1510. https://doi.org/10.1016/j.jma.2022.03.006
9. Roberts CS (1960) Magnesium and its alloys. Wiley, New York
10. Hauser FE, Landon PR, Dorn JE (1956) Deformation and fracture mechanisms of polycrystalline magnesium at low temperatures. Trans Am Soc Metals 48:986–1002

11. Burke EC, Hibbard WR (1952) Plastic deformation of magnesium single crystals. JOM 4:295–303. https://doi.org/10.1007/BF03397694
12. Reed-Hill RE, Robertson WD (1958) Pyramidal slip in magnesium. Trans TMS-AIME 212:256–259
13. Kelley EW, Hosford WF (1968) Plane-strain compression of magnesium and magnesium alloy crystals. Trans Met Soc AIME 242:5–13
14. Lou X, Li M, Boger R, Agnew S, Wagoner R (2007) Hardening evolution of AZ31B Mg sheet. Int J Plast 23:44–86. https://doi.org/10.1016/j.ijplas.2006.03.005
15. Wang H, Wu Y, Wu PD, Neale KW (2010) Numerical analysis of large strain simple shear and fixed-end torsion of HCP polycrystals. Comput Mater Contin 19:255
16. Wang H, Raeisinia B, Wu PD, Agnew SR, Tomé CN (2010) Evaluation of self-consistent polycrystal plasticity models for magnesium alloy AZ31B sheet. Int J Solids Struct 47:2905–2917. https://doi.org/10.1016/j.ijsolstr.2010.06.016
17. Sabbaghian M, Fakhar N, Nagy P, Fekete K, Gubicza J (2021) Investigation of shear and tensile mechanical properties of ZK60 Mg alloy sheet processed by rolling and sheet extrusion. Mater Sci Eng A 828:142098. https://doi.org/10.1016/j.msea.2021.142098
18. Wang H, Wu PD, Wang J (2013) Modeling inelastic behavior of magnesium alloys during cyclic loading–unloading. Int J Plast 47:49–64. https://doi.org/10.1016/j.ijplas.2013.01.007
19. Yu Q, Zhang J, Jiang Y, Li Q (2011) Multiaxial fatigue of extruded AZ61A magnesium alloy. Int J Fatigue 33:437–447. https://doi.org/10.1016/j.ijfatigue.2010.09.020
20. Zhang J, Yu Q, Jiang Y, Li Q (2011) An experimental study of cyclic deformation of extruded AZ61A magnesium alloy. Int J Plast 27:768–787. https://doi.org/10.1016/j.ijplas.2010.09.004
21. Wang L, Huang G, Han T, Mostaed E, Pan F, Vedani M (2015) Effect of twinning and detwinning on the spring-back and shift of neutral layer in AZ31 magnesium alloy sheets during V-bend. Mater Des 68:80–87. https://doi.org/10.1016/j.matdes.2014.12.017
22. Han T, Huang G, Wang Y, Wang G, Zhao Y, Pan F (2016) Enhanced mechanical properties of AZ31 magnesium alloy sheets by continuous bending process after V-bending. Prog Nat Sci Mater Int 26:97–102. https://doi.org/10.1016/j.pnsc.2016.01.005
23. Singh J, Kim M-S, Lee J-H, Guim H, Choi S-H (2019) Microstructure evolution and deformation behaviors of E-form and AZ31 Mg alloys during ex-situ mini-V-bending tests. J Alloy Compd 778:124–133. https://doi.org/10.1016/j.jallcom.2018.11.138
24. Tang D, Zhou K, Tang W, Wu P, Wang H (2022) On the inhomogeneous deformation behavior of magnesium alloy beam subjected to bending. Int J Plast 150:103180. https://doi.org/10.1016/j.ijplas.2021.103180
25. Zhou K, Sun X, Wang H, Zhang X, Tang D, Tang W, Jiang Y, Wu P, Wang P (2023) Texture-dependent bending behaviors of extruded AZ31 magnesium alloy plates. J Magnes Alloy, S2213956723000294. https://doi.org/10.1016/j.jma.2023.02.003
26. Simmons G, Wang H (1971) Single crystal elastic constants and calculated aggregate properties. The MIT Press, p 274

A Numerical Study to Analyze the Effect of Process Parameters on Ring Rolling of Ti-6Al-4V Alloy by Response Surface Methodology

Soumyaranjan Nayak, Abhishek Kumar Singh, Hina Gokhale, M. J. N. V. Prasad, and K. Narasimhan

Abstract Hot ring rolling is a production method to manufacture seamless rings. It is a complex incremental metal-forming process where reduction of cross-section leads to increase in diameter of the ring via circumferential extrusion. High degree of non-linearity and asymmetry is associated with the process. The process results in non-uniform distribution of temperature and plastic strain in the ring cross-section, and this in turn significantly affects the deformation behavior, microstructure, and mechanical properties. Form defect like fishtail defect is also a major concern and incurs loss in terms of labor and machining cost. In this study, rolling of Ti-6Al-4V rings is studied with the help of three-dimensional coupled thermo-mechanical finite element model established using ABAQUS/Explicit environment-based dynamic explicit code. The major parameters taken into consideration for the study are main roll speed (rpm), main roll feed (mm/s), and coefficient of friction. Each parameter was studied at two levels. Twenty simulations with different combinations of major parameters were developed via Central Composite Design (CCD). Coefficient of Variation (CoV) was used as a heterogeneity index to ascertain heterogeneity in equivalent plastic strain (PEEQ) and temperature distribution in the ring. Fishtail defect was quantified using fishtail coefficient as an index. Analysis of variance (ANOVA) was used to ascertain the impact of significant factors and interactions between different parameters affecting the ring rolling process. ANOVA technique requires unrestricted range of $(-\infty, \infty)$ for analysis. Hence, logit transformation is used to transform fishtail coefficient present in the range 0–1 to an unrestricted real number range $(-\infty, \infty)$. Main roll feed rate was found to be the most significant factor affecting CoV (PEEQ), CoV (temperature) and logit transformation of fishtail coefficient and has an inverse correlation and quadratic relationship with all the

S. Nayak · A. K. Singh · H. Gokhale · M. J. N. V. Prasad · K. Narasimhan (✉)
Department of Metallurgical Engineering and Materials Science, Indian Institute of Technology Bombay, Mumbai 400076, India
e-mail: nara@iitb.ac.in

A. K. Singh
Department of Advanced Components and Materials Engineering, Sunchon National University, Suncheon 57922, Republic of Korea

© The Rightsholder, under exclusive licence to [Springer Nature Switzerland AG], part of Springer Nature 2024
J. Kusiak et al. (eds.), *Numerical Methods in Industrial Forming Processes*, Lecture Notes in Mechanical Engineering, https://doi.org/10.1007/978-3-031-58006-2_25

responses. Other sources of variation like main roll speed (rpm) and coefficient of friction (CoF) have minimal impact. Increase in feed rate was found to reduce CoV (PEEQ), CoV (temperature), and logit transformation of fishtail coefficient.

Keywords Ring rolling · Finite element simulation · Response Surface Method

1 Introduction

Ring rolling is a bulk forming process used to manufacture seamless rings [1]. It is highly non-linear process with coupled thermo-mechanical effect [2]. Analyzing the ring rolling process using either experimental or analytical method is a material and time-consuming process. However, the use of finite element (FE) simulations to analyze the process is the most convenient tool [3]. Dynamic boundary contact between the rolls and ring is seen in this process and hence the use of dynamic explicit approach is ideal to simulate the ring rolling process [4]. A detailed study of literature shows that there exists a good number of studies done previously pertaining to ring rolling. Seitz et al. [5] transferred the principle of roll bonding to produce seamless radial composite rings using 3D FE model, which combine the advantages of different materials. S. Guenther et al. [6] further extended the work on composite ring rolling process via 3D FE simulations and experiments on steel rings. Cleaver et al. [7] introduced a novel concept for making profiled L shaped rings without ring growth. Seitz et al. [8] used ring rolling process to produce dish shaped rings using 3D FE simulations. Ring rolling of Ti alloys via FE simulations has also been reported in literature. Taek et al. developed a three-dimensional (3D) finite element (FE) model and used processing map to estimate the locations in the rolled Ti-6Al-4V (Ti-64) ring where defects can be expected to develop during the rolling process [9]. Yang et al. developed 3D FE models of ring rolling process to study the effect of size of rectangular-sectioned Ti-64 blanks on strain and temperature distribution [10]. Liang et al. used Response Surface Methodology (RSM) to analyze the effects of instantaneous ring diameter, ring growth velocity and initial ring temperature and on temperature distribution in TA15 alloy ring [11]. Wang et al. analyzed impact of rolling speed, feed rate and initial temperature on volume fraction and grain size of β phase in Ti-64 alloy using 3D FE model [12]. Yeom et al. developed a 3D FE model and simulated strain and temperature and examined microstructure spread in Ti-64 ring [13].

In addition, there exist some studies pertaining to heterogeneity in temperature and equivalent plastic strain (PEEQ) distribution and formation of fishtail defect via FE analysis for different alloys in literature. Li et al. studied the ring rolling of AA 6061 alloy and established that increase in main roll speed resulted in decrease in the heterogeneity of temperature distribution and increase in PEEQ distribution in the ring [14]. Lee and Kim analyzed ring rolling of AISI 1035 steel and found that a reduction in the heterogeneity in the distribution of temperature and PEEQ when the feed rate was increased [15]. Qian and Peng studied the spread of temperature and PEEQ in AISI 5140 steel ring and demonstrated that increase in feed rate and main roll speed resulted in increase in heterogeneity [16]. But a study by Sun et al. on ring rolling of AISI 5140 steel reported a reverse trend of fall in heterogeneity of temperature and PEEQ distribution with increase in feed rate [17]. Therefore, it becomes evident that alteration in material, process parameters, preform dimensions and roll size can change the heterogeneity patterns. In our recent study [18] on the effect of main roll feed (mm/s) and main roll speed (rpm) on heterogeneity of temperature and PEEQ distribution and fishtail defect during rolling of Ti-64 rings, it was ascertained that feed rate is the only significant process parameter impacting heterogeneity of temperature and PEEQ distribution and formation of fishtail defect. A reduction in heterogeneity index and fishtail defect with increase in feed rate was seen. In the present study, a new process parameter, i.e., coefficient of friction (CoF), has been considered apart from the main roll speed and the feed rate and their impact on heterogeneity of temperature and PEEQ distribution and fishtail defect has been looked into. So, an effort has been made in the present study to determine the effects of process parameters (main roll speed and feed rate and coefficient of friction) on heterogeneity of temperature and PEEQ distribution and form defect like fishtail defect. This aim was achieved by conducting FE simulation runs with process parameters decided as per the three parameters two levels, Central Composite Design (CCD) and detailed statistical analysis.

2 Material and Methodology

2.1 Material

Figure 1a and b shows the forged Ti-64 cylindrical billet obtained from Mishra Dhatu Nigam Limited, Hyderabad, India. Doughnut shaped preform shown in Fig. 1c was machined from the forged billet and used as the starting material. Figure 1d shows the ring after rolling with 30% reduction in cross-section.

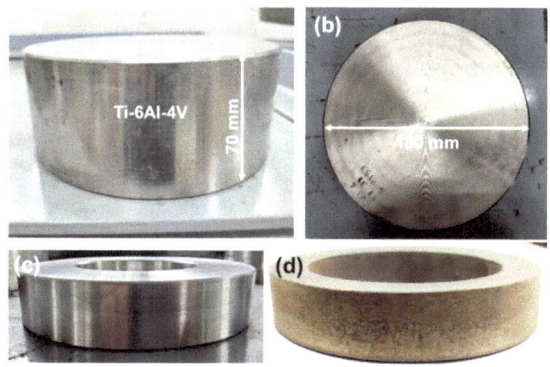

Fig. 1 Dimensions and photographs of forged Ti-64 cylindrical billet captured in two different views **a** side and **b** top. Photographs of the **c** preform and **d** the final ring

2.2 Lab Scale Ring Rolling Facility

The preform was initially soaked in the furnace at 950 °C for 1.5 h. Soaking helps in attaining uniform temperature in the preform cross-section and improves formability [19, 20]. Then the preform was rolled in the lab scale ring rolling facility, when the preform touched 880 °C. The main roll undergoes both radial and linear motion, as shown in Fig. 2. The reduction of cross-section of ring in the deformation process results in circumferential extrusion and increase in ring diameter. The guide roll and mandrel are freely mounted and undriven. Guide roll has the role to maintain circularity of the ring and stabilize the rolling process. Rolling parameters and dimensions of different rollers, preform and ring are provided in Table 1. Table 1 shows that two rings rolled with two different feed rates of 1 and 2.5 mm/s by keeping the other parameters constant. More details of the rolling experiments can be found elsewhere [18].

FLIR T620 thermal imaging camera was used to monitor the temperature in the zone of deformation while the preform was rolled. A cross-section reduction of 30% was achieved in both the rings. After rolling, the rings were left to be cooled in air.

Fig. 2 Lab scale ring rolling facility showing the rotational and translational motion of the main roll

Main roll

Preform
Mandrel

Guide roll

Table 1 Parameters used in lab scale ring rolling experiment

Deformation temperature (°C)	880
Ambient temperature (°C)	25
Inner diameter of preform (mm)	80
Outer diameter of preform (mm)	150
Main roll radius (mm)	266
Mandrel radius (mm)	30
Main roll speed (rpm)	26
Main roll feed (mm/s)	1, 2.5
Final inner diameter of ring (mm)	120
Final outer diameter of ring (mm)	170

2.3   Development of Ring Rolling FE Model

Figure 3a shows the developed 3D coupled thermo-mechanical FE model for the lab scale ring rolling machine using ABAQUS/Explicit software. C3D8RT, the eight-node thermally coupled brick element was used to mesh the ring. Adaptive remeshing was applied to improve the quality of simulation results. Guide roll, mandrel, and main roll are considered analytically rigid. Reduced integration and hourglass control has been implemented. In order to reduce the computational time, mass scaling factor of 1000 was applied to the FE models [21]. Mesh size of 1.85 was chosen, resulting in 56,241 elements, using the mesh sensitivity analysis provided in Fig. 3b. Contact pairs have been provided with coefficient of friction of 0.3 [11]. Convection and radiation are considered on all free surfaces. Taylor-Quinney coefficient is assumed to be 0.9 [22]. Thermo-physical properties of Ti-64 alloy are given in Table 2. The values of specific heat, conductivity, density, and Young's modulus were taken from literature [23–26].

Figure 4 shows the comparison of the experimental results with the FE models developed. Figure 4a and b respectively compares the height of the ring at ID (inner diameter), OD (outer diameter), and center in the final cross sections of the simulated

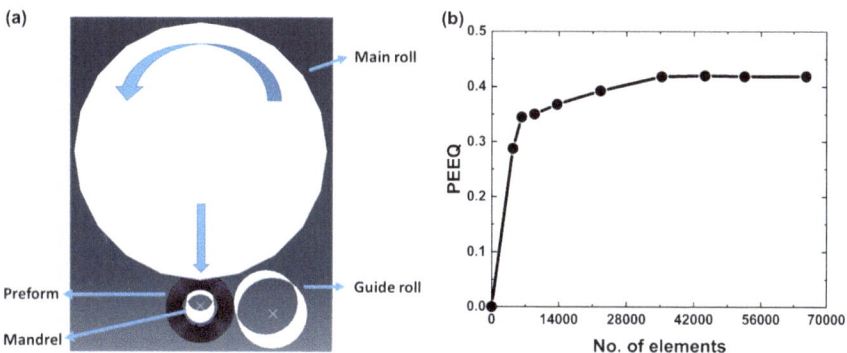

Fig. 3 a FE model of lab scale ring rolling set-up developed using ABAQUS/Explicit platform **b** mesh sensitivity analysis

Table 2 Thermo-physical properties of Ti-64 alloy

Temperature (°C)	Young's modulus (GPa)	Thermal conductivity (W/mK)	Density (kg/m³)	Specific heat (J/K kg)
25	110	7	4420	546
300	94	10	4381	606
600	77	14	4336	673
900	60	20	4294	734
1200	42	22	4252	678

and experimental rings for LFR condition. Similarly, Fig. 4c and d, respectively, shows the final cross-sections of the simulated and experimental rings for HFR condition. Maximum relative error of 13.60% was observed at OD of LFR ring while predicting the height. A comparision of predicted and experimental temperature profile is given in Fig. 4e. Maximum relative error of 3% was observed in temperature prediction in LFR ring. Figure 4f shows the rate of outer diameter growth in both LFR and HFR ring. Maximum relative error of 2.8% in growth predictions was seen in LFR rings.

In order to develop the constitutive equation, Ti-64 cylindrical specimens (15 mm height and 10 mm diameter) were compressed resulting in 60% reduction in height attaining strain level of 0.91 in a strain rate and temperature range of 0.001–10 s^{-1} and 750–950 °C, respectively. The obtained flow curves were adiabatically corrected and subsequently hyperbolic sine equation was developed, which was used as a constitutive equation to run simulations. Details of development of the constitutive equation can be found elsewhere [27]. The general form of hyperbolic sine equation is given in Eq. (1).

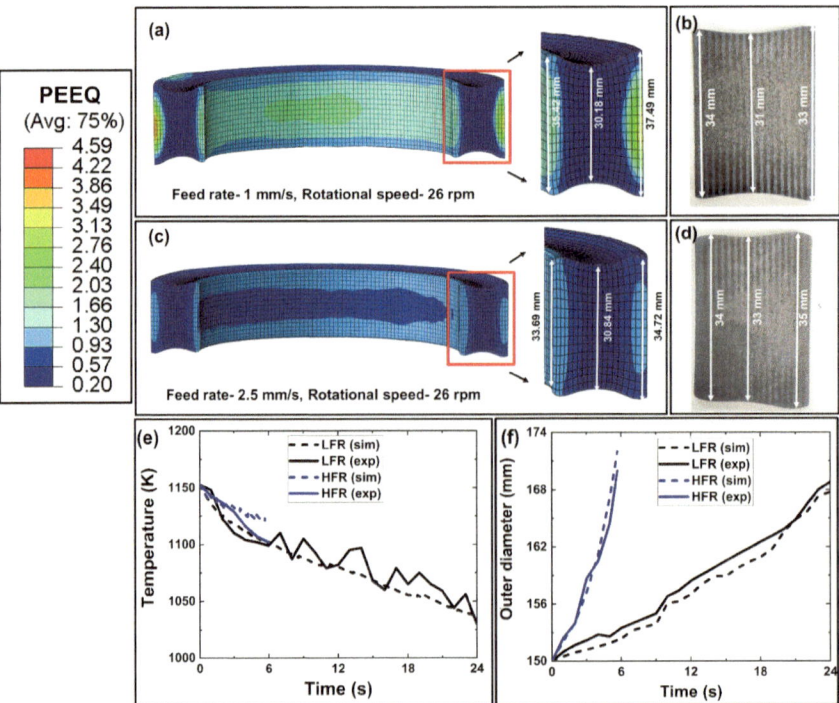

Fig. 4 Cross-sections of **a, c** simulated and **b, d** experimental rings for the LFR and the HFR conditions, respectively, **e** temperature variation in the deformation zone in simulated (sim) and experimental (exp) rings for the LFR and the HFR conditions and **f** outer diameter growth rate for LFR and HFR rings observed in simulation and experiment

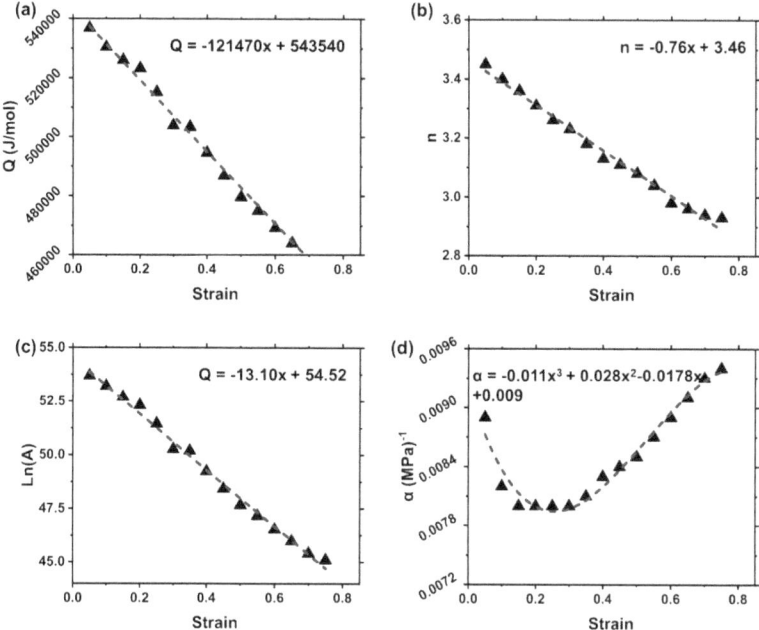

Fig. 5 Plot representing the data as a function of strain with polynomial fit for determining the constants of Arrhenius equation

$$\dot{\varepsilon} = A[\sinh\alpha\sigma]^{n} \cdot \exp\left(-\frac{Q}{RT}\right) \tag{1}$$

where n, α, and A are constants while Q is the activation energy. Equation (1) lacks the strain term. Hence, in order to integrate the strain term, polynomial fit of all the constants (n, α, A and Q) with respect to strain is plotted in Fig. 5. The polynomial equations mentioned are then applied in simulation through a user-defined material subroutine (VUMAT).

2.4 Development of Simulation Matrix

Main roll feed, main roll speed, temperature, and coefficient of friction (μ) are the important process parameters for ring rolling process. In this study, a three-factor (main roll feed (mm/s), main roll speed (rpm) and coefficient of friction (CoF or μ)) and two-level Central Composite Design (CCD) were constituted and are shown in Fig. 6. CCD is a fractional factorial design and is integral part of Response Surface Methodology (RSM). Table 3 shows the minimum and maximum values of process parameters considered for analysis in this study. A range of values chosen for the main roll feed (mm/s) is to ensure that there is no adiabatic damage. The entire functional range of main roll speed (rpm) in the ring rolling set-up is chosen for the study. In case of coefficient of friction a range of 0.1–0.3 was chosen for analysis initially. But, a low coefficient of friction value of 0.1 failed to achieve rolling, as can be seen in Fig. 7. In the process of ring rolling, it is essential that the ring is drawn into the gap between the main roll and mandrel [28]. This process results in gradual reduction in thickness and expansion of the ring. But, in case the friction is less, the drawn-in of the ring to the deformation zone between the main roll and mandrel won't be possible resulting in non-rotation of the ring and compression as shown in Fig. 7. Hence, coefficient of friction range had to be changed to 0.2–0.3. The aim was to see if reduced friction does make any impact. The total deformation level of about 30% was maintained for all the simulations run.

It was found out that there was a noticeable difference between the feed rate that the rolling mill was set to achieve (prescribed feed rate) and the feed rate that it could actually achieve (effective feed rate) during rolling. Hence, the main roll feed range of 0.44–1.76 mm/s was used for simulation, as given in Table 3. Similar difference in prescribed and effective feed rates was also reported in literature [29].

Fig. 6 Central composite design (CCD) showing the scheme of simulations run

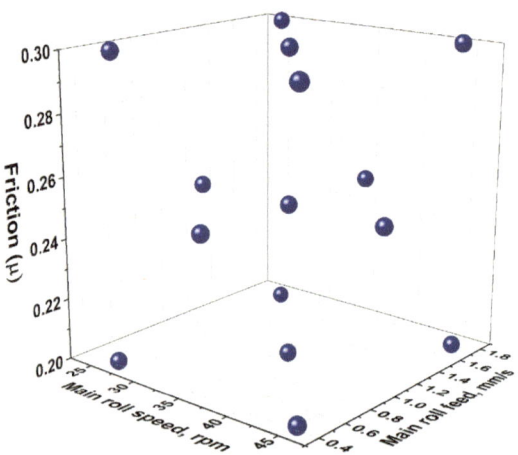

Table 3 Factors and levels of the CCD

Parameters	Notation	Units	Level	
			Minimum	Maximum
Main roll feed	f	mm/s	0.44	1.76
Main roll speed	r	rpm	26	45
Coefficient of friction	μ	–	0.2	0.3

Fig. 7 FE Simulation depicting failure to achieve ring rolling operation at low coefficient of friction ($\mu = 0.1$)

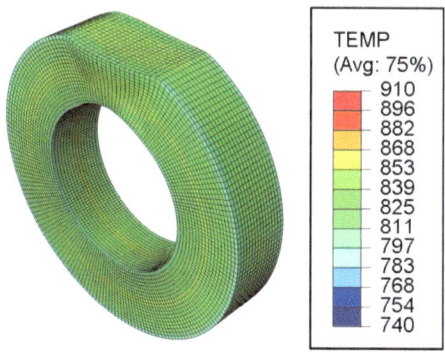

3 Recorded Responses

20 simulations were run following CCD shown in Fig. 6. PEEQ and temperature distribution and fishtail coefficient are recorded in Table 4.

3.1 *Heterogeneity in PEEQ and Temperature Distribution*

Heterogeneity studies of PEEQ and temperature distribution are important for bulk forming processes like ring rolling. The preform in FE simulation was divided into 56,241 elements, and the PEEQ and temperature in each element were measured after rolling process. Coefficient of variation (CoV) was used in the present study as a heterogeneity index to analyze equivalent plastic strain and temperature distribution. It is defined as the ratio of standard deviation and mean [18]. CoV is unit-less and higher is its values suggesting higher is the inhomogeneity in PEEQ and temperature distribution. The formula used to determine the mean (μ) and standard deviation (σ) are given in Eqs. (2) and (3), respectively, for PEEQ. Equation (4) is used to calculate CoV, as presented below.

$$\mu = \sum_{p=1}^{M} PEEQ_p / M \tag{2}$$

$$\sigma = \sqrt{\frac{\sum_{p=1}^{M} \left(PEEQ_p - \mu\right)^2}{M - 1}} \tag{3}$$

$$CoV = \frac{\sigma}{\mu} \tag{4}$$

where, M is the total number of elements into which the ring has been divided. $PEEQ_p$ is the equivalent plastic strain of each element. Similarly, CoV was determined to

Table 4 Simulation runs as per CCD parameters and responses

Serial number	Speed (rpm)	Feed rate (mm/s)	Friction (μ)	CoV (PEEQ)	CoV (Temperature, K) $*10^4$	Logit fishtail
1	35.5	1.1	0.25	0.602	133.333	−0.685
2	35.5	1.1	0.25	0.602	133.333	−0.685
3	35.5	1.1	0.25	0.602	133.333	−0.685
4	35.5	1.76	0.25	0.495	109.890	−0.680
5	35.5	1.1	0.25	0.602	133.333	−0.685
6	26	1.1	0.25	0.545	135.716	−0.692
7	35.5	1.1	0.25	0.602	133.333	−0.685
8	35.5	1.1	0.2	0.631	137.649	−0.589
9	45	0.44	0.3	0.980	172.335	−0.476
10	26	1.76	0.2	0.444	110.462	−0.658
11	45	1.76	0.3	0.527	108.442	−0.683
12	35.5	0.44	0.25	0.926	176.201	−0.414
13	45	1.1	0.25	0.638	133.345	−0.658
14	26	1.76	0.3	0.439	113.339	−0.828
15	45	0.44	0.2	0.946	171.255	−0.487
16	35.5	1.1	0.3	0.530	128.500	−0.529
17	35.5	1.1	0.25	0.602	133.333	−0.685
18	26	0.44	0.2	0.859	175.487	−0.421
19	45	1.76	0.2	0.538	106.876	−0.661
20	26	0.44	0.3	0.868	174.750	−0.492

study the heterogeneity in temperature distribution. CoV calculation for temperature analysis was done in Kelvin scale. Figures 8 and 9, respectively, show the distribution of PEEQ and temperature in the ring cross-section, run as per CCD, after attaining 30% deformation.

3.2 Fishtail Formation

Fishtail is one of the critical form defects encountered in the ring rolling industry where irregular height in the cross-section of the formed ring results because of unrestricted material flow in the axial direction [30], as can be seen in Fig. 10. Fishtail defect is quantified by fishtail coefficient calculated using Eq. (5) [18]. The maximum height of the ring cross-section is subtracted from the minimum height of the ring cross-section and the difference is divided by the preform height. Fishtail coefficient varies between 0 and 1 as it is a normalized value. But, for implementation of the

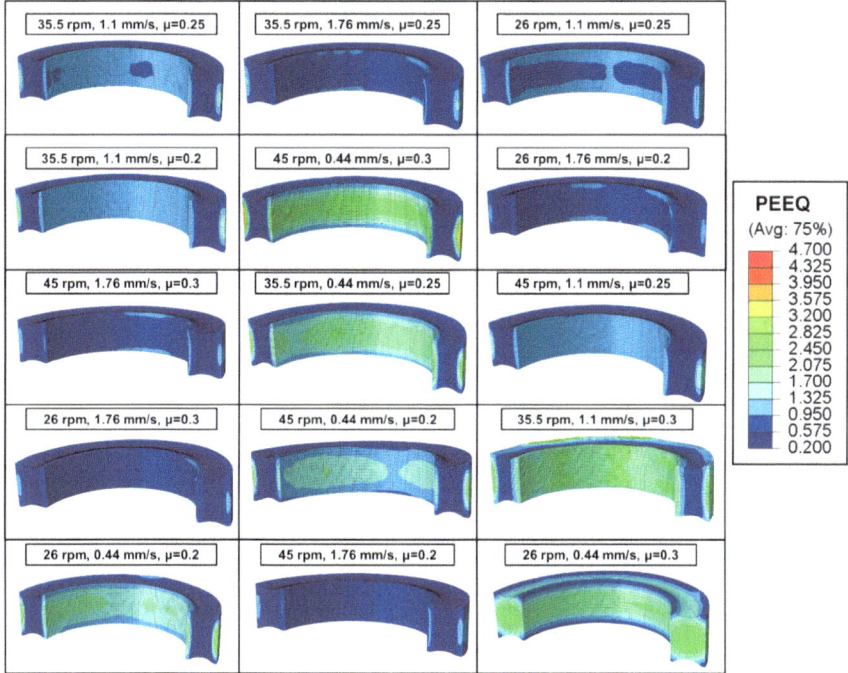

Fig. 8 FE modeling results showing the PEEQ spread after attaining 30% deformation using CCD parameters

ANOVA technique, an unrestricted range of $(-\infty, \infty)$ is required. Hence, for further analysis, logit transformation with log base 10 function has been used, primarily to pull out the ends of the distribution to an unrestricted range of $(-\infty, \infty)$, given by Eq. (6).

$$\text{Fishtail coefficient (x)} = \frac{\text{Max. height of ring cross section} - \text{Min. height of ring cross section}}{\text{Height of the undeformed preform}} \quad (5)$$

$$y = \text{logit}(x) = \log\left(\frac{x}{1-x}\right) \quad (6)$$

4 Regression Equation and Analysis of Variance (ANOVA)

FE simulations yielded CoV (PEEQ), CoV (temperature), and logit transformation of fishtail coefficient as responses. These responses can be denoted as function (f_i) of main roll speed (r), main roll feed (v), and coefficient of friction (μ) [31], as shown below:

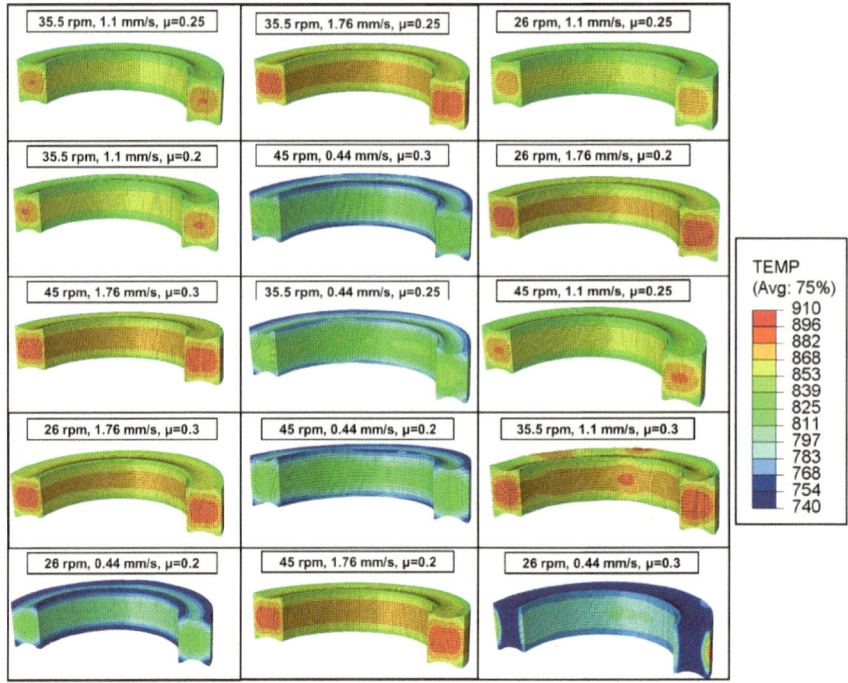

Fig. 9 FE modeling results showing the temperature spread after attaining 30% deformation using CCD parameters

$$\text{CoV (PEEQ)} = f_1(r, v, \mu) \tag{7}$$

$$\text{CoV (Temperature)} = f_2(r, v, \mu) \tag{8}$$

$$\text{logit Fishtail Coefficient} = f_3(r, v, \mu) \tag{9}$$

The polynomial equation can be stated as:

$$f_i(r, v, \mu) = a_0 + a_1 r + a_2 v + a_3 \mu + a_{12} rv + $$
$$a_{13} r\mu + a_{23} v\mu + a_{11} r^2 + a_{22} v^2 + a_{33} \mu^2 + \varepsilon \tag{10}$$

where a_0 is a constant and is estimated as average of the responses and a_1, a_2, a_3, a_{12}, a_{13}, a_{23}, a_{11}, a_{22}, and a_{33} are the regression coefficients for the linear, interaction, and squared terms in Eq. (10). ε is the error term [32].

Fig. 10 Photographs of preform and sectioned ring with fishtail defect

4.1 PEEQ Heterogeneity Analysis

ANOVA test results for PEEQ are presented in Table 5. It shows the various sources of variation and ascertains the p-value for the linear, squared, and interaction terms. Statistically significant terms have p-value less than 0.05. Main roll speed, main roll feed, and square of main roll feed are found to be significant based on the p-values with a contribution of 4.14%, 84.15%, and 6.8%, respectively.

The empirical relationship developed for PEEQ using the coefficients is presented below:

$$\text{CoV (PEEQ)} = 0.732 + 0.0061r - 0.842v + 2.65\mu - 0.000029r^2$$
$$+ 0.2666v^2 - 5.46\mu^2 - 0.00031Rv + 0.0051r\mu - 0.227f\mu \quad (11)$$

The final empirical relationship using coefficients from significant terms in Table 5 is presented below:

$$\text{CoV (PEEQ)} = 0.732 + 0.0061r - 0.842v + 0.2666v^2 \quad (12)$$

The R^2 value was found to be 0.9909, which explains that 99.09% of statistical variation can be explained by the chosen parameters.

4.2 Temperature Heterogeneity Analysis

The empirical relationship representing the impact of factors and interactions is formulated using coefficients as given below:

$$10000 \times \text{CoV (temperature)} = 201.3 - 0.44r - 95.9v + 193\mu + 0.0037r^2$$
$$+ 20.32v^2 - 448\mu^2 - 0.037rv + 0.13r\mu + 15.5v\mu \quad (13)$$

Table 5 ANOVA test results for PEEQ heterogeneity ($R^2 = 0.9909$)

Source	DF	Adj SS	Adj MS	F-value	p-value	Relative contribution (%)
Model	9	0.53644	0.059605	120.40	0.000	99.08
Linear	3	0.478611	0.159537	322.25	0.000	88.40
Main roll speed, rpm	1	0.022448	0.022448	45.34	0.000	4.14
Main roll feed, mm/s	1	0.455603	0.455603	920.28	0.000	84.15
CoF, μ	1	0.000561	0.000561	1.13	0.312	0.10
Square	3	0.057306	0.019102	38.58	0.000	10.5
(Main roll speed, rpm)^2	1	0.000018	0.000018	0.04	0.852	0.00
(Main roll feed, mm/s)^2	1	0.037090	0.037090	74.92	0.000	6.8
(CoF, μ)^2	1	0.000513	0.000513	1.04	0.333	0.00
2-way interaction	3	0.000527	0.000176	0.35	0.787	0.00
Main roll speed * Main roll feed	1	0.000030	0.000030	0.06	0.811	0.00
Main roll speed * Friction	1	0.000048	0.000048	0.10	0.763	0.00
Main roll feed * Friction	1	0.000449	0.000449	0.91	0.363	0.00
Error	10	0.004951	0.000495			0.90
Lack-of-fit	5	0.004951	0.000990	*	*	0.90
Pure error	5	0.000000	0.000000			0.00
Total	19	0.541395				

The significant factors provided in Table 6 contributing to heterogeneity in temperature distribution are the main roll speed, the main roll feed, and square of the main roll feed with contribution of 0.27%, 95.90%, and 2%, respectively.

The final empirical relationship involving significant term is given below:

$$10000 \times \text{CoV (temperature)} = 201.3 - 0.44r - 95.9v + 20.32v^2 \tag{14}$$

The R^2 value was found to be 0.9953.

Table 6 ANOVA test results for temperature heterogeneity ($R^2 = 0.9953$)

Source	DF	Adj SS	Adj MS	F-value	p-value	Relative contribution (%)
Model	9	10695.3	1188.4	237.43	0.000	99.53
Linear	3	10337.9	3446.0	688.50	0.000	96.20
Main roll speed, rpm	1	30.6	30.6	6.12	0.033	0.27
Main roll feed, mm/s	1	10305.3	10305.3	2058.98	0.000	95.90
CoF, μ	1	1.9	1.9	0.38	0.551	0.02
Square	3	354.8	118.3	23.63	0.000	3.30
(Main roll speed, rpm)^2	1	0.3	0.3	0.06	0.808	0.00
(Main roll feed, mm/s)^2	1	215.5	215.5	43.05	0.000	2.00
(CoF, μ)^2	1	3.4	3.4	0.69	0.426	0.03
2-way interaction	3	2.6	0.9	0.17	0.914	0.02
Main roll speed * Main roll feed	1	0.4	0.4	0.08	0.778	0.00
Main roll speed * Friction	1	0.0	0.0	0.01	0.938	0.00
Main roll feed * Friction	1	2.1	2.1	0.42	0.532	0.02
Error	10	50.1	5.0			0.46
Lack-of-fit	5	50.1	10.0	*	*	0.46
Pure error	5	0.0	0.0			0.00
Total	19	10745.3				

4.3 Fishtail Coefficient Analysis

ANOVA results for the fishtail defect are listed in Table 7. The significant terms are found to be the main roll feed rate, square of main roll feed rate, and square of friction factor with contribution of 64.19%, 6.37%, and 4.5%, respectively.

The empirical relationship developed for fishtail coefficient using the coefficients is presented below:

$$
\begin{aligned}
\text{Logit transformation of fishtail coefficient} = {} & 1.126 + 0.0247r - 0.566v - 14.29\mu \\
& - 0.000602r^2 + 0.1680v^2 + 24.6\mu^2 \\
& + 0.00385rv - 0.065r\mu - 0.501v\mu
\end{aligned}
\tag{15}
$$

The final empirical relationship using significant term from Table 7 is given below:

Table 7 ANOVA test results for fishtail defect formation ($R^2 = 0.9101$)

Source	DF	Adj SS	Adj MS	F-Value	p-Value	Relative contribution (%)
Model	9	0.210649	0.023405	11.37	0.000	91.19
Linear	3	0.153620	0.051207	24.88	0.000	66.50
Main roll speed, rpm	1	0.001601	0.001601	0.78	0.398	0.7
Main roll feed, mm/s	1	0.148286	0.148286	72.04	0.000	64.19
CoF, μ	1	0.003733	0.003733	1.81	0.208	1.6
Square	3	0.043574	0.014525	7.06	0.008	18.86
(Main roll speed, rpm)^2	1	0.008128	0.008128	3.95	0.075	3.51
(Main roll feed, mm/s)^2	1	0.014733	0.014733	7.16	0.023	6.37
(CoF, μ)^2	1	0.010413	0.010413	5.06	0.048	4.5
2-way interaction	3	0.013455	0.004485	2.18	0.154	5.82
Main roll speed × Main roll feed	1	0.004656	0.004656	2.26	0.163	2.01
Main roll speed * Friction	1	0.006614	0.006614	3.21	0.103	2.86
Main roll feed * Friction	1	0.002184	0.002184	1.06	0.327	0.94
Error	10	0.020585	0.002058			8.91
Lack-of-fit	5	0.020585	0.004117	*	*	8.91
Pure error	5	0.000000	0.000000			0
Total	19	0.231234				

$$\text{Logit transformation of fishtail coefficient} = 1.126 - 0.566v$$
$$+ 0.1680v^2 + 24.6\mu^2 \quad (16)$$

The R^2 value was of 0.9101.

Hence, it can be clearly stated that the main roll feed is the most significant factor contributing to heterogeneity of PEEQ and temperature and fishtail formation. The second-order polynomial fit of responses (CoV (PEEQ), CoV (Temperature) and logit fishtail) with respect to process parameters are presented in Fig. 11. The results echo the findings of ANOVA in Tables 5, 6 and 7 and affirm that the main roll feed is the most significant factor. There is a negative correlation and quadratic relationship between responses and main roll feed rate.

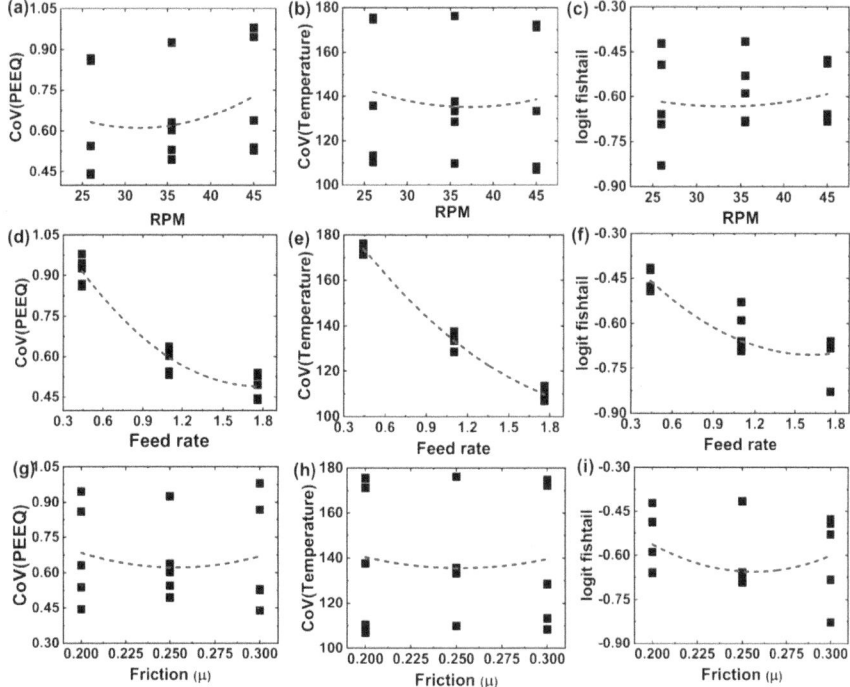

Fig. 11 **a, d, g** CoV (PEEQ), **b, e, h** CoV (temperature) and **c, f, i** fishtail coefficient variation with main roll speed, main roll feed, and coefficient of friction

4.4 Interpretation of DoE Optimization

Relationship, given in Eq. (17), between feed per revolution (Δh_i) and process parameters like main roll speed (n_1) and main roll feed (υ) was proposed Hawkyard et al. [33]:

$$\Delta h_i = \frac{2\pi \upsilon R_i}{n_1 R_1} \tag{17}$$

where the terms R_i and R_1 are the outer radius of the deforming ring and radius of the driver roll, respectively. Equation (17) shows that feed per revolution is directly proportional to main roll feed rate (υ) and inversely proportional to main roll speed (n_1). Higher feed rate makes penetration of plastic deformation easier into the radial thickness, thus creating a more uniform strain distribution.

In Fig. 11, it can be seen that CoV (PEEQ) reduces with increase in feed rate. Increase in feed rate leads to greater penetration in the ring cross-section resulting in greater uniformity in strain distribution and reduced CoV (PEEQ). Increase in feed rate also results in reduced CoV (temperature) as adiabatic heating increases during high feed rate deformation increasing the temperature [34], resulting in internal temperature rise. In case of high feed rate, the desired deformation is attained in less time leading to lesser temperature loss through the surfaces. Hence, a uniform temperature distribution can be resulted in case of high feed rate deformation. Another reason for more uniform deformation in case of high feed rate deformation can be attributed to lower material deformation resistance at higher temperature resulting more uniform deformation [4]. Both the main roll speed (rpm) and coefficient of fraction play little role as suggested both in ANOVA and Fig. 11. It has been reported in literature, a more uniform deformation results in reduction in fishtail coefficient [35], and hence, higher feed rate resulted in reduced fishtail defect.

5   Conclusion

In the present study, the effects of process parameters such as main roll speed, main roll feed and coefficient of friction on heterogeneity of strain (PEEQ) and temperature distribution and form defect like fishtail defect were looked into. A series of 20 3D FE simulations of ring rolling process was run based on a three-parameter two-level Central Composite Design (CCD). Coefficient of Variation (CoV) was used as a heterogeneity index. The extent of Fishtail defect is captured via fishtail coefficient. The main conclusions are as follows:

- FE simulations show that reduced coefficient of friction ($\mu = 0.1$) resulted in failure to achieve the rolling process.
- Feed rate has been established to be the most significant factor affecting the severity of fishtail defect and heterogeneity in temperature and equivalent plastic strain (PEEQ) distribution.
- The effects of other terms like main roll speed, square of feed, and square of coefficient of friction have been found to be minimal.
- ANOVA study shows that CoV (PEEQ), CoV (temperature), and logit fishtail have a quadratic relationship with feed rate (mm/s).
- The responses (CoV (PEEQ), CoV (temperature), and logit of fishtail) shared a negative correlation with feed rate.

References

1. Phalke V, Nayak S, Narasimhan K, Nandedkar VM (2019) 3D coupled thermo-mechanical FE analysis of effect of process parameters in ring rolling process, Вестник Магнитогорского Государственного Технического Университета Им. Г.И. Носова. 17:24–31. https://doi.org/10.18503/1995-2732-2019-17-2-24-31
2. Guo L, Yang H (2011) Towards a steady forming condition for radial-axial ring rolling. Int J Mech Sci 53:286–299. https://doi.org/10.1016/j.ijmecsci.2011.01.010

3. Peng WF, Niu BK, Zhang JH, Hong Z, Shu XD (2017) A 3D-FEM of adaptive movement control of guide and conical rolls in ring rolling process. Int J Adv Manuf Technol 92:3287–3298. https://doi.org/10.1007/s00170-017-0395-x

4. Wang M, Yang H, Sun ZC, Guo LG (2009) Analysis of coupled mechanical and thermal behaviors in hot rolling of large rings of titanium alloy using 3D dynamic explicit FEM. J Mater Process Technol 209:3384–3395. https://doi.org/10.1016/j.jmatprotec.2008.07.054

5. Seitz J, Schwich G, Guenther S, Hirt G (2016) Investigation of a composite ring rolling process by FEM and experiment. In: MATEC web of conferences, vol 80, pp 1–7. https://doi.org/10.1051/matecconf/20168015011

6. Guenther S, Seitz J, Schwich G, Hirt G (2017) Investigation of a composite ring rolling process considering bonding behaviour in FEM and experiment. Procedia Eng 207:1236–1241. https://doi.org/10.1016/j.proeng.2017.10.876

7. Cleaver CJ, Lohmar J, Tamimi S (2021) Limits to making L-shape ring profiles without ring growth. J Mater Process Technol 292:117062. https://doi.org/10.1016/j.jmatprotec.2021.117062

8. Seitz J, Jenkouk V, Hirt G (2013) Manufacturing dish shaped rings on radial-axial ring rolling mills. Prod Eng 7:611–618. https://doi.org/10.1007/s11740-013-0486-y

9. Taek J, Han J, Kwang N, Sik S, Soo C (2007) Ring-rolling design for a large-scale ring product of Ti-6Al-4V alloy. J Mater Process Tech 188:747–751. https://doi.org/10.1016/j.jmatprotec.2006.11.042

10. Yang H, Wang M, Guo LG, Sun ZC (2008) 3D coupled thermo-mechanical FE modeling of blank size effects on the uniformity of strain and temperature distributions during hot rolling of titanium alloy large rings. Comput Mater Sci 44:611–621. https://doi.org/10.1016/j.commatsci.2008.04.026

11. Liang L, Guo L, Wang Y, Li X (2019) Towards an intelligent FE simulation for real-time temperature-controlled radial-axial ring rolling process. J Manuf Process 48:1–11. https://doi.org/10.1016/j.jmapro.2019.09.032

12. Wang M, Yang H, Zhang C, Guo LG (2013) Microstructure evolution modeling of titanium alloy large ring in hot ring rolling. Int J Adv Manuf Technol 66:1427–1437. https://doi.org/10.1007/s00170-012-4420-9

13. Yeom JT, Kim JH, Hong JK, Park NK, Lee CS (2010) FE analysis of microstructure evolution during ring rolling process of a large-scale Ti-6Al-4V alloy ring. Mater Sci Forum 638–642:223–228. https://doi.org/10.4028/www.scientific.net/MSF.638-642.223

14. Li LJ, Luo XD, Zhu YX (2013) Effects of rotational speed of driver roll on hot ring rolling of 6061 aluminum alloy by 3D FE simulation. Adv Mater Res 773:309–315. https://doi.org/10.4028/www.scientific.net/AMR.773.309

15. Lee KH, Kim BM (2013) Advanced feasible forming condition for reducing ring spreads in radial–axial ring rolling. Int J Mech Sci 76:21–32. https://doi.org/10.1016/j.ijmecsci.2013.08.007

16. Qian DS, Peng YY (2015) Effects of forming parameters on coupled thermomechanical behaviours in combined ring rolling. Ironmak Steelmak 42:471–480. https://doi.org/10.1179/1743281214Y.0000000252

17. Sun Z, Yang H, Ou X (2008) Finite element analysis on microstructure evolution of hot ring rolling process. Mater Sci Forum 575–578:1455–1460. https://doi.org/10.4028/www.scientific.net/msf.575-578.1455

18. Nayak S, Kumar Singh A, Gokhale H, Prasad MJNV, Narasimhan K (2023) Optimization of Ti-6Al-4V ring rolling process by FE simulation using RSM. Int J Solids Struct 262–263:112064. https://doi.org/10.1016/j.ijsolstr.2022.112064

19. Ranjan A, Jha JS, Mishra SK (2022) The role of microstructure inhomogeneity in Ti-6Al-4V forging on fracture toughness behavior. J Mater Eng Perform 31:7989–8003. https://doi.org/10.1007/s11665-022-06862-w

20. Ranjan A, Singh A, Jha JS, Mishra SK (2023) Effect of the primary alpha fraction on the dwell fatigue behaviour of Ti-6Al-4V alloy. Int J Fatigue 75:107745. https://doi.org/10.1016/j.ijfatigue.2023.107745

21. Wang ZW, Zeng SQ, Yang XH, Cheng C (2007) The key technology and realization of virtual ring rolling. J Mater Process Technol 182:374–381. https://doi.org/10.1016/j.jmatprotec.2006.08.020

22. Zhu Z, Chen Z, Wang R, Liu C (2022) Forced shear deformation behaviors of annealed pure titanium under quasi-static and dynamic loading. Mater Sci Eng A 839:142872. https://doi.org/10.1016/j.msea.2022.142872

23. Singh AK, Narasimhan K, Singh R (2019) Finite element analysis of thermomechanical behavior and residual stresses in cold flowformed Ti6Al4V alloy. Int J Adv Manuf Technol 103:1257–1277. https://doi.org/10.1007/s00170-019-03609-1

24. Karpat Y (2011) Temperature dependent flow softening of titanium alloy Ti6Al4V: An investigation using finite element simulation of machining. J Mater Process Technol 211:737–749. https://doi.org/10.1016/j.jmatprotec.2010.12.008

25. Mahajan P, Pal M, Kumar R, Agrawal A (2020) Experimental and simulation study of incremental forming for titanium grade 2 sheet. In: ASME 2020 15th international manufacturing science and engineering conference, MSEC 2020, pp 5–10. https://doi.org/10.1115/MSEC2020-8524.

26. Pal M, Pandya V, Agrawal A (2021) Study of formability limit based on ductile damage criteria of incremental sheet forming of titanium grade 2 sheet. In: ASME 2021 16th international manufacturing science and engineering conference, MSEC 2021,. https://doi.org/10.1115/MSEC2021-64005

27. Nayak S, Dhondapure P, Singh AK, Prasad MJNV, Narasimhan K (2020) Assessment of constitutive models to predict high temperature flow behaviour of Ti-6Al-4V preform. Adv Mater Process Technol 6:1–15. https://doi.org/10.1080/2374068X.2020.1731233

28. Lin H, Zhi ZZ (1997) The extremum parameters in ring rolling. Mater Process Technol 69:273–276. https://doi.org/10.1016/S0924-0136(97)00029-0

29. Uchibori T, Matsumoto R, Utsunomiya H (2018) Peripheral speed of steel ring during hot ring rolling. Procedia Manuf 15:89–96. https://doi.org/10.1016/j.promfg.2018.07.174

30. Yang H, Guo L, Zhan M, Sun Z (2006) Research on the influence of material properties on cold ring rolling processes by 3D-FE numerical simulation. J Mater Process Technol 177:634–638. https://doi.org/10.1016/j.jmatprotec.2006.03.209

31. Vairamani G, Kumar TS, Malarvizhi S, Balasubramanian V (2013) Application of response surface methodology to maximize tensile strength and minimize interface hardness of friction welded dissimilar joints of austenitic stainless steel and copper alloy. Trans Nonferrous Met Soc China. English Ed. 23:2250–2259. https://doi.org/10.1016/S1003-6326(13)62725-9

32. Murugan K, Rao TN, Radha K, Gokhale H (2011) Microwave plasma process optimization to produce nano titania through design of experiments. Mater Manuf Process 26:803–812. https://doi.org/10.1080/10426914.2010.496121

33. Hawkyard JB, Johnson W, Kirkland J, Appleton E (1973) Analyses for roll force and torque in ring rolling, with some supporting experiments. Int J Mech Sci 15:873–893. https://doi.org/10.1016/0020-7403(73)90018-0

34. Zheng C, Wang F, Cheng X, Liu J, Liu T, Zhu Z, Yang K, Peng M, Jin D (2016) Capturing of the propagating processes of adiabatic shear band in Ti-6Al-4V alloys under dynamic compression. Mater Sci Eng A 658:60–67. https://doi.org/10.1016/j.msea.2016.01.062

35. Zayadi H, Parvizi A, Farahmand HR, Rahmatabadi D (2021) Investigation of ring rolling key parameters for decreasing geometrical ring defects by 3D finite element and experiments. Arab J Sci Eng 46. https://doi.org/10.1007/s13369-021-05849-4

Comparison of Laser Coordinate Measurements and Hierarchical Multiscale Finite Element Models for the Cup Drawing of Three Commercial Aluminum Alloys

Diego Ricardo Pichardo, Miguel Ángel Ramírez⊙, Leo A. I. Kestens⊙, Albert Van Bael⊙, and Rafael Schouwenaars⊙

Abstract Cup drawing is a benchmark experiment frequently used to validate anisotropic constitutive models and multiscale crystal plasticity codes for yield locus prediction. Earing of the cup rim and thickness variation along the cup wall are sensitive to plastic anisotropy. This test was implemented on an industrial forming press and applied to 85 mm diameter disks of commercial AA1100, AA3103, and AA5005 alloy sheet. Cup geometry was determined using a laser probe coordinate measurement machine (CMM). Finite element models (FEM) were developed with ABAQUS Explicit software and a user-defined subroutine for the anisotropic yield locus based on the hierarchical multiscale model (HMS). As the coordinate cloud produced by the CMM is unrelated to the nodes of the deformed FEM-mesh, both were fitted to a polynomial-Fourier series expansion. After cleaning and correction of the CMM

D. R. Pichardo · M. Á. Ramírez · R. Schouwenaars (✉)
Departamento de Materiales y Manufactura, Facultad de Ingeniería, Edificio O, Universidad Nacional Autónoma de México. Avenida Universidad 3000, Coyoacán 04510, Ciudad de México, México
e-mail: raf_schouwenaars@yahoo.com

M. Á. Ramírez
Centro Tecnológico Aragón, Facultad de Estudios Superiores Aragón, Universidad Nacional Autónoma de México, Av. Rancho Seco S/N, Col. Impulsora, Cd. Nezahualcóyotl 57130, Estado de México, México

L. A. I. Kestens · R. Schouwenaars
Department of Electromechanical, Systems and Metals Engineering, Ghent University, Technologiepark 46, 9052 Ghent, Belgium

L. A. I. Kestens
Department of Materials Science and Engineering, Delft University of Technology, Mekelweg 2, 2628 Delft, CD, Netherlands

A. Van Bael
Department of Materials Engineering, KU Leuven, Kasteelpark Arenberg 44/Box 2450, 3001 Leuven, Belgium

J. Kusiak et al. (eds.), *Numerical Methods in Industrial Forming Processes*, Lecture Notes in Mechanical Engineering, https://doi.org/10.1007/978-3-031-58006-2_26

data, point-by-point comparison can be performed between model and experiment. For AA1100, the position of the ears was correctly identified but their magnitude was underestimated. Excellent coincidence was found for AA5005, with strong ears at 0 and 90°. The small ears at -30° and 30° and secondary ears at 90° were correctly predicted for AA3013.

Keywords Cup drawing · Aluminum alloy · Coordinate measurement · Yield locus · Multiscale modeling

1 Introduction

Modeling of metal forming processes has become an indispensable instrument in industry. It allows for drastic reductions in the lead-up times for process development and in the costs associated with tool building and retooling [1]. While bulk forming processes can often be modeled with the isotropic von Mises yield criterion, this is not the case for sheet forming, where anisotropy plays an important role [2, 3].

An extensive overview of the different yield loci, physical and virtual test methods and the crystal plasticity models used in virtual experiments was given in the framework of ESAFORM [4], where 11 teams tested the accuracy of finite element models for the cup-drawing of a 6016-T4 aluminum sheet. Elements for the validation were the prediction of yield stress and r-values in different directions in the sheet, cup height, earing amplitude, and the force–displacement curve during the process.

Cup drawing has been popular to test the accuracy of proposed yield loci because it provides a simple example of industrial deep-drawing processes. Early models [5, 6] introduced anisotropy in existing continuum models for cup drawing [7] and validated by manually measured cup heights at determined positions of the cup rim [8]. Full constraints and relaxed constraints' Taylor models [9] were tested; Engler followed a similar approach based on the VPSC model [10, 11]. Neither of these studies involved the formal definition of an anisotropic yield locus nor the use of FEM. Generally, the global shape of the cup profile was predicted in a satisfactory manner, but the absolute cup height was not studied in detail.

Early efforts to incorporate an anisotropic yield locus into FEM based on a polynomial fit to virtual experiments were provided by Van Houtte et al. [12] and Munhoven et al. [13]. These papers focused more on the technical aspects and proof of concept than on the precision of the results. Consecutive versions of Barlat's yield criteria [3, 4, 14] have been incorporated into FEM and applied to cup drawing by Yoon et al. [15–17] and Han et al. [18], who incorporated virtual experiments based on the CPFEM [19] to analyze the effect of texture evolution.

A hierarchical multiscale (HMS) [20] approach was used by Ramírez et al. [21] to model cup drawing on two AA6016 alloys with different sheet thicknesses. HMS uses the ALAMEL crystal plasticity model [9] to execute the virtual experiments required to fit the FACET [22] yield locus. Good coincidence was found in terms of earing profile, but average cup height depended critically on the friction coefficient.

The latter value also affects the punch force, although to a lesser extent. A significant effect was found by modeling the tools as elastic solids, contrary to the generally used approximation of perfectly rigid tools. This work expands the latter study by analyzing three commercial Al-alloy sheets with significantly different textures.

2 Experimental Set-Up

The alloys used are commercial AA1100-O, AA3103-O, and AA5005-O. Tensile curves were measured according to ISO 6892–1:2016. Only the results in the rolling direction (RD) are required in HMS. Tensile curves were fitted to a Voce hardening law. X-ray texture measurements were performed on a Siemens D500 texture goniometer. The orientation distribution function (ODF) was determined using four incomplete (111), (200), (220), and (311) pole figures measured with the reflection method [23]. The ODFs were calculated using the series expansion method [24] with $L_{max} = 22$, considering the orthorhombic sample symmetry of the sheet. Representative ODF sections are shown in Figs. 1, 2 and 3, using the Bunge convention for Euler angles.

Cup drawing experiments were performed on a 75 T mechanical press (Meister, Germany). The press and tool assembly are shown in Fig. 4. The punch velocity, as measured from the load cell signal, is 50 mm/s. The tool geometry follows ISO 11531:2015. One important difference with a hydraulic system is that the blank holder force is imposed by springs and is not constant during the operation. The geometry of the cups was measured with a FARO laser line probe mounted on a FARO-arm (FARO Inc., Rochester, NY).

The finite element model was described in Ref. [21], but wall thickness was rescaled from 1.1 to 0.85 mm, which is the blank thickness for the present samples. The mesh consists of linear 8-node hexagonal (brick) reduced-integration elements with hourglass control. The yield locus uses a 6th-degree FACET formulation based on the ALAMEL or FC Taylor model with 5000 orientations.

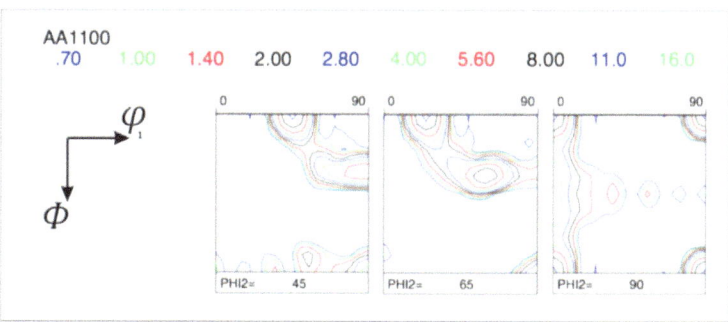

Fig. 1 $\varphi_2 = $ constant sections of the ODF of the AA1100-alloy

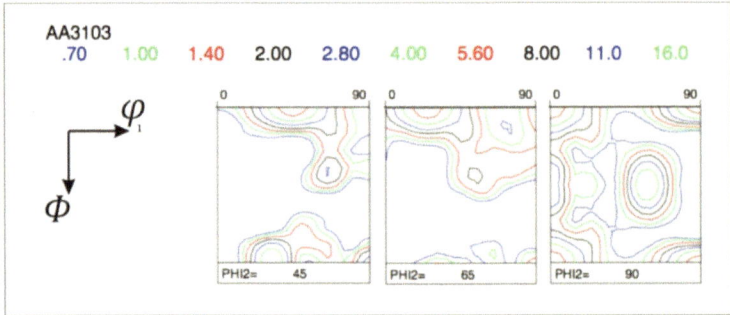

Fig. 2 $\varphi_2 =$ constant sections of the ODF of the AA3103-alloy

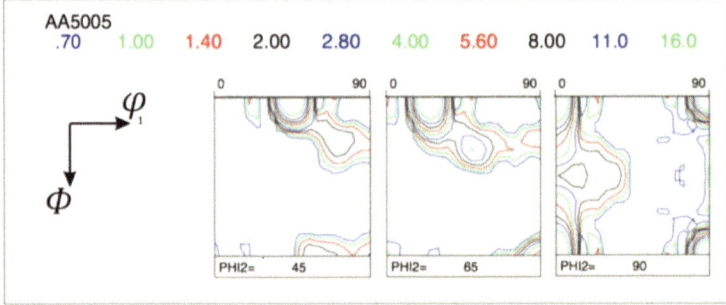

Fig. 3 $\varphi_2 =$ constant sections of the ODF of the AA5005-alloy

Fig. 4 Press and tool assembly. **a** Punch; **b** Blank holder; **c** Die; **d** Springs; **e** Load cell

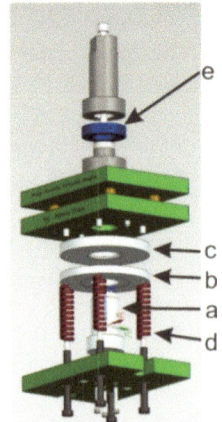

3 Data Analysis

The measurement cloud obtained with the CMM does not coincide with the nodes of the deformed FEM-mesh (Fig. 5). To analyze the measurements, a cylindrical coordinate system (r,θ,z) is used. The minimum of the Z-coordinate, at the bottom of the cup, is set equal to 0 by a rigid body translation; points below a threshold height h_0 (typically 2.2 mm) are removed, i.e., only the cup walls are analyzed. Outliers are removed, and the measurement cloud is rotated to move its central axis into vertical position.

The corrected dataset is sliced vertically into sections of width $\Delta\theta$ (typically 1°). A polynomial fit of the radial distance between wall and center line is made for each slice using the following equation:

$$r(\theta_m, z) = \sum_{n=0}^{N} a_{mn} z^n \tag{1}$$

where θ_m is the central angle of the mth segment and N the degree of the polynomial. In the first step, all points in the slice are used, which defines the central line of the wall segment. Then, the fit is made separately to the points inside and outside this central line, to fit the inner and outer surfaces producing a set of $360/\Delta\theta$ values a_{mn}^{ext} and a_{mn}^{int}. The values of the polynomial coefficients in Eq. (1) are fitted by a Fourier series:

$$a_n^p(\theta) = \sum_{l=0}^{L} c_{ln}\cos(l\theta) + \sum_{l=1}^{L} d_{ln}\sin(l\theta) \tag{2}$$

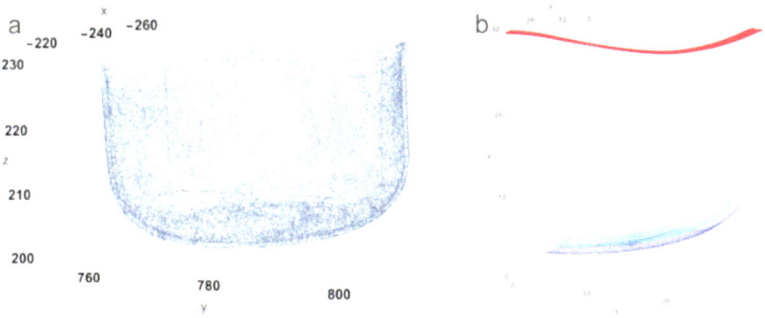

Fig. 5 a Measurement cloud obtained from CMM (only 1/10th of 482 269 points are plotted). **b** Surface nodes of the FEM model of one quarter of the cup, assuming orthotropic symmetry

The superscript p stands for the inner ($p = int$) and outer ($p = ext$) surface. The rim is detected as the highest point in each vertical slice on inner and outer surface. The average of both values is then fitted by a Fourier series for the rim height $h(\theta)$:

$$h(\theta) = \sum_{l=0}^{L} c_l \cos(l\theta) + \sum_{l=1}^{L} d_l \sin(l\theta) \tag{3}$$

Note that c_0 gives the average cup height. The cup is described by:

$$\begin{cases} r^p(\theta, z) = \sum_{n=0}^{N} a_n^p(\theta) r^n \\ h_0 < z < h(\theta) \end{cases} \tag{4}$$

The second line in Eq. 4 is essential to avoid extrapolation. Too low values of N and L produce a poor fit, too high values will introduce spurious undulations. Here, $N = L = 6$. The same fitting procedure is used for the nodes of the FEM-mesh, with orthotropic symmetry imposed in Eq. 3 by including only terms in $\cos(2lq)$.

Two corrections are made to the measured data. Asymmetry due to an eccentric positioning of the blank is mostly eliminated by setting the terms for $l = 1$ equal to 0. Small alignment errors between the rolling direction of the blank and the $\theta = 0$ direction in the CMM can be removed by adding a correction $\delta\theta$ to θ. However, even after these corrections, deviations of orthotropic symmetry are sometimes still observed. The measured rim height is symmetrized by the following operation [8]:

$$h_{Crr}(\theta) = \frac{h(\theta + \delta\theta) + h(-\theta) + h(\pi + \theta + \delta\theta) + h(\pi - \theta + \delta\theta)}{4} \tag{5}$$

4   Results and Discussion

The measured and modeled cups for AA5005 are shown in Fig. 6. The morphology of the rim, both in terms of radius and height, is very satisfactory. A more detailed comparison of the rim height profile is found in Fig. 7 for AA1100, Fig. 8 for AA3103, and Fig. 9 for AA5005. A first observation is that the FEM results overestimate the cup height by approximately 3mm. This may be due to the consideration of a constant blank holder force, as opposed to the variable force exercised by the springs, inaccuracies of the model in the bending zone at the punch radius at the bottom of the cup or an underestimation of the friction coefficient.

Strain rate sensitivity is not considered in the simulations. The deformation speed in the industrial press is much higher than on a laboratory rig for the Swift test. In turn, the strain rate in the laboratory test is higher than the one defined in ISO 6892–1:2016 for the tensile test. In aluminum alloys, this effect may be significant. Other factors which may affect the earing height in AA1100 and the small deviations in

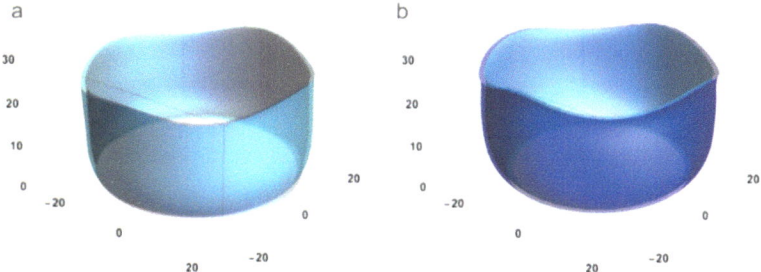

Fig. 6 AA5005. **a** CMM result, **b** FEM result. Good qualitative similarity is observed

Fig. 7 Cup profile as predicted by FEM and height-corrected measured profile for AA1100

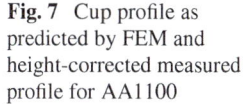
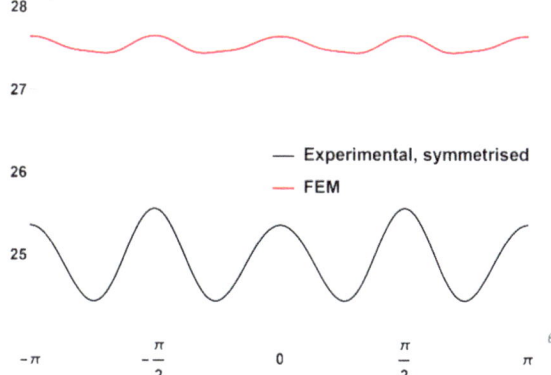

Fig. 8 Cup profile as predicted by FEM and height-corrected measured profile for AA3103

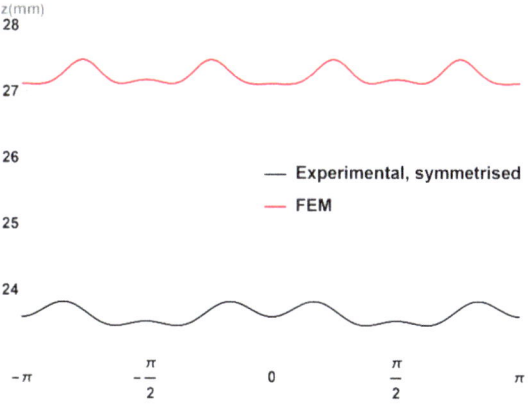

peak positions for AA3103 are the discretization procedure used to select the 5000 orientations used in the simulations, or fundamental effects of the crystal plasticity approach used.

Fig. 9 Cup profile as predicted by FEM and height-corrected measured profile for AA5005

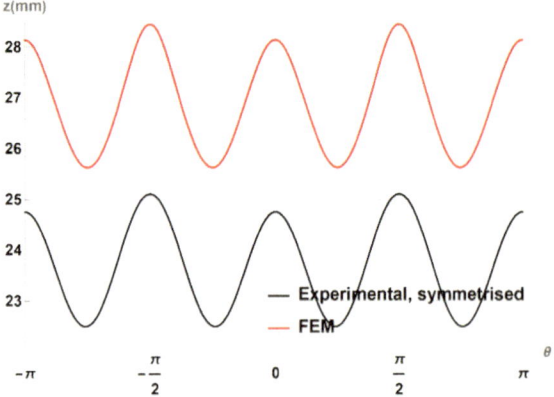

5   Conclusions

A cup drawing test was successfully developed on an industrial mechanical drawing press and full-field measurements of the deformed geometry were obtained with a mechanical-optical coordinate measurement machine. A Fourier-series fitting technique was developed to compare the measured data to FEM results.

Differences in strain rate, blank holder force and friction coefficient between the mechanical press and existing laboratory testing equipment induce challenges, which are unrelated to the theoretical approaches in the anisotropic plasticity codes. Further research on the effect of non-texture-related modeling parameters is required to enable a more precise evaluation of the theoretical models used in the FEM analysis.

The strong overestimation of the earing profile for the AA1100-alloy probably is an effect of the crystal plasticity models used or may be an effect of texture discretization, although the detailed reason for this discrepancy is difficult to identify. For the AA3103 and AA5005, coincidence between experiment and model is excellent, indicating that the modeling approach used here is highly reliable in most cases.

Finally, for AA1100 and AA3103, the earing amplitude is small. This imposes the need for a statistical evaluation of the measurement precision, in terms of eccentricity and misalignment of samples, to provide an estimation of the relative importance of modeling assumptions and statistical spread of experimental results.

Acknowledgements D.R.P., M.A.R., and R.S. acknowledge support from DGAPA grant IN113123 during the final stages of this project. D.R.P. acknowledges his scholarship by CONACYT for his master's studies.

References

1. Dixit US (2020) Modeling of metal forming: a review. Mechanics of materials in modern manufacturing methods and processing techniques, pp 1–30
2. Van Houtte P, Gawad J, Eyckens P, Van Bael A, Samaey G, Roose D (2011) A full-field strategy to take texture induced anisotropy into account during FE simulations of metal forming processes. J Metals 63:37–43
3. Banabic D, Barlat F, Cazacu O, Kuwabara T (2020) Advances in anisotropy of plastic behaviour and formability of sheet metals. IntJ Mater Form 13:749–787
4. Habraken AM, Aksen TA, Alves JL et al (2022) Analysis of ESAFORM 2021 cup drawing benchmark of an Al alloy, critical factors for accuracy and efficiency of FE simulations. IntJ Mater Form 15(5):61
5. Clarke AP, Van Houtte P, Saimoto S (1994) Mater Sci Forum 157–162:1953
6. Van Houtte P, Clarke P, Saimoto S (1993) A quantitative analysis of earing during deep drawing. In: Morris JG et al. (eds), Aluminum alloys for packaging, TMS, 261
7. Hosford WF, Caddell RM (2011) Metal forming: mechanics and metallurgy. Cambridge University Press
8. Schouwenaars R, Van Houtte P, Van Bael A, Winters J, Mols K (1996) Analysis and prediction of the earing behaviour of low carbon steel sheet. Textures Microst 27:553–570
9. Van Houtte P, Li S, Seefeldt M, Delannay L (2005) Deformation texture prediction: from the Taylor model to the Advanced Lamel model. Int J Plast 21:589–624
10. Engler O, Kalz S (2004) Simulation of earing profiles from texture data by means of a visco-plastic self-consistent polycrystal plasticity approach. Mater Sci Eng A, A 373:350–362
11. Engler O, Hirsch J (2007) Polycrystal-plasticity simulation of six and eight ears in deep-drawn aluminum cups. Mater Sci Eng A, A 452:640–651
12. Van Houtte P, Van Bael A, Winters J (1995) The incorporation of texture-based yield loci into elasto-plastic finite element programs. Textures Microstr 24(4):255–272
13. Munhoven S, Habraken AM, Winters J, Schouwenaars R, Van Houtte P (1995) Application of an anisotropic yield locus based on texture to a deep drawing simulation. In: Proceedings Numiform 1995, 5th international conference on numerical methods in industrial forming processes, pp 767–772. Balkema, Rotterdam
14. Barlat F, Aretz H, Yoon JW, Karabin M, Brem JC, Dick R (2005) Linear transfomation-based anisotropic yield functions. Int J Plast 21(5):1009–1039
15. Yoon JW, Barlat F, Chung K, Pourboghrat F, Yang DY (2000) Earing predictions based on asymmetric nonquadratic yield function. Int J Plast 16:1075–1104
16. Yoon JW, Barlat F, Dick RE, Karabin ME (2006) Prediction of six or eight ears in a drawn cup based on a new anisotropic yield function. Int J Plast 22:174–193
17. Yoon JH, Cazacu O, Yoon JW, Dick RE (2010) Earing predictions for strongly textured aluminum sheets. Int J Mech Sci 52(12):1563–1578
18. Han F, Diehl M, Roters F, Raabe D (2020) Using spectral-based representative volume element crystal plasticity simulations to predict yield surface evolution during large scale forming simulations. J Mater Process Technol 277:116449
19. Roters F, Diehl M, Shanthraj P, Eisenlohr P, Reuber C, Wong SL, Maiti T, Ebrahimi A, Hochrainer T, Fabritius HO, Nikolov S (2019) DAMASK–The Düsseldorf Advanced Material Simulation Kit for modeling multi-physics crystal plasticity, thermal, and damage phenomena from the single crystal up to the component scale. Comput Mater Sci 158:420–478
20. Gawad J, Van Bael A, Eyckens P, Samaey G, Van Houtte P, Roose D (2013) Hierarchical multi-scale modeling of texture induced plastic anisotropy in sheet forming. Comput Mater Sci 66:65–83
21. Ramírez MA, Schouwenaars R, Eyckens P, Gawad J, Kestens L, Van Bael A, Van Houtte P (2016) Experimental validation and effect of modelling assumptions in the hierarchical multi-scale simulation of the cup drawing of AA6016 sheets. Modell Simul Mater Sci Eng 25(1):015002

22. Van Houtte P, Yerra SK, Van Bael A (2009) The Facet method: a hierarchical multilevel modelling scheme for anisotropic convex plastic potentials. Int J Plast 25:332–360
23. Van Houtte P (1984) A new method for the determination of texture functions from incomplete pole figures—comparisons with older methods. Textures Microst 6:137–162
24. Bunge HJ (1982) Texture analysis in materials science. Butterworth's, London

Numerical Simulation and Experimental Validation of Electromagnetic-Impacted Micro-forming Processes of Aluminum Alloy Thin Sheet

Wei Liu, Tao Cheng, Zhenghua Meng, Jiaqi Li, and Shangyu Huang

Abstract Electromagnetic forming is a kind of high-speed forming technology for improving the formability of aluminum alloy sheets; however, the inherent non-uniform electromagnetic forces always restrict its application into the micro-forming field. Electromagnetic-impacted micro-forming processes were adopted by using flexible mediums for force transmission. The processes of electromagnetic-impacted rubber micro-forming, electromagnetic-impacted micro-hydroforming, and electromagnetic-impacted pneumatic micro-forming were compared to inspect the influences of flexible mediums. The effects of discharge voltages on the electromagnetic-impacted micro-forming processes of an AA5052-O aluminum alloy thin sheet were investigated by the numerical simulation of electromagnetic and mechanical coupled fields. The numerical simulations were validated via comparisons with the experimental results; these showed that the forming accuracy and thickness uniformity of micro-channels were the greatest by the electromagnetic-impacted micro-hydroforming process.

Keywords Electromagnetic forming · Micro-forming · Micro-hydroforming

W. Liu (✉) · T. Cheng · J. Li · S. Huang
School of Materials Science and Engineering, Wuhan University of Technology, Wuhan 430070, China
e-mail: weiliu@whut.edu.cn

W. Liu · S. Huang
Hubei Engineering Research Center for Green Precision Material Forming, Wuhan University of Technology, Wuhan 430070, China

Z. Meng
School of Automotive Engineering, Wuhan University of Technology, Wuhan 430070, China

347
J. Kusiak et al. (eds.), *Numerical Methods in Industrial Forming Processes*, Lecture Notes in Mechanical Engineering, https://doi.org/10.1007/978-3-031-58006-2_27

1 Introduction

With the rapid development of micro-electromechanical systems (MEMS), fuel cells, biomedical devices, and other fields, the advanced manufacturing technology of micro-parts is becoming urgently needed. Nowadays, a number of non-conventional energy-assisted forming technologies have been studied for manufacturing these micro-parts. Electromagnetic forming (EMF) has attracted wide attention because of its ability to improve the formability and deformation uniformity of difficult-to-form metallic sheets at room temperature and high strain rates.

Langstädtler et al. [1] performed the electromagnetic micro-embossing of a 1050 aluminum alloy thin sheet. Triangular cross-section micro V-grooves with a width of 86.6 μm and structure angle of 30° were achieved. Kamal et al. [2] developed a uniform pressure actuator (UPA) for generating uniform electromagnetic force distribution on such a sheet. The electromagnetic micro-forming of a 5052-H32 aluminum alloy thin sheet with UPA was performed to fabricate a micro-part with optical diffraction grating features. Zhao et al. [3] performed the electromagnetic micro-punching experiments of a 20-μm-thick T2 copper thin sheet with a UPA. Micro holes with a diameter of 0.4 mm were successfully punched. Dong et al. [4] performed the electromagnetic micro-forming (EMMF) of titanium bipolar plates with a UPA. The simulated and experimental results showed that EMMF can significantly improve the thickness uniformity and the uniformity of the channel depth; however, a UPA has the limitations of coil winding intervals, complex structures, and electrical contacts. Long et al. [5] carried out the electromagnetic micro-forming of a 0.2-mm-thick 1060 aluminum alloy thin sheet with a plane runway coil. An eddy current must be induced in order to generate electromagnetic forces.

Shang et al. [6] proposed the EMF process with a compliant layer for fabricating bipolar plates; a non-uniform electromagnetic force was transmitted by the compliant layer to deform the sheet. Zhu et al. [7] developed a hybrid forming process that combined EMF and stamping. The dimensional precision and thickness uniformity of the micro-channel were significantly higher under this hybrid process than that of single EMF or stamping. Yan et al. [8] experimentally and numerically investigated the electromagnetic hydraulic forming process of an AA5052 aluminum alloy sheet. This not only obtained good fittability with the die but also improved the formability of the materials. Thus, electromagnetic hybrid micro-forming processes exhibit enormous potential for the fabrication of micro-parts.

Herein, electromagnetic-impacted micro-forming (EMIMF) processes were adopted to solve the problem of coil-winding intervals and non-uniform electromagnetic forces. Numerical simulations of electromagnetic-impacted rubber micro-forming (EMIRMF), electromagnetic-impacted micro-hydroforming (EMIMHF), and electromagnetic-impacted pneumatic micro-forming (EMIPMF) were performed in order to compare the influences of a flexible medium. By comparing these with the experimental results, the numerical simulations were validated.

2 Numerical Simulation

2.1 Numerical Model

The numerical simulations of the EMIMF processes were fulfilled by using LS-DYNA software. Figure 1a shows half of a numerical model of an EMIRMF simulation, which consisted of a coil, drive plate, piston, rubber, pressure chamber, aluminum alloy thin sheet, and die. The electromagnetic fields of the coil and drive plate were calculated by using the finite element method (FEM), while their surrounding air field was analyzed by the boundary element method (BEM). The coil, drive plate, and piston were defined to be rigid bodies, those numerical models were divided into hexahedral entity elements. The electromagnetic and mechanical fields were sequentially coupled. The time steps of the electromagnetic and structural fields were both set to be 2 μs. The mechanical fields of the other components were simulated by the Lagrange algorithm. The nonlinear hyper-elastic behavior of the rubber was described by the Mooney-Rivlin model [9]. AA5052-O aluminum alloy thin sheets with a thickness of 0.1 mm were used. The sheet was discretized with shell elements, and the effect of the high strain rate on the mechanical behavior of the sheet was described by the Johnson–Cook model ($\sigma = (35.4\text{MPa} + 114.1\text{MPa}\varepsilon_p^{0.3246})(1 + 0.0028\ln\dot{\bar{\varepsilon}}/\dot{\bar{\varepsilon}}_0)$). A surface-to-surface contact with a friction coefficient of 0.2 was considered for the sheet and the tools. The cross-section sizes of the die are detailed in Fig. 1a. The meshes for the deformation zone of the sheet and its surrounding die zone were locally refined to reduce the computational cost. The longitudinal direction (LD) and transverse direction (TD) sections of the micro-channel were inspected.

Half of a numerical model of the EMIMHF/EMIPMF simulation was established (as shown in Fig. 1b). Instead of rubber for EMIRMF, the flexible mediums of EMIMHF and EMIPMF were substituted by liquid and forming gas, respectively. The fluid–structure interaction (FSI) of the aluminum alloy thin sheet and fluid (water, forming gas, and air) was simulated by the arbitrary Lagrangian–Eulerian (ALE) algorithm in order to avoid the problem of mesh distortion due to the massive deformation of the fluid. The state equation of the air was defined by the linear polynomial equation, and the state equations of the water and forming gas were described by the Gruneisen equation [10].

2.2 Simulated Results

To study the effects of the discharge voltages on the deformation profile and thickness distribution of the micro-channel, numerical simulations of EMIRMF were performed at different discharge voltage values (1.4, 1.6, 1.8, 2.0, and 2.2 kV). For EMIMHF, these voltages were 0.8, 1.0, 1.2, 1.4, and 1.6 kV, while for EMIPMF, they were 1.6, 1.7, 1.8, 1.9, and 2.0 kV.

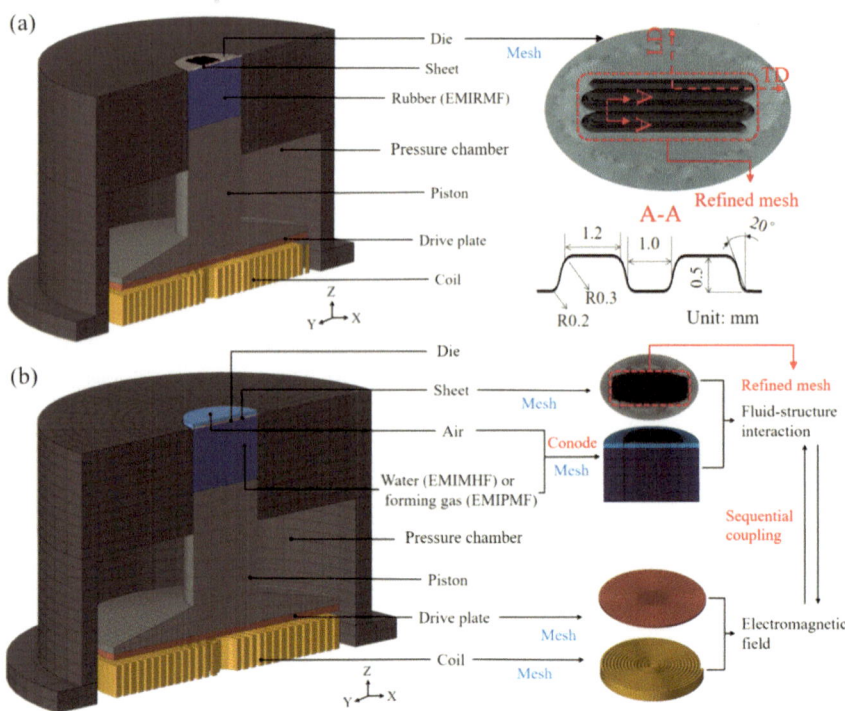

Fig. 1 Numerical model: **a** EMIRMF; **b** EMIMHF/EMIPMF

With increase in the discharge voltage, the forming accuracy and thickness distribution of the micro-channel were gradually improved and decreased, respectively (as shown in Figs. 2, 3, and 4). As the discharge voltage further increased to the maximum value of each process, the minimum thickness of the micro-channel was greatly reduced, and the local thinning became serious. Thus, the optimum discharge voltages of the EMIRMF, EMIMHF, and EMIPMF processes for fabricating the micro-channel were chosen to be 2.0, 1.4, and 1.9 kV, respectively.

The thinning rate was used to analyze the formability (which is defined as the change of the sheet thickness divided by the sheet's initial thickness). Figure 5 shows the thinning-rate contours of the micro-channel at the optimum discharge voltages for each process. The maximum thinning-rate values of the micro-channel for the EMIRMF, EMIMHF, and EMIPMF processes were 28.6, 25.4, and 26.2%, respectively.

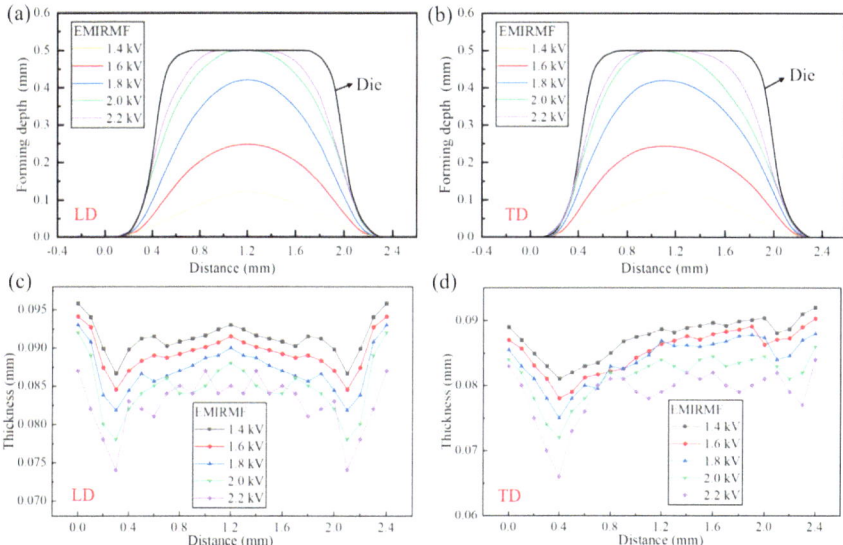

Fig. 2 Effects of discharge voltage on simulated results of micro-channel during EMIRMF: **a** profile along LD; **b** profile along TD; **c** thickness along LD; **d** thickness along TD

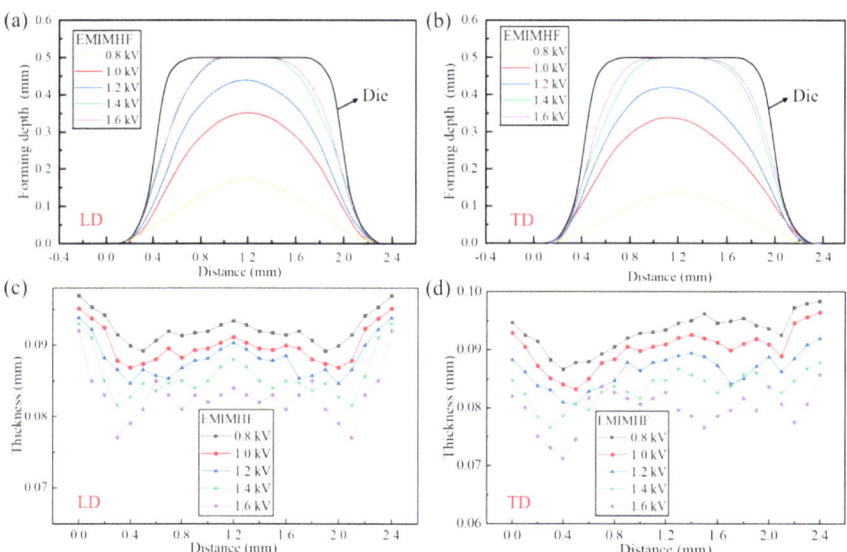

Fig. 3 Effects of discharge voltage on simulated results of micro-channel during EMIMHF: **a** profile along LD; **b** profile along TD; **c** thickness along LD; **d** thickness along TD

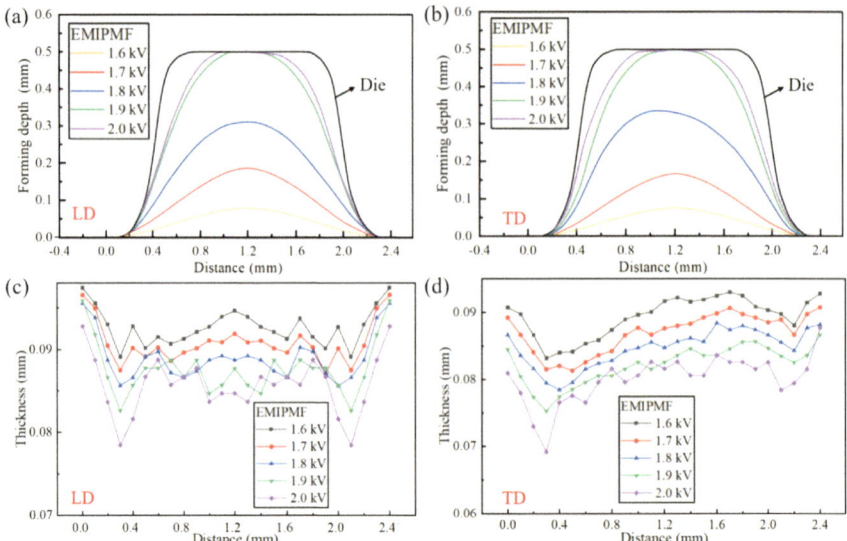

Fig. 4 Effects of discharge voltage on simulated results of micro-channel during EMIPMF: **a** profile along LD; **b** profile along TD; **c** thickness along LD; **d** thickness along TD

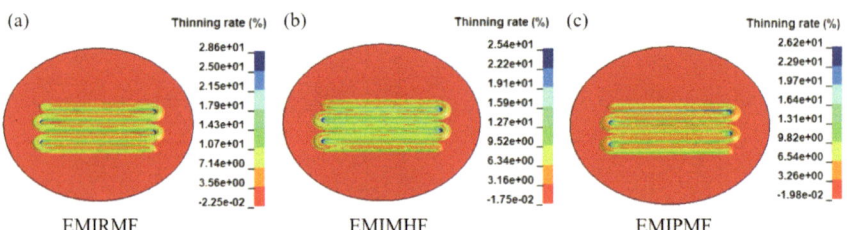

Fig. 5 Thinning rate contours of micro-channel

3 Experimental Validation

The experimental setup of the EMIMF processes is shown in Fig. 6a. The forming coil was selected as a flat spiral coil. The coil and drive plate were made of copper due to the property of high electrical conductivity. When the electrical current flowed along the coil, a transient magnetic field was created. Subsequently, an eddy current was evoked between the drive plate, and non-uniform Lorentz forces were stimulated between the coil and the drive plate (as shown in Fig. 6b). When the drive plate impacted the piston, the flexible medium made the aluminum alloy thin sheet fit with the die. Seal rings were installed in the pressure chamber to prevent fluid leakage. The specimens were obtained by performing the EMIMF processes at the optimum discharge voltages for each process (as shown in Fig. 7). These specimens were cut along the LD and TD used wire cut electrical discharge machining. An optical microscope with a micron resolution was used to measure the deformation profiles and thickness distributions of the specimens. The relative error of measurement was less than 1.5%.

Figure 8 compares the simulated and experimental deformation profiles of the specimens for each process. The simulated and experimental deformation profiles agreed well. Figure 9 compares the simulated and experimental thicknesses of the specimens. Similarly, the simulated thickness values nearly matched those of the experimental ones; thus, the numerical simulations of the EMIMF processes were validated. The experimental maximum thickness reduction η_{max} is also compared in Fig. 9. The specimen that was obtained by the EMIMHF process exhibited more thickness uniformity due to the smaller maximum thinning rate. To directly assess the forming accuracy of the specimen, the experimental filling rate values of the specimens by each process were compared (as shown in Fig. 10). The filling rate values of the specimen that was obtained by the EMIMHF process were the greatest;

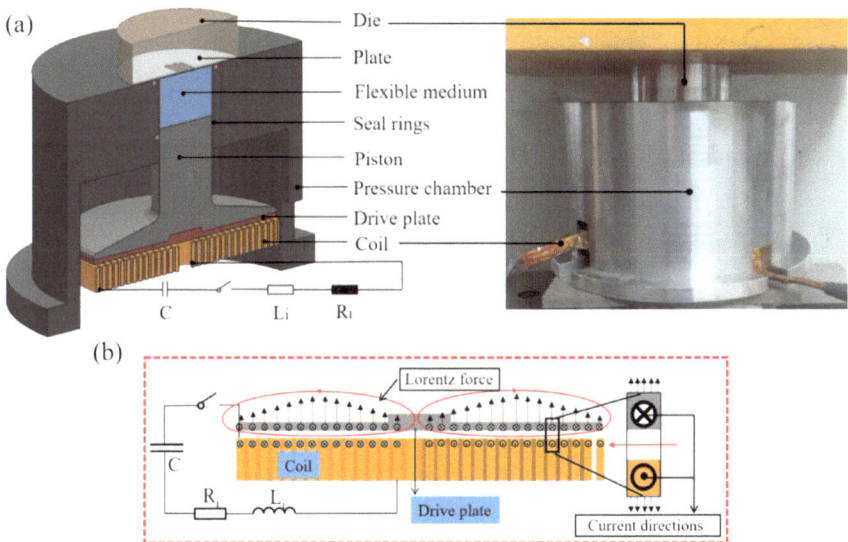

Fig. 6 EMIMF processes: **a** experimental setup; **b** schematic diagram

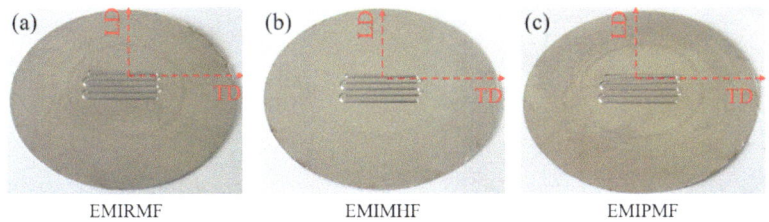

Fig. 7 Specimens via different processes

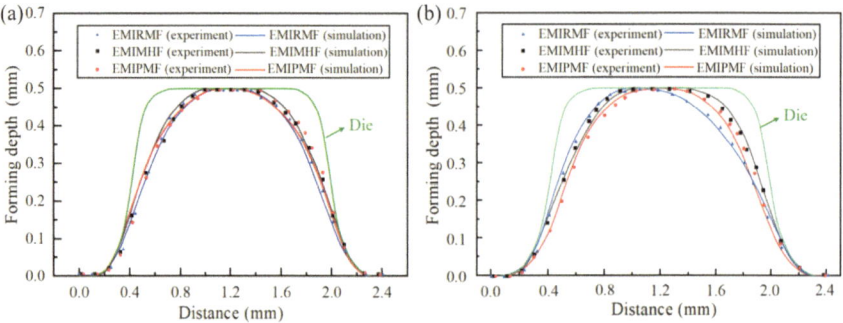

Fig. 8 Comparisons of simulated and experimental deformation profiles: **a** LD; **b** TD

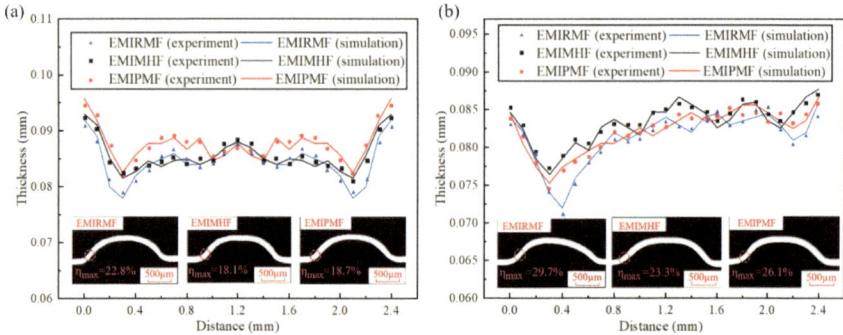

Fig. 9 Comparisons of simulated and experimental thicknesses: **a** LD; **b** TD

therefore, the forming accuracy and thickness uniformity of the micro-channel were the best via the EMIMHF process.

Fig. 10 Comparisons of filling rates of different processes

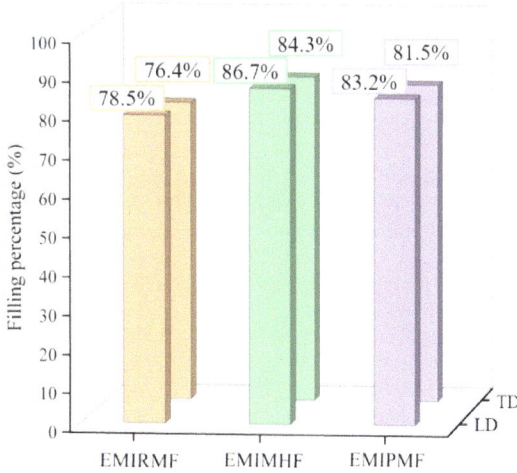

4 Conclusion

The EMIMF processes that used flexible mediums were adopted in order to avoid the problem of the coil-winding interval and non-uniform electromagnetic forces. The EMIRMF, EMIMHF, and EMIPMF processes were applied to fabrications of micro-channels in order to compare the influences of a flexible medium. The numerical simulations of the EMIMF processes were performed with the sequential coupling method of the electromagnetic and mechanical fields. The discharge voltage could significantly affect the deformation profile and thickness distribution of the micro-channel of an AA5052-O aluminum alloy thin sheet. By comparing the simulated and experiment deformation profiles and thicknesses, the numerical simulations of the EMIMF processes were deemed to be accurate. The forming accuracy and thickness uniformity of the micro-channel were the best via the EMIMHF process.

Acknowledgements The authors are grateful for the financial support of the National Natural Science Foundation of China (Grant Number 52005374), the Key Research and Development Program of Hubei Province, China (Grant Number 2021BAA174), and the open research fund of the State Key Laboratory of High-Performance Complex Manufacturing, Central South University, China (Grant Number Kfkt2021-04).

References

1. Langstädtler L, Schönemann L, Schenck C et al (2016) Electromagnetic embossing of optical microstructures. J Micro Nano-Manuf 4(2):021001
2. Kamal M, Daehn GS (2007) A uniform pressure electromagnetic actuator for forming flat sheets. J Manuf Sci Eng 129(2):369–379

3. Zhao QJ, Xu J, Wang CJ et al (2015) Electromagnetic micro-punching process of T2 copper foil. Adv Mater Res 1120–1121(1):1220–1225
4. Dong P, Li Z, Feng S et al (2021) Fabrication of titanium bipolar plates for proton exchange membrane fuel cells by uniform pressure electromagnetic forming. Int J Hydrogen Energy 46(78):38768–38781
5. Long Z, Liu H, Yan Z et al (2022) Electromagnetic micro-forming using flat spiral coil. Int J Adv Manuf Technol 121(1):1161–1171
6. Shang J, Wilkerson L, Hatkevich S et al (2010) Commercialization of fuel cell bipolar plate manufacturing by electromagnetic forming. Int Conf High Speed Forming 4:47–56
7. Zhu C, Xu J, Yu H et al (2022) Hybrid forming process combining electromagnetic and quasi-static forming of ultra-thin titanium sheets: formability and mechanism. Int J Mach Tools Manuf 180:103–929
8. Yan Z, Xiao A, Zhao P et al (2022) Deformation behavior of 5052 aluminum alloy sheets during electromagnetic hydraulic forming. Int J Mach Tools Manuf 179:103–916
9. Koubaa S, Belhassen L, Wali M et al (2017) Numerical investigation of the forming capability of bulge process by using rubber as a forming medium. Int J Adv Manuf Technol 92:1839–1848
10. Varas D, Zaera R, López-Puente J (2012) Numerical modelling of partially filled aircraft fuel tanks submitted to Hydrodynamic Ram. Aerosp Sci Technol 16(1):19–28

Gotoh's 1977 Yield Stress Function with Kinematic Hardening for Modeling Strength Differential Yielding of Orthotropic Sheet Metals

Jie Sheng, Seung-Yong Yang, and Wei Tong

Abstract When a sheet metal is subjected to both tensile and compressive stresses in a forming process, there is a need to formulate a yield stress function that can accurately account for its strength differential effect in anisotropic yielding. The earliest classical approach is to combine Hill's 1948 quadratic yield stress function with Prager's kinematic hardening concept. Consistent with the requirement that a polynomial stress function admits only even-order shear stress components for an orthotropic sheet metal, the resulting quadratic yield stress function in plane stress has up to five material parameters for on-axis yielding but only one material parameter for off-axis yielding. The latter feature limits its modeling capabilities in general sheet metal forming simulations. In this paper, we present a user-friendly approach of formulating a non-quadratic yield stress function with tension-compression asymmetry by combining Gotoh's 1977 quartic yield stress function with kinematic hardening. The new fourth-order yield stress function in plane stress has up to a total of eleven material constants: seven for on-axis yielding and four for off-axis yielding. The nonlinear parameter identification by least-square minimization with positivity and convexity constraints on the yield stress function is detailed for various sheet metals exhibiting strength differential effects. The results show that the new Gotoh-Prager yield stress function has adequate capabilities for modeling both on-axis and off-axis asymmetric yielding of many orthotropic sheet metals investigated over the years.

Keywords Tension-compression asymmetry · Fourth-order yield function · Least-square minimization · Anisotropic plasticity

J. Sheng · W. Tong (✉)
Southern Methodist University, Dallas, TX 75275, USA
e-mail: wtong@smu.edu
URL: https://s2.smu.edu/~wtong

S.-Y. Yang
SFTC, Columbus, OH 43235, USA

© The Rightsholder, under exclusive licence to [Springer Nature Switzerland AG], part of Springer Nature 2024
J. Kusiak et al. (eds.), *Numerical Methods in Industrial Forming Processes*, Lecture Notes in Mechanical Engineering, https://doi.org/10.1007/978-3-031-58006-2_28

357

1   Introduction

Plastic yielding and flow behavior of a metal may be noticeably different in tension and compression. One well-known example of such behavior is the so-called Bauschinger effect in metals deformed plastically upon a reversal of loading directions. The strength differential effect has also often been observed to be rather strong in some rolled sheet metals [1–4]. An advanced anisotropic yield stress function with tension-compression asymmetry is thus sought for accurately modeling those sheet metals as they are increasingly used in various engineering applications [5–8].

For an orthotropic sheet metal under a state of plane stress $\boldsymbol{\sigma} = (\sigma_x, \sigma_y, \tau_{xy})$, the well-known Hill's 1948 quadratic yield stress function [9] has the following compact form [10]

$$\Phi_2(\boldsymbol{\sigma}) = \Phi_2(\sigma_x, \sigma_y, \tau_{xy}) = A_1\sigma_x^2 + A_2\sigma_x\sigma_y + A_3\sigma_y^2 + A_4\tau_{xy}^2, \tag{1}$$

where (A_1, A_2, A_3, A_4) are its four non-dimensional material constants that are often determined using three uniaxial and one equal biaxial tensile yield stresses. Hill's 1948 quadratic yield stress function $\Phi_2(\boldsymbol{\sigma})$ is however symmetric for tension and compression, i.e., $\Phi_2(-\sigma_x, -\sigma_y, \tau_{xy}) = \Phi_2(\sigma_x, \sigma_y, \tau_{xy})$. To account for the strength differential effect in anisotropic yielding, the simplest approach has been to incorporate the back stress concept in Prager's kinematic hardening model [11] into Hill's quadratic yield stress function, namely [2, 12]

$$\Phi_{hp}(\boldsymbol{\sigma}) = A_1(\sigma_x - \alpha_x)^2 + A_2(\sigma_x - \alpha_x)(\sigma_y - \alpha_y) + A_3(\sigma_y - \alpha_y)^2 + A_4\tau_{xy}^2, \tag{2}$$

where (α_x, α_y) are two additional material constants.[1] The so-called Hill-Prager yield stress function $\Phi_{hp}(\boldsymbol{\sigma})$ can model reasonably well the tension-compression asymmetric yielding of a sheet metal in on-axis or coaxial loading ($\tau_{xy} = 0$) with a total of five material constants $(A_1, A_2, A_3, \alpha_x, \alpha_y)$. However, its ability to model strength differential effect of the sheet metal under off-axis or non-coaxial loading is very limited as there is only one material constant A_4 uniquely related to off-axis yield stresses.

In this study, we presented a user-friendly asymmetric quartic yield stress function with enhanced capabilities for modeling the strength differential effect in anisotropic yielding of sheet metals. Following the similar development of quadratic yield stress functions above, we proposed to add the two axial back stress components per kinematic hardening as two additional material constants to Gotoh's 1977 quartic yield stress function [13–15]. A brief description of the new Gotoh-Prager quartic yield stress function with seven on-axis material constants and four off-axis material constants will be given first. A novel nonlinear optimization algorithm will then be detailed to simultaneously identify its total 11 material constants and guarantee the convexity of any as-calibrated Gotoh-Prager quartic yield stress function. Finally,

---

[1] They are often called back stresses. The shear back stress component α_{xy} is set to be zero due to the orthotropic symmetry of a rolled sheet, i.e., $\Phi_{hp}(\sigma_x, \sigma_y, -\tau_{xy}) = \Phi_{hp}(\sigma_x, \sigma_y, +\tau_{xy})$.

the capabilities of the new yield stress function will be assessed and demonstrated through numerical results of modeling various orthotropic sheet metals with asymmetric yielding.

2 Quartic Yield Stress Function with Kinematic Hardening and Its Calibration

The new asymmetric yield stress function in terms of Gotoh's quartic yield stress function with two axial back stress components (β_x, β_y) has the following form:

$$
\begin{aligned}
\Phi_{gp}(\boldsymbol{\sigma}) = {} & B_1(\sigma_x - \beta_x)^4 + B_2(\sigma_x - \beta_x)^3(\sigma_y - \beta_y) + B_3(\sigma_x - \beta_x)^2(\sigma_y - \beta_y)^2 \\
& + B_4(\sigma_x - \beta_x)(\sigma_y - \beta_y)^3 + B_5(\sigma_y - \beta_y)^4 + B_6(\sigma_x - \beta_x)^2\tau_{xy}^2 \\
& + B_7(\sigma_x - \beta_x)(\sigma_y - \beta_y)\tau_{xy}^2 + B_8(\sigma_y - \beta_y)^2\tau_{xy}^2 + B_9\tau_{xy}^4,
\end{aligned}
\tag{3}
$$

where $(B_1, B_2, ..., B_9, \beta_x, \beta_y)$ are its 11 material constants. As the new yield stress function $\Phi_{gp}(\boldsymbol{\sigma})$ has seven on-axis material constants $(B_1, B_2, B_3, B_4, B_5, \beta_x, \beta_y)$ and four off-axis material constants (B_6, B_7, B_8, B_9), its modeling capabilities will be superior in comparison with the Hill-Prager yield stress function $\Phi_{hp}(\boldsymbol{\sigma})$.

To apply the Gotoh-Prager yield stress function $\Phi_{gp}(\boldsymbol{\sigma})$ to model a sheet metal with asymmetric yielding, one needs to identify its material constants from available measured yield stress values and to ensure that the calibrated yield stress function is positive and convex. The second task is rather easy: if Gotoh's quartic yield stress function itself (i.e., both back stress components β_x and β_y are zero in Eq. 3) is known to be positive and convex, then the Gotoh-Prager yield stress function $\Phi_{gp}(\boldsymbol{\sigma})$ is also positive and convex.[2] Several practical methods have already been established for certifying the convexity of a calibrated Gotoh's quartic yield stress function using either sufficient conditions only or sufficient and necessary conditions [14, 18, 19].

The task of identifying the material constants of the Gotoh-Prager yield stress function $\Phi_{gp}(\boldsymbol{\sigma})$ involves numerically solving 11 nonlinear equations per the yield condition

$$
\Phi_{gp}(\boldsymbol{\sigma}^{(1)}) - \sigma_{gp}^4 = 0, \ \ \Phi_{gp}(\boldsymbol{\sigma}^{(2)}) - \sigma_{gp}^4 = 0, \ \ ..., \ \ \Phi_{gp}(\boldsymbol{\sigma}^{(11)}) - \sigma_{gp}^4 = 0, \tag{4}
$$

where $(\boldsymbol{\sigma}^{(1)}, \boldsymbol{\sigma}^{(2)}, ..., \boldsymbol{\sigma}^{(11)})$ are measured yield stresses at 11 unique and suitable plane stress states and $\sigma_{gp}(\xi)$ is the flow strength at the current hardening state (in terms of a scalar internal state variable ξ) of the sheet metal appeared in the yield condition $\Phi_{gp}(\boldsymbol{\sigma}) = \sigma_{gp}^4(\xi)$. Ideally, the first seven yield stresses shall be from on-axis loading (uniaxial and biaxial) with $\tau_{xy} = 0$ so one can solve the first seven equations in Eq. (4) for the seven on-axis material constants $(B_1, B_2, B_3, B_4, B_5, \beta_x, \beta_y)$.

---

[2] Any linear transformation of stress preserves the convexity of a stress function [16, 17]. The convexity of a quartic homogeneous polynomial is also positive [18].

Afterward, one can find the four off-axis material constants (B_6, B_7, B_8, B_9) from the remaining four equations in Eq. (4) where four off-axis yield stresses are used.

When there are more than 11 measured yield stresses available for a given sheet metal, one can use a nonlinear optimization algorithm to solve for the 11 material constants by minimizing a sum of squared errors in the yield condition. Here we proposed to add the sum-of-squares (SOS) convexity constraints [18] directly to the least-square minimization used for parameter identification. The SOS-convexity constraints are the algebraic sufficient conditions to ensure a calibrated Gotoh's quartic yield stress function is indeed convex and they are given as $G_{3\times3} \geq 0$ and $G_{6\times6} \geq 0$ [18–20], where

$$
G_{3\times3} = \begin{pmatrix} 12B_1 & 3B_2 & 2B_3 \\ 3B_2 & 2B_3 & 3B_4 \\ 2B_3 & 3B_4 & 12B_5 \end{pmatrix}, \quad G_{6\times6} = \begin{pmatrix} 12B_1 & 3B_2 & 0 & 2B_3 & 0 & 2B_6 \\ 3B_2 & 2B_3 & 0 & 3B_4 & 0 & B_7 \\ 0 & 0 & 2B_6 & 0 & B_7 & 0 \\ 2B_3 & 3B_4 & 0 & 12B_5 & 0 & 2B_8 \\ 0 & 0 & B_7 & 0 & 2B_8 & 0 \\ 2B_6 & B_7 & 0 & 2B_8 & 0 & 12B_9 \end{pmatrix}. \quad (5)
$$

Consequently, calibration of the Gotoh-Prager yield stress function $\Phi_{gp}(\sigma)$ for a given sheet metal with some measured yield stresses is to solve the following constrained optimization problem with respect to its 11 material constants:

$$
\min \sum_{k=1}^{K} w_k [\Phi_{gp}(\sigma^{(k)}) - \sigma_p^4]^2, \quad \text{subject to } G_{3\times3} \geq 0 \text{ or } G_{6\times6} \geq 0, \quad (6)
$$

where the SOS-convexity condition $G_{3\times3} \geq 0$ is to be used first for on-axis yield stresses (with $K_{on} \geq 7$) in identifying the seven on-axis material constants $(B_1, B_2, B_3, B_4, B_5, \beta_x, \beta_y)$ and the SOS-convexity condition $G_{6\times6} \geq 0$ is then to be used for off-axis yield stresses (with $K_{off} \geq 4$) in identifying the remaining four material constants (B_6, B_7, B_8, B_9).[3] The weight coefficient w_k is set between 0 and 1 for each measured yield stress depending on its accuracy and precision (unless it is explicitly changed, it is by default set to 1 in this study).

When there are less than 11 measured yield stresses available for a given sheet metal, one can impose some additional conditions (equations) of reduced plastic anisotropy [15] to the above-constrained optimization problem. Alternatively, one can first calibrate the Hill-Prager yield stress function $\Phi_{hp}(\sigma)$ of Eq. (2) using the similarly constrained optimization with six or more yield stress inputs ($11 > K \geq 6$) with respect to its six material constants $(A_1, A_2, A_3, A_4, \alpha_x, \alpha_y)$

---

[3] One can also simply use the latter condition for identifying all 11 constant materials together using all available 11 or more (i.e., $K = K_{on} + K_{off} \geq 11$) measured yield stresses.

$$\min \sum_{k=1}^{K} w_k [\Phi_{hp}(\boldsymbol{\sigma}^{(k)}) - \sigma_{hp}^2]^2, \text{ subject to } A_1 > 0, \ 4A_1 A_3 > A_2^2, \ A_4 > 0,$$

(7)

where the three inequalities here are necessary and sufficient conditions to ensure the strict convexity of Hill's 1948 yield stress function $\Phi_2(\boldsymbol{\sigma})$ of Eq. (1) [10]. One can then use the calibrated Hill-Prager yield stress function $\Phi_{hp}(\boldsymbol{\sigma})$ to estimate the missing yield stresses needed for calibrating the Gotoh-Prager yield stress function $\Phi_{gp}(\boldsymbol{\sigma})$. One additional benefit of having a calibrated Hill-Prager yield stress function $\Phi_{hp}(\boldsymbol{\sigma})$ on hand first is to provide the initial guesses for the nonlinear optimization problem of Eq. (6), namely (noting $\Phi_{hp}^2(\boldsymbol{\sigma}) \approx \Phi_{gp}(\boldsymbol{\sigma})$)

$$B_1^{(0)} = A_1^2, \ B_2^{(0)} = 2A_1 A_2, \ B_3^{(0)} = (A_2^2 + 2A_1 A_3), \ B_4^{(0)} = 2A_2 A_3, \ B_5^{(0)} = A_3^2,$$
$$B_6^{(0)} = 2A_1 A_4, \ B_7^{(0)} = 2A_2 A_4, \ B_8^{(0)} = 2A_3 A_4, \ B_9^{(0)} = A_4^2, \beta_x^{(0)} = \alpha_x, \ \beta_y^{(0)} = \alpha_y.$$

(8)

3 Modeling Results of Selected Sheet Metals

Here application examples of using the Gotoh-Prager yield stress function $\Phi_{gp}(\boldsymbol{\sigma})$ for describing tension-compression asymmetry in anisotropic yielding are presented for some nine HCP, FCC and BCC sheet metals, respectively. Experimental values of their yield stresses have already been reported in the literature [1–7] and their strength differential behaviors have motivated some very recent research on developing advanced asymmetric yield stress functions [5–8].

In the following, modeling results were grouped according to the availability of numbers and types of measured yield stress data for each of the nine representative sheet metals. The calibrated values of material constants of both Hill-Prager and Gotoh-Prager yield criteria are listed in Table 1. The unit for the flow strength σ_{hp} or σ_{gp} that appeared in Hill-Prager and Gotoh-Prager yield criteria is MPa. For the first four materials listed in Table 1, only on-axis material constants are given as there are no actual off-axis yield stresses reported for those materials in [1, 2].

HCP metals with only five on-axis yield stresses reported. The well-known 1968 study by Kelley and Hosford [1] reported in their tabulated data only five on-axis yield stresses for three HCP metals, namely ($\sigma_{t0}, \sigma_{t90}, \sigma_{c0}, \sigma_{c90}, \sigma_{tb}$) at three deformed states. These yield stresses happen to match the total number of on-axis material constants of a biaxial Hill-Prager yield stress function $\Phi_{hp}(\sigma_x, \sigma_y, 0)$. One might expect that the Hill-Prager yield stress function would be good enough for modeling their biaxial yielding behavior and $\Phi_{gp}(\sigma_x, \sigma_y, 0) = \Phi_{hp}^2(\sigma_x, \sigma_y, 0)$. This is indeed the case as shown in Fig. 1a, b for two magnesium alloys Mg-0.5%Th and Mg-4%Li: the five solid symbols in each figure correspond to the five reported yield stresses that are well predicted by the corresponding Hill-Prager yield criterion. The insertion of dashed-line yield surfaces given by the symmetric Hill's 1948 yield function

Table 1 List of material constants of two yield stress functions

	Mg-0.5%Th	Mg-4%Li	Zircaloy 2J,350C	Zircaloy 2K,25C	CP Ti Grade 4	AZ31	AA2008-T4	AA2090	DP780-T3
A_2	−0.1958	−1.0431	−1.1633	−1.5131	−1.3388	−0.6579	−0.9265	−1.1276	−1.0289
A_3	1.0869	1.1803	0.6833	1.0556	1.2232	0.8324	1.1008	1.0224	1.0126
A_4	–	–	–	–	3.9761	2.6862	3.0777	4.6679	3.6146
α_x/σ_{hp}	0.3242	0.2520	−0.1054	0.1479	0.0106	0.3254	−0.0372	0.0683	−0.0609
α_y/σ_{hp}	0.2534	0.1751	−0.0949	0.1445	0.1212	0.3795	−0.0696	0.0149	−0.0949
σ_{hp}	147.44	81.544	281.61	520.12	677.37	146.92	213.21	263.82	744.03
B_2	−0.3915	−2.0861	−2.0765	−3.9624	−2.7241	−0.3213	−1.8531	−2.2553	−3.0818
B_3	2.2122	3.4487	1.6570	6.3318	2.8408	0.3562	3.0600	3.3162	3.7018
B_4	−0.4255	−2.4623	−0.7742	−4.1789	−1.6993	−0.7911	−2.0398	−2.3057	−1.7488
B_5	1.1814	1.3932	0.4531	1.1137	1.3091	0.7075	1.2117	1.0452	0.9966
B_6	–	–	–	–	6.7610	7.7519	4.7139	4.1822	10.000
B_7	–	–	–	–	−10.019	−4.8478	1.0704	−11.390	−11.786
B_8	–	–	–	–	7.8067	7.0311	5.6394	7.7545	10.614
B_9	–	–	–	–	16.297	3.3734	2.6302	24.408	9.1169
β_x/σ_{gp}	0.3242	0.2520	−0.1166	0.1877	−0.0643	0.2973	−0.0372	0.0683	−0.0929
β_y/σ_{gp}	0.2534	0.1751	−0.0751	0.1827	0.0816	0.3976	−0.0696	0.0149	−0.1050
σ_{gp}	147.44	81.544	279.55	520.27	663.65	136.55	213.21	263.82	741.83

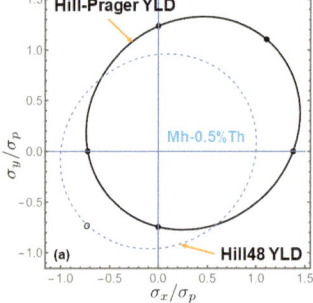 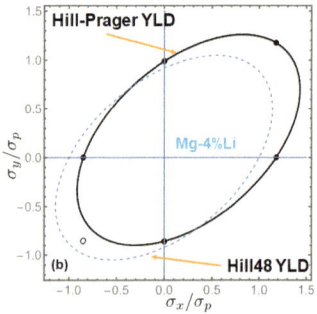

Fig. 1 **a** Biaxial yield surfaces of Mg-0.5%Th at a plastic strain of 1% as given by the calibrated Hill-Prager (solid lines) and Hill 1948 (dashed lines) yield criteria; **b** Biaxial yield surfaces of Mg-4%Li at a plastic strain of 1% as given by the calibrated Hill-Prager (solid lines) and Hill 1948 (dashed lines) yield criteria. Solid and open circular symbols are reported [1] and calculated yield stresses, respectively, where $2\sigma_{cb} = \sigma_{c0} + 2\sigma_{c45} + \sigma_{c90}$ was used. Results of Gotoh-Prager yield surfaces are omitted here as they are identical to Hill-Prager ones

$\Phi_2(\sigma_x, \sigma_y, 0)$ (using the same A_1, A_2, A_3 as in $\Phi_{hp}(\sigma_x, \sigma_y, 0)$) in each figure is to highlight visually the asymmetric nature of the yield surfaces given by the Hill-Prager yield criterion.

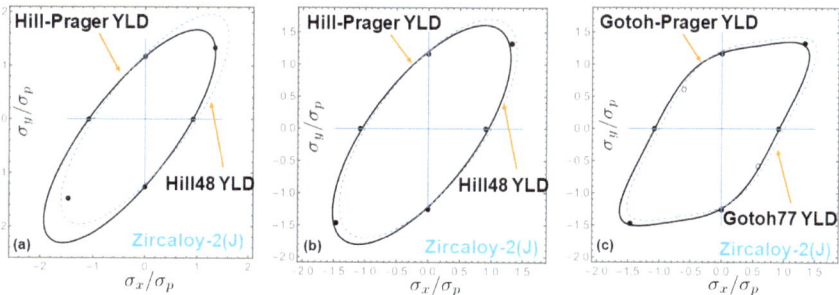

Fig. 2 **a, b** Biaxial yield surfaces of Zicaloy-2(J) at a plastic strain of 0.002% and 350° as given by the calibrated Hill-Prager (solid lines) and Hill 1948 (dashed lines) yield criteria; **c** Biaxial yield surfaces of the same material at the same conditions as given by the calibrated Gotoh-Prager (solid lines) and Gotoh 1977 (dashed lines) yield criteria. Solid symbols are reported yield stresses [2] while open circles are calculated ones in pure shear from the best-fit Hill-Prager yield stress function in **b**

HCP metals with six on-axis yield stresses reported. On the other hand, a total of six on-axis yield stresses (σ_{t0}, σ_{t90}, σ_{c0}, σ_{c90}, σ_{tb}, σ_{cb}) are provided by Shih and Lee [2] in their Table 1 for three HCP metals. If only the first five yield stresses are used to determine the material constants of the Hill-Prager yield stress function, its predicted equal biaxial compression yield stress σ_{cb} will be way off from the measured one as shown in Figs. 2a and 3a. The yield surface defined by the least-square best-fit Hill-Prager yield stress function via Eq. (7) is much better as shown in Figs. 2b and 3b. Finally, the six reported on-axis yield stresses plus two calculated on-axis pure shear yield stresses (σ_{ss0}, σ_{ss90}) from the best-fit biaxial $\Phi_{hp}(\sigma_x, \sigma_y, 0)$ were used to find the best-fit biaxial $\Phi_{gp}(\sigma_x, \sigma_y, 0)$ according to Eq. (6). The weights in the equation were set to be 0.1 instead of the default value of 1 for yield conditions based on

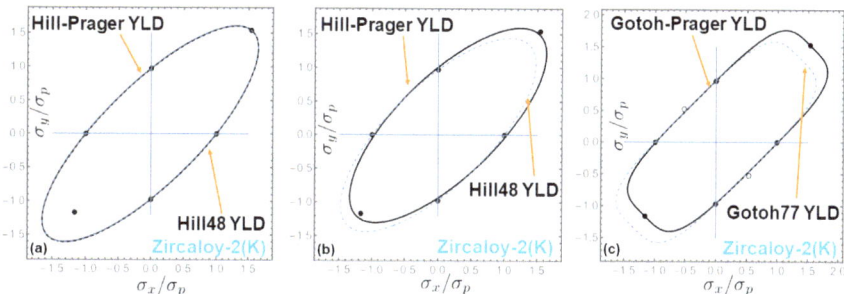

Fig. 3 **a, b** Biaxial yield surfaces of Zicaloy-2(K) at a plastic strain of 0.002% and 25° as given by the calibrated Hill-Prager (solid lines) and Hill 1948 (dashed lines) yield criteria; **c** Biaxial yield surfaces of the same material at the same conditions as given by the calibrated Gotoh-Prager (solid lines) and Gotoh 1977 (dashed lines) yield criteria. Solid symbols are reported yield stresses [2] while open circles are calculated ones in pure shear from the best-fit Hill-Prager yield stress function in **b**

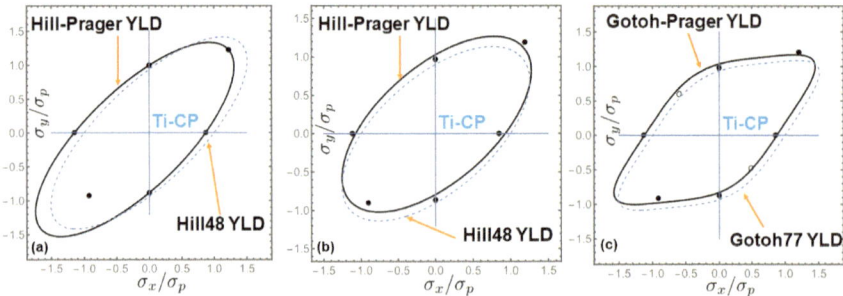

Fig. 4 a, b Biaxial yield surfaces of CP Ti as given by the calibrated Hill-Prager (solid lines) and Hill 1948 (dashed lines) yield criteria; **c** Biaxial yield surfaces of the same material at the same conditions as given by the calibrated Gotoh-Prager (solid lines) and Gotoh 1977 (dashed lines) yield criteria. Solid symbols are reported yield stresses [5] while open circles are calculated ones in pure shear from the best-fit Hill-Prager yield stress function in **b**

the calculated pure shear yield stresses. The yield surfaces given by such calibrated Gotoh-Prager yield criteria are shown in Figs. 2c and 3c, respectively.

HCP metals with six on-axis and two off-axis yield stresses reported. A total of eight such yield stresses have been provided by Raemy et al. [5] for CP titanium and by Hu and Yoon [6] for magnesium alloy AZ31. The best-fit on-axis material constants of both yield criteria were identified in a similar way as described above. The resulting biaxial yield surfaces are given in Fig. 4 for CP Ti and Fig. 5 for AZ31, respectively. With both σ_{t45} and σ_{c45} yield stresses available, one obtained the best-fit A_4 of $\Phi_{hp}(\boldsymbol{\sigma})$ per Eq. (7). For the case of $\Phi_{gp}(\boldsymbol{\sigma})$, one first added to the measured two off-axis yield stresses $(\sigma_{t45}, \sigma_{c45})$ four additional estimated off-axis yield stresses $2\sigma_{t225} = \sigma_{t0} + \sigma_{t45}$, $2\sigma_{t675} = \sigma_{t45} + \sigma_{t90}$, $2\sigma_{c225} = \sigma_{c0} + \sigma_{c45}$ and $2\sigma_{c675} = \sigma_{c45} + \sigma_{c90}$. The best-fit off-axis material constants of the Gotoh-Prager yield stress function were

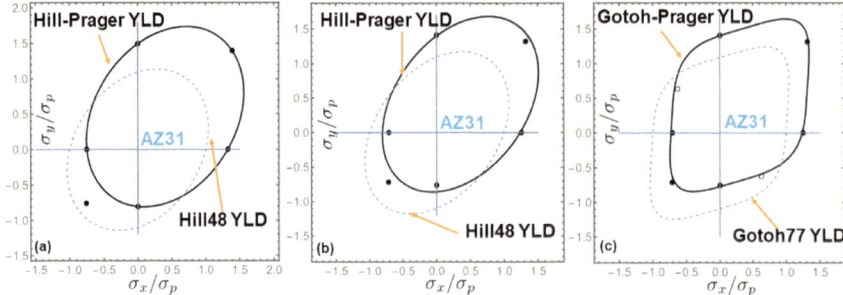

Fig. 5 a, b Biaxial yield surfaces of AZ31 as given by the calibrated Hill-Prager (solid lines) and Hill 1948 (dashed lines) yield criteria; **c** Biaxial yield surfaces of the same material at the same conditions as given by the calibrated Gotoh-Prager (solid lines) and Gotoh 1977 (dashed lines) yield criteria. Solid symbols are reported yield stresses [6] while open circles are calculated ones in pure shear from the best-fit Hill-Prager yield stress function in **b**

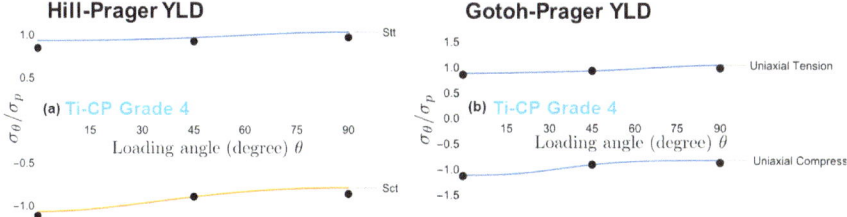

Fig. 6 **a** Dependence of uniaxial tensile and compressive yield stresses $\sigma_{t\theta}$ and $\sigma_{c\theta}$ of CP Ti on the loading angle θ as given by the best-fit Hill-Prager (solid lines) yield criterion; **b** The same dependence as given by the best-fit Gotoh-Prager (solid lines) yield criterion. Solid symbols are reported yield stresses [5]

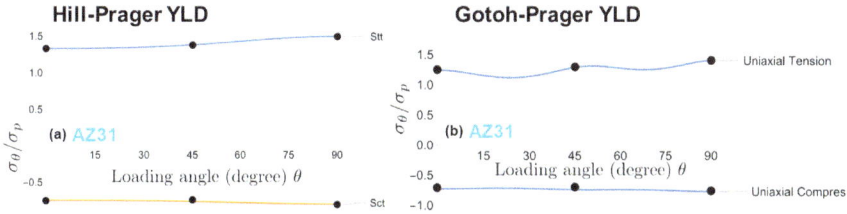

Fig. 7 **a** Dependence of uniaxial tensile and compressive yield stresses $\sigma_{t\theta}$ and $\sigma_{c\theta}$ of AZ31 on the loading angle θ as given by the best-fit Hill-Prager (solid lines) yield criterion; **b** The same dependence as given by the best-fit Gotoh-Prager (solid lines) yield criterion. Solid symbols are reported yield stresses [6]

identified using these six off-axis yield stresses per Eq. (6). The weights for those four estimated yield stresses were set to 0.1 instead of the default value of 1. The resulting dependence of uniaxial yield stresses on the loading angle θ is given in Fig. 6 for CP Ti and Fig. 7 for AZ31, respectively.

FCC aluminum alloys with five on-axis and ten off-axis yield stresses reported. Such a set of yield stress data has been provided in [7] for the two aluminum alloy sheets AA2008-T4 [3] and AA2090-T3 [4]. As there are only five on-axis yield stresses available for both aluminum sheets, one obtained biaxial Gotoh-Prager yield function from biaxial Hill-Prager yield stress function as before to (i.e., getting $B_1, …, \beta_y$ per Eq. (8) directly). Then the best-fit four off-axis material constants (B_6, B_7, B_8, B_9) were identified using those 10 reported off-axis yield stresses per Eq. (6). The resulting biaxial yield surfaces are given in Fig. 8a for AA2008-T4 and Fig. 8b for AA2090-T3, respectively. The resulting dependence of uniaxial yield stresses on the loading angle θ is given in Fig. 9 for AA2008-T4 and Fig. 10 for AA2090-T3, respectively.

BCC Steel with seven on-axis inputs and three off-axis yield stresses reported. Such a set of yield stress data has been provided in [7] for a DP780 steel sheet. The seven on-axis yield stresses $(\sigma_{t0}, \sigma_{t90}, \sigma_{c0}, \sigma_{c90}, \sigma_{tb}, \sigma_{s0}, \sigma_{s90})$ provided happen to be sufficient in calibrating the on-axis material constants of both Hill-Prager

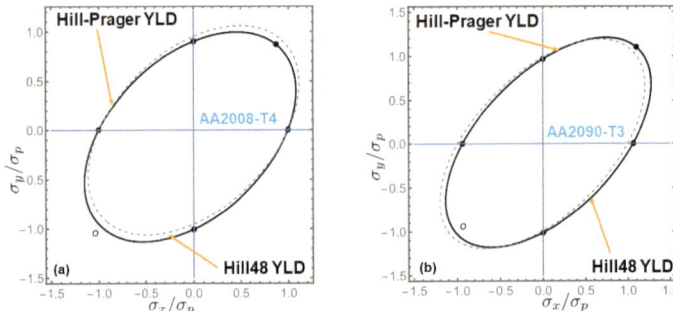

Fig. 8 **a** Biaxial yield surfaces of AA2008-T4 as given by the calibrated Hill-Prager (solid lines) and Hill 1948 (dashed lines) yield criteria; **b** Biaxial yield surfaces of AA2090-T3 as given by the calibrated Hill-Prager (solid lines) and Hill 1948 (dashed lines) yield criteria. Solid and open circular symbols are reported [7] and calculated yield stresses, respectively, where $2\sigma_{cb} = \sigma_{c0} + 2\sigma_{c45} + \sigma_{c90}$ was used

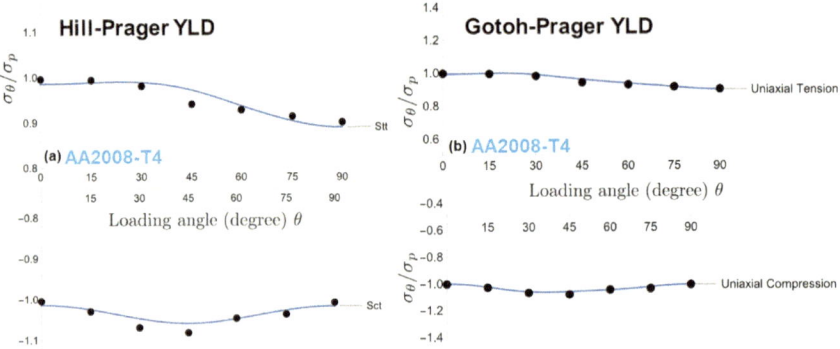

Fig. 9 **a** Dependence of uniaxial tensile and compressive yield stresses $\sigma_{t\theta}$ and $\sigma_{c\theta}$ of AA2008-T4 on the loading angle θ as given by the best-fit Hill-Prager (solid lines) yield criterion; **b** The same dependence as given by the best-fit Gotoh-Prager (solid lines) yield criterion. Solid symbols are reported yield stresses [7]

and Gotoh-Prager yield functions. For the case of $\Phi_{gp}(\boldsymbol{\sigma})$, one first added to the three measured off-axis yield stresses ($\sigma_{t45}, \sigma_{c45}, \sigma_{s45}$) four additional estimated off-axis yield stresses $2\sigma_{t225} = \sigma_{t0} + \sigma_{t45}, 2\sigma_{t675} = \sigma_{t45} + \sigma_{t90}, 2\sigma_{c225} = \sigma_{c0} + \sigma_{c45}$ and $2\sigma_{c675} = \sigma_{c45} + \sigma_{c90}$. The best-fit off-axis material constants of Gotoh-Prager yield stress function were identified using these seven off-axis yield stresses per Eq. (6). The weights for those four estimated yield stresses were set to 0.1 instead of the default value of 1. The resulting biaxial yield surfaces and dependence of uniaxial yield stresses on the loading angle θ are shown in Figs. 11 and 12, respectively.

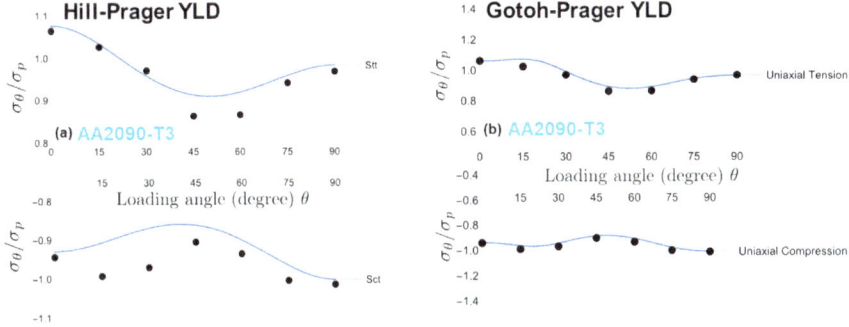

Fig. 10 **a** Dependence of uniaxial tensile and compressive yield stresses $\sigma_{t\theta}$ and $\sigma_{c\theta}$ of AA2090-T3 on the loading angle θ as given by the best-fit Hill-Prager (solid lines) yield criterion; **b** The same dependence as given by the best-fit Gotoh-Prager (solid lines) yield criterion. Solid symbols are reported yield stresses [7]

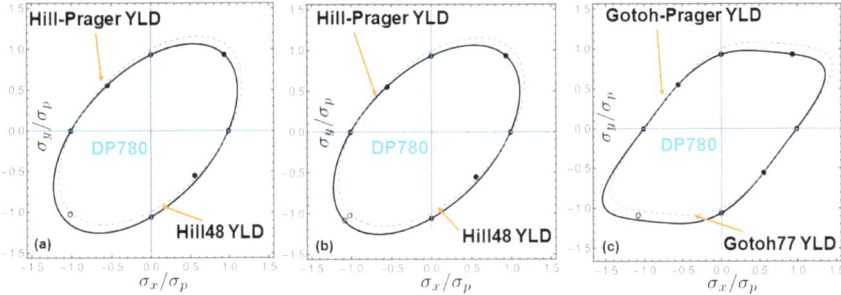

Fig. 11 **a, b** Biaxial yield surfaces of DP780 steel as given by the calibrated Hill-Prager (solid lines) and Hill 1948 (dashed lines) yield criteria; **c** Biaxial yield surfaces of the same material at the same conditions as given by the calibrated Gotoh-Prager (solid lines) and Gotoh 1977 (dashed lines) yield criteria. Solid symbols are reported yield stresses at a plastic strain of 1.5% [7] while the open circle is the calculated equal biaxial compression yield stress $4\sigma_{cb} = \sigma_{c0} + 2\sigma_{c45} + \sigma_{c90}$

4 Discussion and Conclusions

In practice, slightly stronger SOS-convexity conditions $\boldsymbol{G}_{3\times3} > 0$ and $\boldsymbol{G}_{6\times6} > 0$ (both matrices are positive define) instead of $\boldsymbol{G}_{3\times3} \geq 0$ and $\boldsymbol{G}_{6\times6} \geq 0$ (both matrices are positive semi-define) were used here. One reason to do so is due to the fact that leading principal minors of those two matrices are somewhat simpler to obtain than their principal minors [18, 20]. The strong SOS-convexity conditions were actually implemented in numerical calculations given here, i.e., all eigenvalues of both $\boldsymbol{G}_{3\times3}$ and $\boldsymbol{G}_{6\times6}$ were required to be equal or larger than a small positive value (say, 10^{-4}). This proves to be effective in avoiding any possible violation of convexity conditions when only five significant digits or up to four decimal points were kept for the calibrated material constants as given in Table 1.

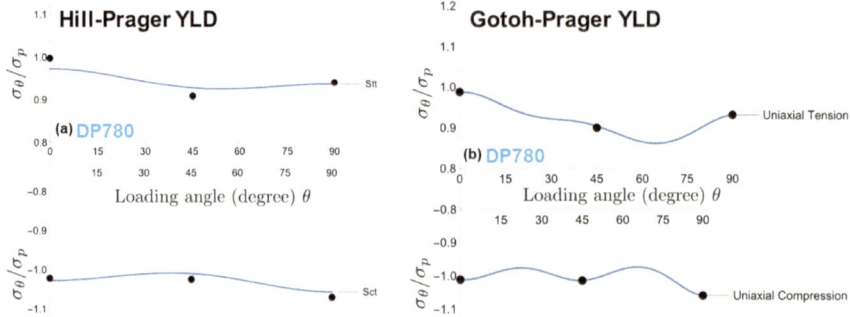

Fig. 12 a Dependence of uniaxial tensile and compressive yield stresses $\sigma_{t\theta}$ and $\sigma_{c\theta}$ of DP780 steel on the loading angle θ as given by the best-fit Hill-Prager (solid lines) yield criterion; **b** The same dependence as given by the best-fit Gotoh-Prager (solid lines) yield criterion. Solid symbols are reported yield stresses at a plastic strain of 1.5% [7]

As shown through numerical results for modeling tension-compression asymmetry of all representative sheet metals considered in the previous section, the SOS-convexity constrained least-square minimization per Eq. (6) was proved to be effective for parameter identification of the non-quadratic Gotoh-Prager yield function. This is similar to the use of any yield function based on a convex isotropic function of linearly transformed stresses (such as Yld2000-2d): any calibrated Gotoh-Prager yield function with a total of 11 material constants is also guaranteed to be convex automatically without a need for further certification. For sheet metals with purely isotropic-kinematic hardening behaviors with evolving (β_x, β_y) but fixed $(B_1, ..., B_9)$, the convexity of Gotoh-Prager yield function is preserved. On the other hand, for sheet metals with anisotropic-kinematic hardening behaviors with evolving (β_x, β_y) and $(B_1, ..., B_9)$, the convexity of Gotoh-Prager yield function has to be verified continuously during the plastic deformation. This can be easily done in either the parameter identification stage or the post-calibration certification stage by using the algebraic SOS-convexity conditions $G_{3\times3} > 0$ and $G_{6\times6} > 0$.

In conclusion, the first effort in developing and applying the Gotoh-Prager yield function as presented here has shown that such a simpler asymmetric yield function can be an attractive alternative for modeling strength differential effects in sheet metals.

References

1. Kelley EW, Hosfrod WF Jr (1968) The deformation characteristics of textured magnesium. Trans Metall Soc AIME 242:654–661

2. Shih CF, Lee D (1978) Further developments in anisotropic plasticity. ASME J Eng Mater Tech 100(3):294–302
3. Yoon J-W, Song IS, Yang D-Y, Chung K, Barlat F (1995) Finite element method for sheet forming based on an anisotropic strain-rate potential and the convected coordinate system. Int J Mech Sci 37:733–752
4. Yoon J-W, Barlat F, Chung K, Pourboghrat F, Yang DY (2000) Earing predictions based on asymmetric nonquadratic yield function. Int J Plast 16:1075–1104
5. Raemy C, Manopulo N, Hora P (2017) On the modelling of plastic anisotropy, asymmetry and directional hardening of commercially pure titanium: a planar Fourier series based approach. Int J Plast 91:182–204
6. Hu Q, Yoon JW (2021) Analytical description of an asymmetric yield function (Yoon2014) by considering anisotropic hardening under non-associated flow rule. Int J Plast 140:102978
7. Hou Y, Min J, Lin J, Lee M-G (2022) Modeling stress anisotropy, strength differential, and anisotropic hardening by coupling quadratic and stress-invariant-based yield functions under non-associated flow rule. Mech Mater 174:104458
8. Soare SC (2023) Bezier5YS and SHYqp: a general framework for generating data and for modeling symmetric and asymmetric orthotropic yield surfaces. Euro J Mech A Solids 97:104781
9. Hill R (1948) A theory of the yielding and plastic flow of anisotropic metals. Proc R Soc Lond A193:281–297
10. Tong W, Alharbi M (2017) Comparative evaluation of non-associated quadratic and associated quartic plasticity models for orthotropic sheet metals. Int J Solids Struct 128:133–148
11. Prager W (1955) The theory of plasticity: a survey of recent achievements (James Clayton Lecture). Proc Inst Mech Eng 169:41–57
12. Wu H-C, Hong H-K, Shiao Y-P (1999) Anisotropic plasticity with application to sheet metals. Int J Mech Sci 41:703–724
13. Gotoh M (1977) A theory of plastic anisotropy based on a yield function of fourth order (plane stress state). Int J Mech Sci 19:505–520
14. Tong W (2016) Application of Gotoh's orthotropic yield function for modeling advanced high-strength steel sheets. ASME J Manu Sci Eng 138:094502-1
15. Tong W (2018) An improved method of determining Gotoh's nine material constants for a sheet metal with only seven or less experimental inputs. Int J Mech Sci 140:394–406
16. Rockafellar RT (1970) Convex analysis. Princeton University Press, Princeton, NY
17. Boyd SP, Vandenberghe L (2004) Convex optimization. Cambridge University Press. ISBN 978-0-521-83378-3
18. Tong W (2018) Algebraic convexity conditions for Gotoh's non-quadratic yield function. ASME J Appl Mech 85:074501-1
19. Tong W (2018) On the certification of positive and convex Gotoh's fourth-order yield function. In: Journal of physics: conference series, vol 1063, p 012093
20. Yang S-Y, Sheng J, Tong W (2023) Unpublished research

Analysis of Forming Limits During Cold Forging of Aluminum Hybrid Billets

Karl C. Grötzinger, Roman Kulagin, Dmitry Gerasimov, and Mathias Liewald

Abstract Metal forming processes for manufacture of hybrid components with complex geometry is one of the most promising trends in materials science. At the same time, such processes pose serious challenges regarding the interface properties between the different materials. Thus far, models are existing to describe cold welding of hybrid components and their plasticity during following deformation processes. However, these models have been developed and validated mainly for processes, where stress and strain states are recognized as simple (e.g., drawing, roll bonding). In this contribution, a flange upsetting process using hybrid billets made of AA-6060 and AA-7075 is subjected, where the deformation conditions significantly differ (non-monotonous and complex strain history). The main aim of this paper is to investigate such process with special emphasis on acting stress conditions in the interface with respect to delamination effects occurring during cold forging experiments. Also, the influence of an initial bond between the aluminum components, created by compound hot extrusion, is investigated both experimentally and numerically. In this context, three geometrical models of the hybrid billet (variation of diameter and wall thickness of the reinforcing component) were used. Analyses show that even if bonding in extrusion process is not ideal and depends on various influencing parameters, it makes subsequent cold forging process more stable and reliable regarding material flow compared to billets without initial bond.

Keywords Hybrid materials · Cold forging · Numerical simulation

K. C. Grötzinger (✉) · M. Liewald
Institute for Metal Forming Technology, University of Stuttgart, Holzgartenstr. 17, 70174 Stuttgart, Germany
e-mail: karl.groetzinger@ifu.uni-stuttgart.de

R. Kulagin
Institute of Nanotechnology (INT), Karlsruhe Institute of Technology, Hermann-Von-Helmholtz-Platz 1, 76344 Eggenstein-Leopoldshafen, Germany

D. Gerasimov
Micas Simulations, Temple Court, 107 Oxford Road, Oxford OX4 2ER, UK

J. Kusiak et al. (eds.), *Numerical Methods in Industrial Forming Processes*, Lecture Notes in Mechanical Engineering, https://doi.org/10.1007/978-3-031-58006-2_29

371

1 Introduction

The manufacture of hybrid components by metal forming poses serious challenges to process design and thus pre-determines numerous decisive trends in materials science [1]. Such components allow for combining of different materials to achieve a significant potential for lightweight design and integration of special functions. The property of the individual, separate parts that make up the hybrid, differs from the one of the joined assembly. Joining by forming therefore increases productivity and makes additional joining processes and connecting elements needless. As a result, energy and material can be used more efficiently. There are, however, several challenges to produce hybrid components, such as the inevitable presence of an interface between the parts, where property gradients and thus stress peaks are prevailing. Various process routes for the manufacture of hybrid components of different shape, such as sheet and bulk material, have been focused in literature. Cold forging is one such process for joining two or more components by forming [2]. The joining principle between involved components can be either a form fit, force fit or even a metal bond [3]. Comprehensive studies on the influence of component geometry on the interfacial contact during upsetting of bi-metallic billets were shown in [4]. In recent times, analytical approaches for estimating process load in hybrid forming processes were developed [5]. The kind of joint being created during deformation highly depends on the billet preparation, such as surface treatment, roughness, tolerances, and surface expansion in the interface exposed to prevailing normal and shear stresses [6]. Joining processes by cold forging of tubular components, however, pose several challenges such as buckling and contact loss as described in [4]. To overcome these challenges, the use of a two-step sequence seems promising: A first process includes the manufacture of the hybrid billet by extrusion, and a second one targets the deformation of component by cold forging. Such process route, as subjected in this publication, at first was introduced in [7]. It consists of a compound hot extrusion process of AA-6060 aluminum alloy featuring a continuously embedded tubular reinforcing component consisting of AA-7075.

In this publication, a comparison between different initial bond characteristics (unbond and bonded by compound hot extrusion) and their effect on subsequent cold flange upsetting process is subjected. Especially the stress distribution internally in the workpiece volume during flange upsetting process and suitable modeling of different bond characteristics belong to the main challenges in terms of estimating process limits. Cold flange upsetting in this study is investigated experimentally and numerically using three different geometrical billet variations. Gained results show that a certain initial bond, created by compound extrusion, leads to more stable process conditions and higher possible deformation in cold forging prior to delamination. Process simulations including stress analyses for different billet configurations were performed using QForm UK 10.2.4 commercial FE code. An approach for modeling opening and closing voids emerging in the interstice between both components is presented objecting a qualitative prediction of delamination effects during cold flange upsetting of extruded billets.

2 Materials and Methods

2.1 System Description

Compound hot extrusion, performed by the project partner Institute for Metal Forming and Lightweight Technology (IUL) Dortmund, was used to manufacture a hybrid aluminum extrudate (Ø 20 mm). Extruded hybrid rod consisted of a matrix from AA-6060 with a tubular reinforcing element from AA-7075 (see Fig. 1a). An extrusion ratio $R = 10.9$ was realized for the compound extrusion process to create hybrids with different inner diameter d_i and outer diameter d_a as depicted in Fig. 1c. The dimensions of the billet components prior to extrusion were chosen in such a way, that the process was feasible without fracture of the tubular reinforcing element. Main reason for choosing mentioned hot extrusion process parameters was to create a proper metal bond between the hybrid components during extrusion. For subsequent cold forging process, billets were cut from the continuous hybrid extrudate (length L_0, diameter d_0). Cold flange upsetting was chosen as a suitable process as it poses challenges to the bond between the materials during deformation. Depending on the diameter ratio $\overline{d_i}$ and $\overline{d_a}$ of the three different reinforcing element dimensions I-III, particular stress conditions are prevailing, that provoke delamination effects between the material layers.

As shown in literature, compound hot extrusion may result into metallic bond between the components [8]. Qualitative estimation of bond strength for such process was performed in [9]. Numerical modeling of the compound extrusion process showed large deformation (effective strain up to $\varepsilon_{eff} = 3$) in the interface between

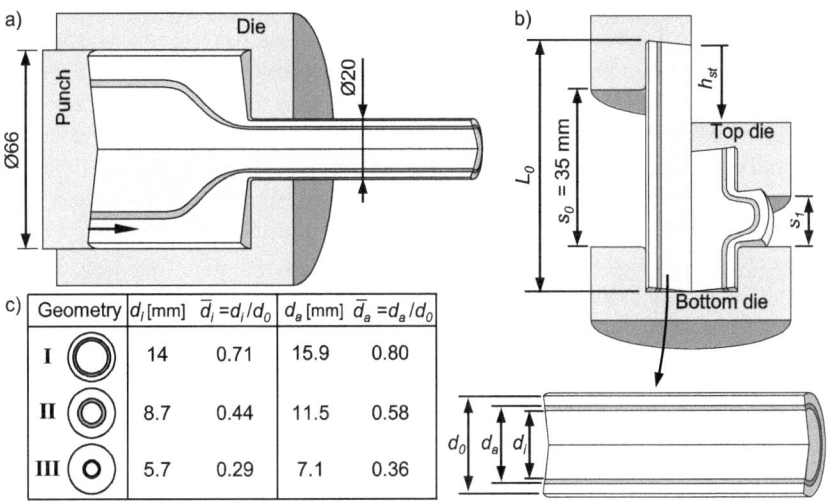

<table>
<tr><th>c) Geometry</th><th></th><th>d_i [mm]</th><th>$\overline{d_i} = d_i/d_0$</th><th>d_a [mm]</th><th>$\overline{d_a} = d_a/d_0$</th></tr>
<tr><td>I</td><td>◯</td><td>14</td><td>0.71</td><td>15.9</td><td>0.80</td></tr>
<tr><td>II</td><td>◉</td><td>8.7</td><td>0.44</td><td>11.5</td><td>0.58</td></tr>
<tr><td>III</td><td>○</td><td>5.7</td><td>0.29</td><td>7.1</td><td>0.36</td></tr>
</table>

Fig. 1 **a** Compound hot extrusion, **b** cold flange upsetting, **c** billet dimensions

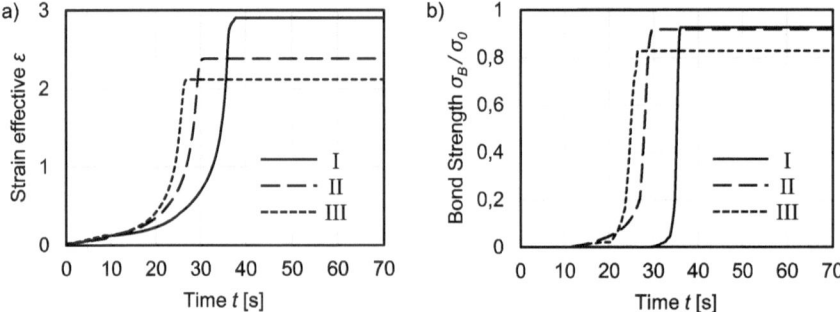

Fig. 2 Process parameters during compound hot extrusion **a** effective strain **b** relative bond strength according to Bay welding model [10]

the components and thus can lead to a metal bond due to the resulting high surface exposure in combination with the hydrostatic pressure during extrusion.

A qualitative bond strength calculation was examined using the Bay pressure welding criterion as published in [10]. Bond strengths for the three different geometrical variants I-III are depicted in Fig. 2b. Geometrical changes of the reinforcing element position lead to bond strength variations. For further effect analysis of bond characteristics on subsequent cold flange upsetting, billets with a clearance fit between the components were prepared by machining process.

2.2 Experimental Investigations

The billets for subsequent cold flange upsetting process were prepared using two different processing methods leading to different interface characteristics between the aluminum alloys. To investigate the effect of a clearance fit between the components and thus no bonding ($\sigma_B = 0$), samples were machined from round bars featuring a diametral clearance of 0.09 mm between the corresponding inner and outer part, see Fig. 3a. After machining, the three parts, consisting of core, tube, and shell, were manually assembled to make up the hybrid billet for cold forging. The hybrid billets with a certain bond due to compound hot extrusion process (assuming $\sigma_B \sim 1$) were cut from the hybrid extrudate as shown in Fig. 3b.

After the manufacturing process, the billets had been annealed, maintaining a holding time $t = 90$ min at $T = 380\,°C$, followed by controlled cooling with a rate of 30 K/s down to $T = 250\,°C$. Lubricant zinc stearate was applied in a tumbling process with a specific mass of 5.4 g/m^2 on the billet surface. Cold flange upsetting tests were performed on a 1000 kN hydraulic programmable press with position control featuring a ram speed of 10 mm/s. The experimental setup and the tool rack are shown in Fig. 4a. During forming tests, both tool load and ram position were measured in order to investigate the critical upsetting deformation, by which delamination effects between the material layers initially occur. The upsetting deformation is defined by φ

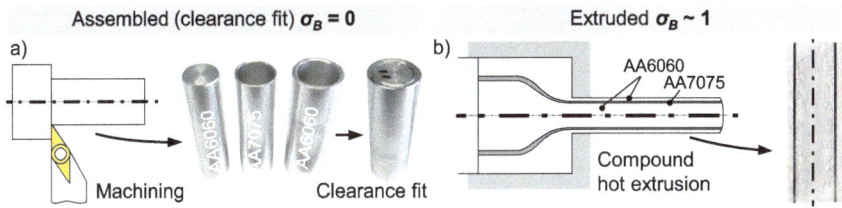

Fig. 3 Billet preparation by **a** machining with clearance fit and **b** extrusion with bonding

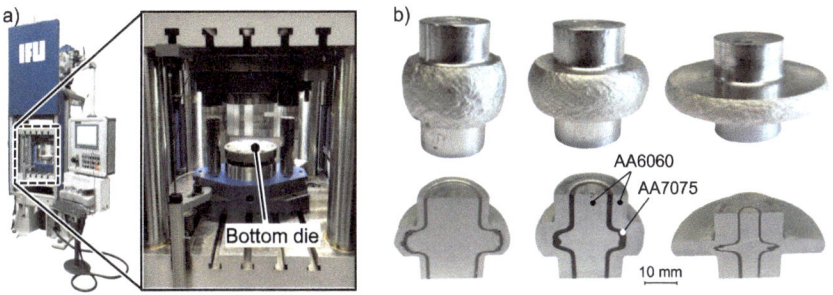

Fig. 4 Experimental setup for cold flange upsetting **a** hydraulic programmable press, **b** deformed hybrid components before and after cutting and etching

$= \ln(s_0/s_1)$, where $s_0 = 35$ mm is the initial distance between the dies and s_1 the final flange height at the end of the process according to the schematic in Fig. 1b. Delamination effects were detected by means of microscope analysis in the cross sections of the considered flange component. For better visibility of different alloys, sodium hydroxide was used to make the AA-7075 component appear in dark color, see Fig. 4b.

2.3 Numerical Investigation of Cold Flange Upsetting Using Hybrid Billets

Commercial FE code QForm UK 10.2.4 was used for numerical modeling of hybrid cold flange upsetting process. A 2D axisymmetric model with plastic workpiece definition and rigid tools was created. Initial workpiece temperature was 20 °C. Material characterization tests for both AA-6060 and AA-7075 in soft annealed state were conducted on a thermomechanical testing system Gleeble 3800 C in a temperature range of $20\,°C \leq T \leq 200\,°C$ applying strain rates between $0.1\,s^{-1} \leq \dot{\varphi} \leq 10\,s^{-1}$. The resulting flow curves for AA-6060 are depicted in Fig. 5a and for AA-7075 in Fig. 5b.

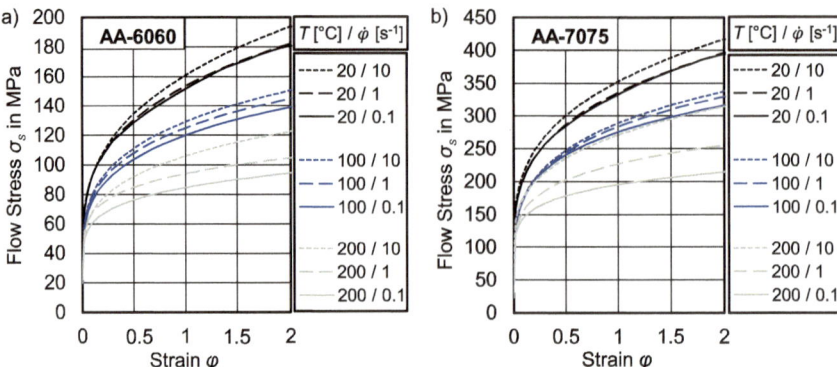

Fig. 5 Flow curves of **a** AA-6060 and **b** AA-7075 in annealed state based on table function

For the two proposed billet variants ($\sigma_B = 0$ and $\sigma_B \sim 1$), different modeling approaches available in QForm UK 10.2.4 were applied. In case of billets with clearance fit ($\sigma_B = 0$), a three-body design with interface contact definition showed similar deformation behavior as to experiments (see Fig. 6a). Friction factor $m = 0.4$ according to shear friction model in the FE code was used to define prevailing contact between the single workpiece bodies, where separation between the surfaces was allowed. Flange deformation geometry and process force were used to identify a suitable friction coefficient between tools and workpiece of $m = 0.6$. Top die velocity was set to 10 mm/s and heat transfer coefficient between workpiece and tools was defined to 30,000 W/(m²K).

In case of an initial bond between the workpieces due to billet manufacture by composite hot extrusion ($\sigma_B \sim 1$), a one-body model featuring local areas with different material properties was suitable (see Fig. 6b). As a result, one homogeneous calculation mesh was used and contact definition between the area of different material properties was neglected. The comparison between simulation and experiments resulted in friction factor $m = 0.3$ in the interface of tool and workpiece. Point tracking as well as subroutines were used to gain the required components of stress tensor in the interface area of the hybrid for further analysis.

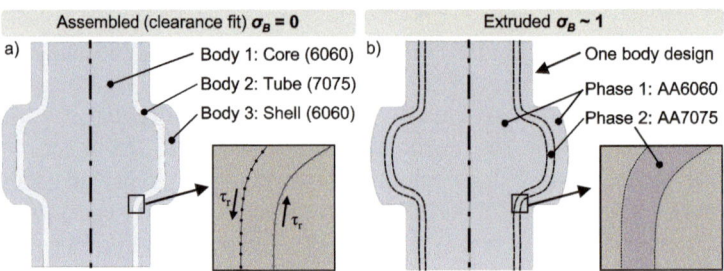

Fig. 6 Simulation models for **a** no initial bond between components **b** extruded billet

3   Results and Discussion

3.1   Experimental Results

Representative cross sections of all three geometry variants I-III and for both billet conditions ($\sigma_B = 0$ and $\sigma_B \sim 1$) are depicted in Fig. 7 as an axisymmetric view with a magnification next to the corresponding image. For each of the variants, multiple tests were conducted by increasing ram stoke and thus by decreasing flange height s_1, which is shown here by the upsetting logarithmic strain $\varphi = \ln(s_0/s_1)$. In case of billets with a clearance fit ($\sigma_B = 0$), an undefined and irregular deformation of the tubular reinforcing component made of AA-7075 was observed for multiple repetitions. In fact, buckling occured due to the clearance fit, which led eventually in some cases to vortex formation as can be seen for variants I and III. Obvious delamination, however, could be detected only for geometrical variants I and II, where the tubular reinforcing element remained under radial compression for variant III. In each case, delamination effects occurred on the outside boundary of the reinforcement.

For compound hot extruded billets ($\sigma_B \sim 1$), the deformation of reinforcing element was much more stable and homogeneous (no buckling, no vortex formation)

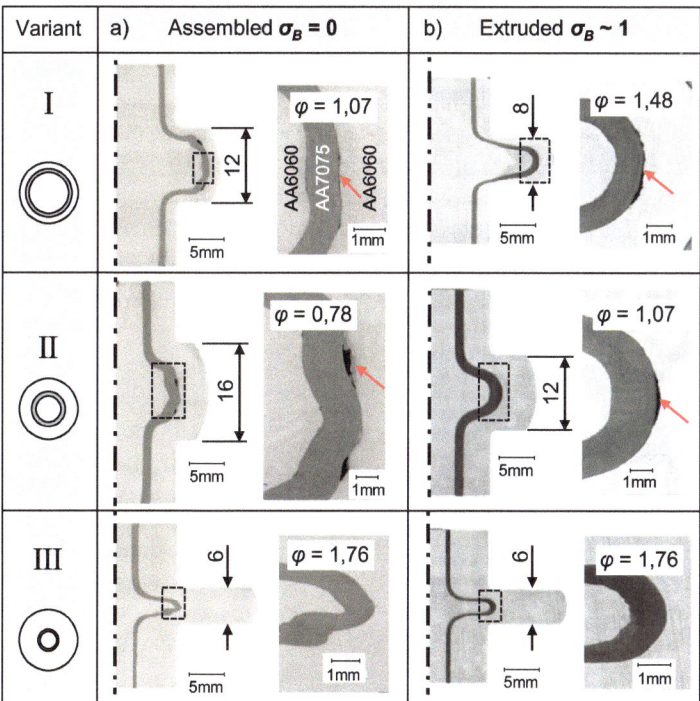

Fig. 7 Flange upsetting experimental results of different geometrical variants **a** without initial bond and **b** with extruded billets

than without bond. Regardless of which bond strength and which characteristics of bond actually prevailed in the extruded billets, bonding disclosed a positive effect on process stability and homogeneous strain distribution during cold flange upsetting. Also in this case, delamination effects appeared only at the outside boundary and if so, at approx. 35% higher deformation than without bond effect (see Fig. 7b, red arrows). For variant III, in both cases without bond and extruded billets, high deformations could be achieved without delamination effects.

3.2 Simulation Results

Numerical results according to the FE-model in Fig. 6a without initial bonding between the hybrid billet components are shown for the three geometry variants I-III and representative deformation states in Fig. 8. Radial stress σ_{xx} was used as a suitable state variable for analysis of possible bond detachment areas. For both variants I and II, experiments showed a noticeable separation between the outside shell and the reinforcing component at a certain amount of logarithmic strain. Using the proposed simulation model, loss of contact could be obtained in the outside boundary at similar deformation as to the experiment. Utilized simulation model, however, did not predict buckling of the reinforcing element as it could be obtained in experiments.

In case I, tensile radial stresses of approx. 30 MPa were detected close to the boundary in the outside shell from AA-6060, where detachment was actually prevailing also in experiments (see right column of first row in Fig. 8). When decreasing the outer diameter of the reinforcing component, the radial stress state close to the interface became more compressive. For geometry II, low compressive stress values had been calculated in the outer shell in the horizontal symmetry plane of the flange. Tensile radial stresses in the corners of the flange, together with the increased stiffness of the reinforcing element due to thick walls led to a visible loss of contact in the simulation model. Small outer diameter of reinforcement (III) resulted into a highly compressive radial stress state, which prevented delamination effects in both simulation and experiment.

Considering initial bonding between the billet components ($\sigma_B \sim 1$), the one-body simulation model presented in Fig. 6b showed areas with possible bond separation using the hydrostatic stress state. In both cases I and II, a positive hydrostatic stress state of approx. 25 MPa was obtained in the area of largest reinforcing element diameter, in which delamination also occurred in experiments. Considering geometry variant III with smallest outside diameter, more or less similar conditions were recognized as shown in Fig. 9, including large compressive hydrostatic stress level on the outside boundary, thus not allowing separation between the bodies.

Cold flange upsetting process showed highly non-monotonous stress profiles, which are depicted in Fig. 10 as triaxiality $\eta = \sigma_m/\sigma_{eff}$ over effective strain ε_{eff}, where σ_m is the hydrostatic stress and σ_{eff} effective stress. The triaxiality profiles were analyzed for two points P1 and P2 located representatively in the flange area, where delamination could be obtained in experiments. P1 (see Fig. 10a) was located

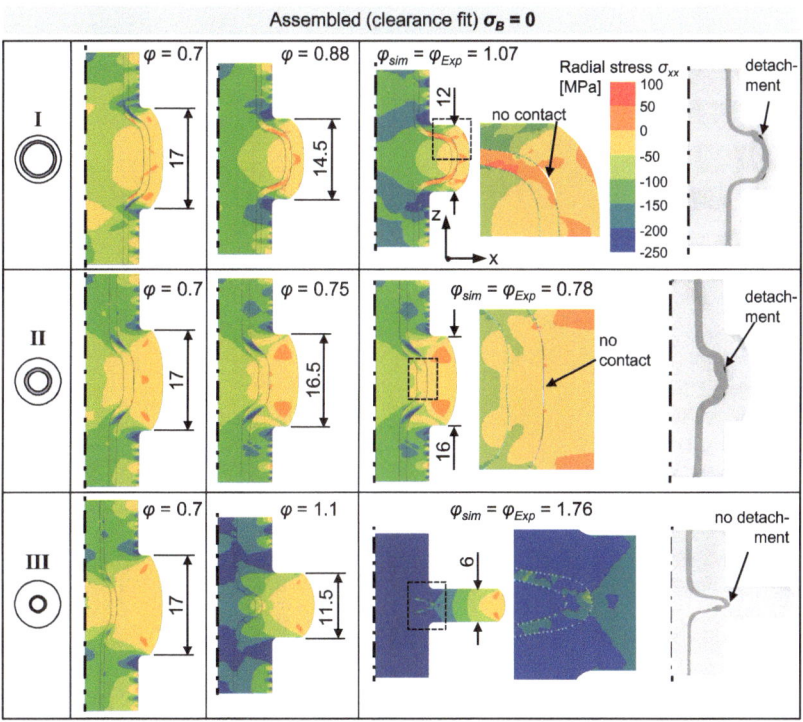

Fig. 8 Simulation results overview for different geometrical variants I-III with no initial bond between the materials $\sigma_B = 0$ for different deformation increments

at the corner of the bonding boundary in the transition zone towards the horizontal part of the reinforcing element and P2 was located right at the horizontal symmetry (see Fig. 10b). The characteristics in both points were similar, where geometrical variant III showed in each case negative stress triaxiality while variants I and II showed partly positive values. After a local maximum, triaxiality decrease could be obtained for all three variants. Due to the non-monotonous stress conditions in the interface, the porosity model, first introduced by Beygelzimer [11] was used to predict detachment of bond. The model describes porosity Θ as a function of triaxiality η, effective strain ε_{eff} and two constants A and B. Using this model, alternating triaxiality as present in this process, can be considered by increasing and decreasing porosity:

$$\frac{d\Theta}{d\varepsilon_{eff}} = A + B \times \Theta \frac{\sigma_m}{\sigma_{eff}} [11] \tag{1}$$

As thoroughly investigated in [11], the critical porosity to failure for several monolithic materials is $\Theta_{crit} \approx 1\%$. In a first approach of application, constants of this model were chosen to $A = 0.01$ and $B = 2.5$. The porosity calculation results for both points and all three geometrical variants are shown in Fig. 10c and d.

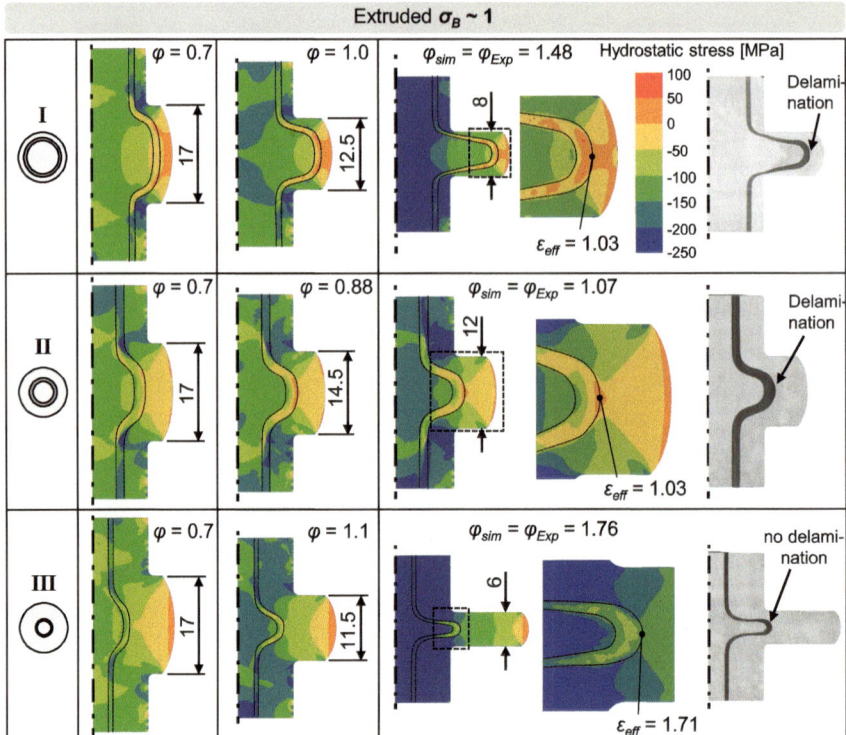

Fig. 9 Simulation results overview for different geometrical variants I-III with initial bond between the materials ($\sigma_B \sim 1$) for different deformation increments

For P1 and P2, observed tendencies appear comparable, so porosity was calculated for variant I to $\varepsilon_{eff,I}$ $(\Theta_{crit}) \approx 1.0$ and for variant II to $\varepsilon_{eff,II}$ $(\Theta_{crit}) \approx 1.2$. Variant III, however, did not exceed critical porosity value and even showed a decrease, which confirmed the finding of experiments as there was no detachment of bond found. Especially for P1 (see Fig. 10c), porosity decreased for all variants I-III, thus separation of interface was more likely to be reversed after certain degree of deformation. Comparing the critical effective strains $\varepsilon_{eff,I}$ (Θ_{crit}) and $\varepsilon_{eff,II}$ (Θ_{crit}) with the simulation results in Fig. 9, a good agreement could be found for variant I. For variant II, the strain to critical porosity was estimated approx. 16% higher than in experiments.

4 Summary and Conclusion

This study has shown that a certain bond strength between the components of a bi-metallic billet, manufactured by composite hot extrusion, in fact makes subsequent cold forging process more stable regarding material flow compared to billets, that

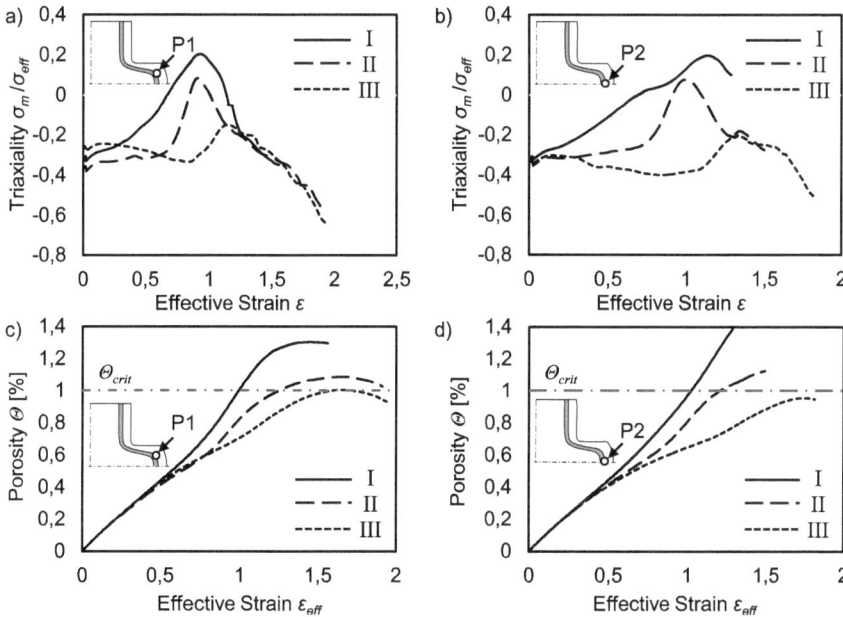

Fig. 10 Triaxiality profiles during deformation for **a** point P1 and **b** point P2 in interface and porosity for **c** P1 and **d** P2 for all geometrical variants I-III according to [11]

were manually assembled without initial bond due to clearance fit. Billets without initial bond were more likely to show several defects, such as detachment of contact surfaces, loss in stability during compression, and undesired strain localization within the reinforcing component. The formability during cold flange upsetting of extruded AA-6060/AA-7075 billets was found at least 35% higher than without bond, before delamination effects had been initiated. Two simulation approaches had been evaluated for different initial billet bonding conditions. A 3-body design with separable shear friction contact definition between joint components was suitable to model hybrid cold forging process without initial bond between the parts, for which detachment was detected by contact loss. Analysis of radial stress acting in outside shell of flange showed that increasing tensile stresses led to detachment between the components. A 1-body design with different local material properties was used for modeling of extruded billet deformation. The used porosity model approach allowed a qualitative prediction of delamination between the initially bonded billets. For quantitative estimation, further experiments should be performed in order to investigate more deeply prevailing relationships between stress triaxiality, surface enlargement during forming, and other parameters defining gained effectiveness of bond.

References

1. Estrin Y, Beygelzimer Y, Kulagin R, Gumbsch P, Fratzl P, Zhu Y, Hahn H (2021) Architecturing materials at mesoscale: some current trends. Mater Res Lett 9(10):399–421
2. Yoshida Y, Matsubara T, Yasui K, Ishikawa T, Suganuma T (2012) Influence of processing parameters on bonding conditions in backward extrusion forged bonding. Key Eng Mater 504–506:387–392
3. Wohletz S, Groche P (2014) Temperature influence on bond formation in multi-material joining by forging. Procedia Eng 81:2000–2005
4. Essa K, Kacmarcik I, Hartley P, Plancak M, Vilotic D (2012) Upsetting of bi-metallic ring billets. J Mater Process Technol 212(4):817–824
5. Plancak M, Kacmarcik I, Vilotic D, Krsulja M (2012) Compression of bimetallic components—analytical and experimental investigation. Int J Eng, 157–160
6. Ossenkemper S, Dahnke C, Tekkaya AE (2019) Analytical and experimental bond strength investigation of cold forged composite shafts. J Mater Process Technol 264:190–199
7. Grötzinger KC, Benko K, Kotzyba P, Hering O, Liewald M, Tekkaya AE (2021) Manufacture of hybrid components with tubular reinforcement by composite hot extrusion and subsequent cold forging. In: MEFORM, pp 93–98
8. Kolpak F, Schulze A, Dahnke C, Tekkaya AE (2019) Predicting weld-quality in direct hot extrusion of aluminium chips. J Mater Process Technol 274:116294
9. Kotzyba P, Grötzinger KC, Hering O, Liewald M, Tekkaya AE (2020) Introduction of composite hot extrusion with tubular reinforcements for subsequent cold forging. In: Proceedings of the 10th congress of the German academic association for production technology, pp 193–201
10. Bay N (1979) Cold pressure welding—The mechanisms governing bonding. J Manuf Sci Eng, Trans ASME 101(2):121–127
11. Beigelzimer JE, Efros BM, Varyukhin VN, Khokhlov AV (1994) Continuum model of the structural-inhomogeneous porous body and its application for the study of stability and viscous fracture of materials deformed under pressure. Eng Fract Mech 48(5):629–640

Experimental and Numerical Simulation on Formability and Failure Behavior of Thermoplastic Carbon Fiber/AL Composite Laminates

Chen Sun, Minghua Dai, Liang Ying, Kai Du, Zhigang Chen, and Ping Hu

Abstract Carbon fiber reinforced thermoplastic/aluminum alloy (CFRTP/AL) composite laminates have the advantages of low density, high specific strength, and good fatigue resistance, which is a new type of engineering composite material to realize lightweight vehicle body. Heterogeneous interface delamination failure occurs in the forming process of the fiber metal laminates (FMLs). It is necessary to establish an effective finite element simulation strategy to accurately predict the delamination failure behavior of FMLs. In this work, thermoplastic PA6 continuous carbon fiber/AL FMLs were taken as the research object, and the double cantilever beam (DCB) and the end-notched flexure (ENF) experiments were carried out to determine the basic mechanical parameters between the interlayer interfaces of CFRTP/AL. Furthermore, a numerical simulation model based on ABAQUS software was developed to describe the progressive damage failure behavior of the CRFTP/AL in the forming process by using the equivalent modeling strategy of discontinuous micro-shear, which realized the effective prediction of ply directional damage failure of FMLs on the basis of the S-beam model. The results show that the established damage constitutive model and numerical method coupled with cohesive zone model (CZM) can effectively predict the ply directional damage failure behavior of CFRTP/AL composites during the large deformation forming.

Keywords Thermoplastic carbon fiber · Fiber metal laminates · Interlaminar damage failure · Formability · CZM

C. Sun · M. Dai · L. Ying (✉)
School of Mechanical Engineering, Dalian University of Technology, Dalian 116024, China
e-mail: yingliang@dlut.edu.cn

C. Sun · L. Ying · Z. Chen · P. Hu
School of Automotive Engineering, Dalian University of Technology, Dalian 116024, China

K. Du
School of Materials Science and Engineering, Shenyang University of Technology, Shenyang 110870, China

J. Kusiak et al. (eds.), *Numerical Methods in Industrial Forming Processes*, Lecture Notes in Mechanical Engineering, https://doi.org/10.1007/978-3-031-58006-2_30

383

1   Introduction

Automobile lightweight can reduce the weight of the body and improve fuel efficiency, which is considered to be an important means to solve current energy and environmental issues [1–3]. For the ever-growing new energy vehicles (NEVs), lightweight automobile can improve the mileage of NEVs and solve the problem of "mileage anxiety" to a certain extent. Automobile lightweight has a positive effect on both traditional fuel vehicles and NEVs [4]. FMLs are a new type of lightweight engineering material that combines the respective advantages of metal and fiber-reinforced composite materials, which has been widely focused by the automotive industry for its excellent mechanical properties such as low density, corrosion resistance, and fatigue resistance [5–7].

However, FMLs is a hyper-hybrid composite material with multiple interfaces, which may have various failure behaviors such as fiber fracture and matrix damage during the stamping forming process, especially the delamination failure between heterogeneous interfaces would seriously affect the mechanical properties of the component. Vieille [8] found that the softening of the resin matrix at high temperature would lead to serious degradation of the fiber/resin bonding interface. Hirsch [9] confirmed that the interface roughness has a positive impact on the bonding performance of the FMLs heterogeneous interface. How to accurately quantify the mechanical properties of FMLs heterogeneous interfaces and accurately predict the interlayer damage failure is an urgent problem to be solved.

In this paper, the preparation of CFRTP/AL laminates and the stamping of S-beams were firstly carried out. Then DCB test and ENF test were executed according to ASTM test standard, which accurately quantified the mechanical properties of FMLs prepared by the hot stamping process and obtained the fracture toughness of type I and type II that characterize the interfacial bonding strength. Furthermore, the finite element simulation method coupled with CZM was introduced to simulate the experimental process of DCB and ENF to verify the correctness of the model. Finally, the developed numerical modeling strategy of CFRTP/AL was applied to the stamping simulation of S-beams successfully, which realized the accurate prediction of layered failure of heterogeneous interfaces in the process of large deformation of FMLs.

2   Fabrication of CFRTP/AL Laminates

FMLs is an interlayer hyper-hybrid composite material, which is obtained by stacking metal sheets and fiber-reinforced composite materials at a certain angle and curing under a certain temperature and pressure conditions [10]. In this study, FMLs were prepared by a HPFM-100D comprehensive compression molding test platform. AA5754 aluminum alloy produced by Southwest Aluminum is selected as the surface metal, and its material composition and performance parameters are shown in

Table 1 Constituent elements and compositions of AA5754 aluminum alloy

Element content	Si	Cu	Mg	Zn	Mu	Cr	Fe
Max (%)	0.40	0.10	3.6	0.20	0.50	0.30	0.40
Min (%)	–	–	2.6	–	–	–	–

Table 2 Parameters of AA5754 aluminum alloy

Parameters	Density (g·cm^{-3})	Elasticity modulus (GPa)	Yield strength (MPa)	Tensile strength (MPa)	Poisson ratio
Value	2.7	69	278	472	0.33

Table 3 Parameters of thermoplastic PA6 carbon fiber reinforced composites

Material parameter	Value
Carbon fiber content/(%)	53
Prepreg density/(g·cm^{-3})	1.393
Prepreg average thickness/(mm)	0.2
Longitudinal elasticity modulus/(GPa)	86.6
Longitudinal tensile strength/(GPa)	1036
Poisson ratio	0.2

Tables 1 and 2. The fiber-reinforced composite material is the unidirectional prepreg of the thermoplastic PA6 carbon fiber produced by Zhongfu Shenying Carbon Fiber Co., Ltd. The relevant material parameters are shown in Table 3.

After the preparation of CFRTP/AL laminate, a 40 T hydraulic press was utilized to stamp the S-beam. The mold gap of the S-beam was 1.6 mm. Therefore, when preparing the laminate, two layers of carbon fiber prepreg were stacked between the upper and lower surface aluminum alloys. An adhesive film was added between the heterogeneous interfaces to further enhance the interlayer bonding ability. In order to obtain better formability, the CFRTP/AL laminate was heated to 150 °C and quickly transferred to the mold for stamping, and then cooled to room temperature with the furnace to obtain the S-beam formed part The layering method of CFRTP/AL and related hot molding process parameters are shown in Fig. 1c.

3 Interfacial Fracture Toughness Test

During the forming process of CFRTP/AL, the failure of delamination cracking between the heterogeneous interfaces of laminates is a key issue of concern. In order to accurately measure the interfacial fracture toughness of CFRTP/AL laminates prepared by hot molding process, DCB and ENF tests were carried out on the basis of ASTM D5528 and ASTM D7905 standards [11, 12]. The total thickness of the

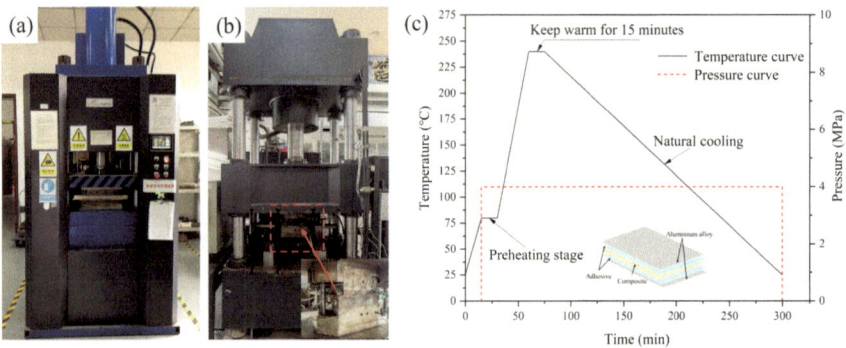

Fig. 1 Preparation and forming of CFRTP/AL laminates **a**: HPFM-100D comprehensive molding test platform **b**: S-beam forming press **c**: The temperature–pressure curve of hot molding

specimen is 4 mm. In order to ensure that the single component material of the specimen deforms harmoniously under the action of the test force, the carbon fiber prepregs are stacked in the orthogonal lay-up method of 0°/90° (fiber direction is specified as 0°) for 10 layers, so that the thickness of the composite material and the thickness of the aluminum alloy are both 2 mm. To prefabricate the crack, a layer of 0.05 mm polyimide release paper was added between the carbon fiber prepregs and the aluminum alloy. The release paper should not be too thick, as excessively thick release paper will lead to a lipid-rich area at the binding tip of the heterogeneous interface, resulting in inaccurate experimental results.

The DCB/ENF test was conducted on the WDW-100 universal tensile testing machine, and the test process is shown in Fig. 2. In the DCB test, the carbon fiber bundles were pulled out of the resin matrix, showing the "bridging" phenomenon in Fig. 2a. At this point, the bearing capacity had been greatly reduced. The final calculated mode I fracture toughness data G_{IC} is 753.91 J/m^2. In the ENF specimen, the heterogeneous interface was mainly subjected to shear stress. Due to the large longitudinal stiffness of CFRTP, it cannot coordinate with the aluminum alloy layer to deform when subjected to a large bending displacement load. The glue layer on the left side of the punch mainly bore this incongruous deformation and suffered shear stress, resulting in shear failure. At the same time, the carbon fiber layer gradually broke and failed from the outside to the inside, and finally the carbon fiber layer and the aluminum alloy layer were completely cracked, as shown in Fig. 2b. The final calculated mode II fracture toughness data G_{IIC} is 3.59 kJ/m^2.

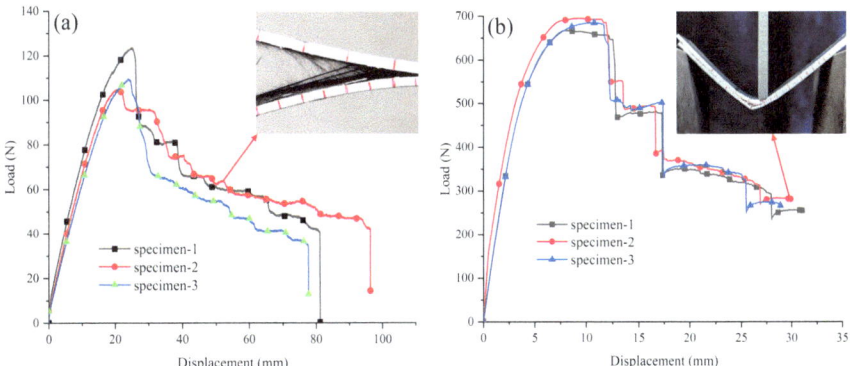

Fig. 2 Interface mechanical properties test **a**: DCB **b**: ENF

4 Numerical Simulation of Forming Process for CFRTP/AL

4.1 Cohesive Zone Model

CZM is widely used in the field of continuum fracture simulation and crack propagation, and has been derived into bilinear, trapezoidal, exponential, and other forms. Figure 3 shows the damage propagation process of the heterogeneous interface adhesive layer and the exponential CZM. It can be seen that the exponential traction–separation relationship assumes that the adhesive layer is initially in the linear elastic stage. In ABAQUS, both cohesive element and cohesive contact can be used to simulate the bonding behavior of heterogeneous interfaces. When the cohesive contact method is used for simulation, the slope of the elastic stage is the stiffness of adhesive layer. When the traction force reaches the initial damage, the damage evolves exponentially. The area enclosed by the traction force-separation curve and the coordinate axis represents the fracture energy G_C. The damage initiation criteria commonly used in CZM include the maximum nominal stress criterion and the quadratic nominal stress criterion. In order to accurately describe the delamination failure between heterogeneous interfaces, the quadratic nominal stress criterion is introduced, which means that when the sum of the squares of the nominal stress and the maximum nominal stress ratio in the three directions is 1, it is determined that the adhesive layer fails, as shown in Eq. 1.

$$\left(\frac{\langle t_n \rangle}{t_n^{\max}} \right)^2 + \left(\frac{t_s}{t_s^{\max}} \right)^2 + \left(\frac{t_t}{t_t^{\max}} \right)^2 = 1 \tag{1}$$

where t_n is the normal traction stress component, t_s and t_t are the tangential traction stress components, and " $< >$ " indicates that the failure of the interface would not be caused by the compressive deformation or stress in the normal direction.

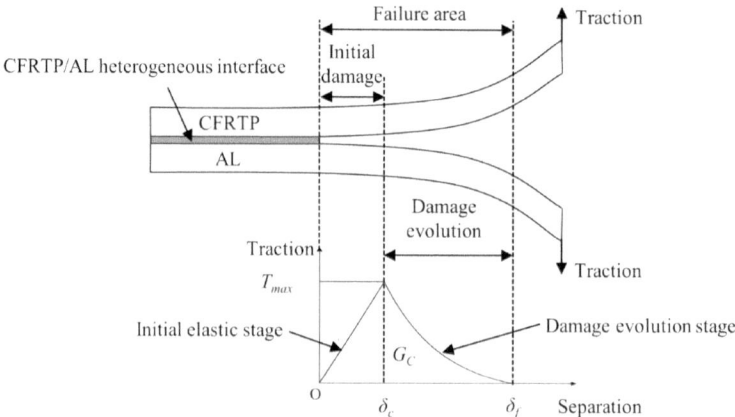

Fig. 3 Crack growth process and exponential CZM of heterogeneous interface

When the adhesive layer reaches the damage initiation, the damage evolution is carried out. Considering that the heterogeneous interface of the CFRTP/AL laminate is a mixed failure model of type I and type II fracture during the forming process, it is necessary to consider the combined action of normal traction and tangential traction at the same time. In this paper, the damage propagation of the adhesive layer is simulated based on the second-order power law, as shown in Eq. 2.

$$\left(\frac{G_I}{G_{IC}}\right)^2 + \left(\frac{G_{II}}{G_{IIC}}\right)^2 = 1 \tag{2}$$

where G_{IC} is the energy released by the normal traction force, and G_{IIC} is the energy released by the tangential traction force, namely, the type I fracture energy and the type II fracture energy.

4.2 DCB/ENF FE Simulation

A 3D finite element model is established in the commercial software ABAQUS, and the mechanical behavior of the DCB specimen under type I loading conditions is simulated using cohesive element. The model can be divided into three parts: the aluminum alloy layer, the CFRTP layer, and the middle adhesive interface. In order to accurately simulate the actual stress of the DCB specimen during the test, two reference points are established 5 mm above the end of the model, and the motion constraints are imposed on them respectively to simulate the role of the hinge in the test. Boundary conditions are shown in Fig. 4a. In ABAQUS, the simulation result of scalar stiffness degradation (SDEG) is expressed as the damage of the adhesive layer. Larger SDEG value indicates more serious damage of the adhesive layer.

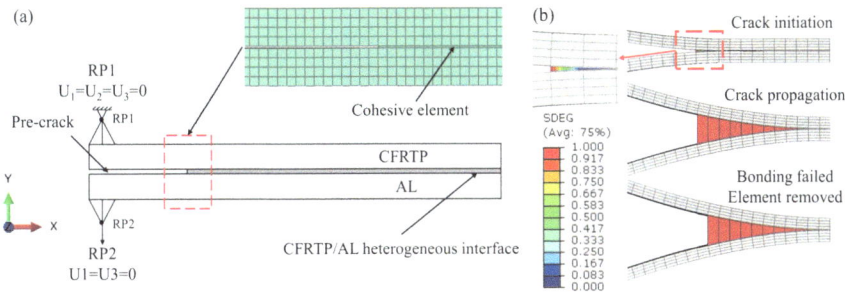

Fig. 4 DCB finite element simulation **a**: boundary conditions and motion constraints **b**: simulation results

When its value is 1, it indicates that the adhesive layer has completely failed and the corresponding elements are deleted. The DCB simulation results are shown in Fig. 4b. It can be seen from the figure that with the increase of the tip displacement load, the heterogeneous interface experiences the continuous increase of traction stress, damage initiation, damage evolution, and complete failure elements deletion in four stages, which is consistent with the test results.

ENF finite element model and boundary conditions are shown in Fig. 5a. The radius of the punch and the supporting roller is 5 mm, which is set as the rigid body through reference point. Combined with the ENF experimental process, the degrees of freedom of the two rollers are set as complete fixed, and a downward displacement load is applied to the punch. The simulation result CSQUADSCRT represents the quadratic traction stress damage initiation criterion. When using cohesive contact calculation method, this parameter can be used to judge whether the heterogeneous interface is damaged. When its value is 1, it means that the damage has been reached. The ENF simulation results are shown in Fig. 5b. From the simulation results, it can be seen that in the early stage of simulation, the adhesive layer generates shear traction at the tip and first appeared damage failure. As the punch continues to move down, the failure gradually expands and finally gathers at the position of the punch. The adhesive layer failure at both ends of the punch is earlier than other positions. It can be concluded from the DCB/ENF simulation results that the damage simulation of the failure behavior of the CFRTP/AL heterogeneous interface adhesive layer can be realized based on both the cohesive element and the cohesive contact.

4.3 Simulation of S-beam CFRTP/AL Stamping Forming

Furthermore, CZM is introduced into the stamping forming simulation of the S-beam for automobile parts. The finite element model is shown in Fig. 6. Among them, punch and die are meshed with R3D4 shell elements, which are constrained as discrete rigid bodies by reference points. The sheet material is deformable CFRTP/

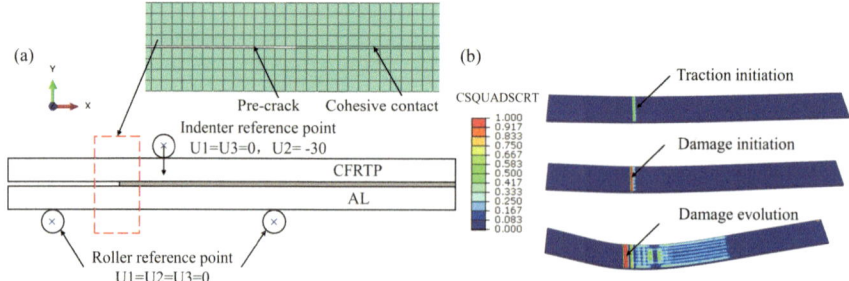

Fig. 5 ENF finite element simulation **a**: boundary conditions and motion constraints **b**: simulation results

AL composite material, and the mesh type of surface aluminum alloy is C3D8R solid element, and the core composite material mesh type is SC8R continuous shell element. Considering that the central part of the sheet material is the main deformation area, this section is remeshed by transition mesh division method, which gradually densifies from both sides to the middle. Thickness direction is divided into three layers to prevent warping in the forming process. The adhesive ability between the heterogeneous interfaces of FMLs is the focus of attention in the stamping process. During the setting of contact properties, the contact type between the upper and lower surface metals and the core composite is specially designated as cohesive contact. Based on the second-order power law, the delamination failure between the heterogeneous interfaces is effectively predicted by using the type I and type II fracture toughness. The rest of the parts adopt the general contact properties and the contact law is penalty function method. Tangential friction coefficient is set as 0.125. Moreover, the cohesive element has high requirements on the quality of the mesh. The cohesive element is meshed at 0.5 mm to prevent the simulation from failing to converge.

The results of finite element simulation are shown in Fig. 7. It can be seen from the results that S-beam has good formability and there is no delamination cracking between the heterogeneous interfaces. The S-beam part obtained by stamping with the 40T press is shown in Fig. 7b. The experimental results are consistent with the simulation results, which verifies that the modeling strategy proposed in this paper

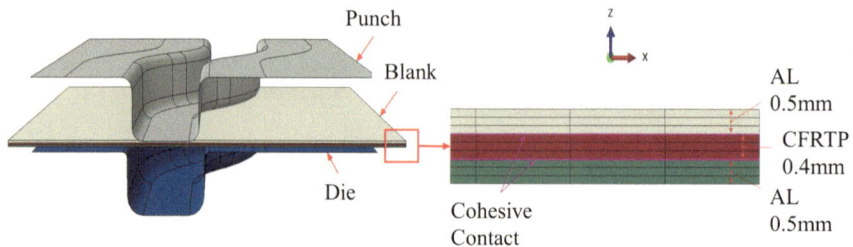

Fig. 6 Finite element model of stamping forming of CFRTP/AL S-beam laminate

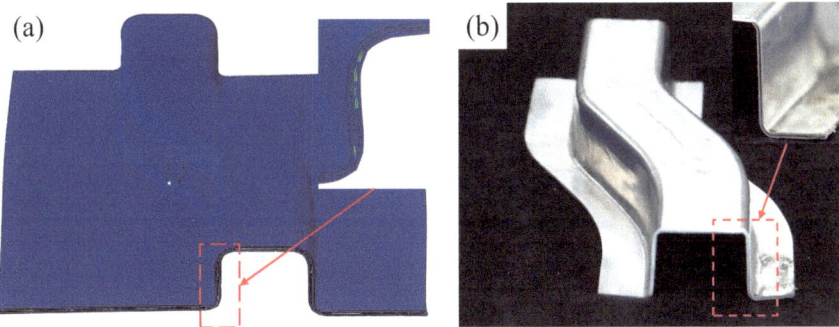

Fig. 7 Finite element simulation results of CFRTP/AL laminated S-beam **a**: Coupled CZM finite element simulation results **b**: Experimental results

can effectively predict the ply directional failure of CFRTP/AL laminates during the forming process.

The simulation result CSDMG is the stiffness degradation coefficient of cohesive surface, which can be used to describe the combination of heterogeneous interfaces of CFRTP/AL, and its value of 1 indicates that the interface is completely invalid. In order to observe the change history of the ply directional adhesive properties of heterogeneous interface during the forming process of S-beam, the bottom aluminum alloy is taken as an example. Its upper surface is in viscous contact with the surface of composite material. The simulation results are shown in Fig. 8. It can be seen that the initial damage between CFRTP/AL heterogeneous interface occurs at the contact area between the sheet metal and the die fillet. As the punch continues to move down, the damage gradually expands to both sides, and the maximum stiffness degradation coefficient is 0.829, which does not completely fail. However, due to the large damage, the interface delamination may occur under the influence of alternating stress during the service. The figure also shows the fiber tensile failure simulation results of the core composite material. From the simulation results, some fibers have broken, and the position is mainly concentrated in the punch corner.

5 Conclusions

In this paper, CFRTP/AL laminates were firstly fabricated based on the hot molding process, and the stamping test of S-beams was completed with the 40T press. The formability of the parts was good, and no delamination crack occurred. According to ASTM D5528 and ASTM D7905 test standards, DCB and ENF tests were performed. The fracture toughness of type I was calculated as GIC = 753.91 J/m^2, and the fracture toughness of type II was calculated as GIIC = 3.59 kJ/m^2. FE simulation of DCB/ENF test process was carried out based on the damage constitutive model of coupled CZM, which confirmed that CZM can be used to describe the bonding mode

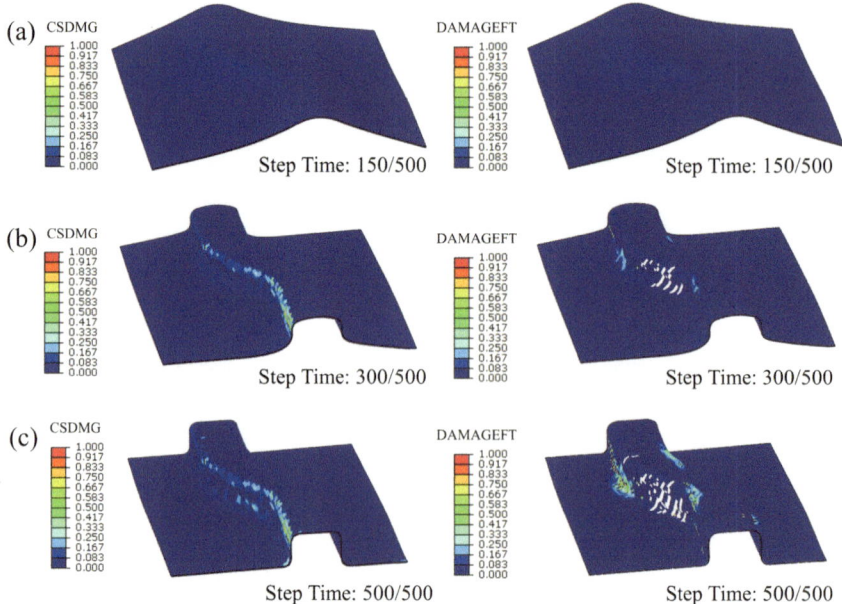

Fig. 8 Damage history of adhesive layer during CFRTP/AL forming process

between heterogeneous interfaces and accurately predict the damage and failure. By using the equivalent modeling strategy of discontinuous micro-shear, the simulation model constructed in this paper is introduced into the stamping process of S-beam of typical automobile parts. The results show that damage of heterogeneous interface mainly occurs in the rounded part of the die, and expands to both sides with the downward movement of the punch. The maximum stiffness degradation coefficient CSDMG of the adhesive layer is 0.829, and no stratification phenomenon occurs.

Acknowledgements This work was supported by the National Natural Science Foundation of China (52375310,52305396) and Aeronautical Science Foundation of China (20230036063002).

References

1. Ming JJ (2020) Research status of automobile lightweight materials and manufacturing technology. Mod Manuf Technol Equip 56(10):146–147
2. Liu Q, Lin Y, Zong Z et al (2013) Lightweight design of carbon twill weave fabric composite body structure for electric vehicle. Compos Struct 97:231–238
3. Han C (2020) Research on the development and application of lightweight automotive materials. J Phys: Conf Ser 1676(1):12085
4. Pan ZF, Li Y, Fu L et al (2021) Application of lightweight technology in automobile. Automob Technol Mater 2021(5):1–8

5. Shamloo A, Fathi B, Elkoun S et al (2018) Impact of compression molding conditions on the thermal and mechanical properties of polyethylene. J Appl Polym Sci 135(15):46176
6. Dawei Z, Qi Z, Xiaoguang F et al (2018) Review on joining process of carbon fiber-reinforced polymer and metal: methods and joining process. Rare Met Mater Eng 47(12):3686–3696
7. Sim K, Baek D, Shin J et al (2020) Enhanced surface properties of carbon fiber reinforced plastic by epoxy modified primer with plasma for automotive applications. Polymers 12(3):556
8. Vieille B, Aucher J, Taleb L (2011) Carbon fiber fabric reinforced PPS laminates: influence of temperature on mechanical properties and behavior. Adv Polym Technol 30(2):80–95
9. Hirsch F, Natkowski E, Kästner M (2021) Modeling and simulation of interface failure in metal-composite hybrids. Compos Sci Technol 214:108965
10. Wu XT, Zhan LH, Li SJ et al (2018) Research progress on forming process of fiber metal laminates. Chin J Nonferrous Metals 28(01):12–20
11. American Society for Testing and Materials (2021) Standard Test Method for Mode I Interlaminar Fracture Toughness of Unidirectional Fiber-Reinforced Polymer Matrix Composites: ASTM D5528–2021[S]. West Conshohocken: ASTM
12. American Society for Testing and Materials (2019) Standard Test Method for Determination of the Mode II Interlaminar Fracture Toughness of Unidirectional Fiber-Reinforced Polymer Matrix Composites: ASTM-D7905–2019[S]. West Conshohocken: ASTM